The Diffusion of Military Technology and Ideas

The Diffusion of Military Technology and Ideas

Edited by

Emily O. Goldman and Leslie C. Eliason

Stanford University Press, Stanford, California, 2003

Stanford University Press
Stanford, California
© 2003 by the Board of Trustees of the
Leland Stanford Junior University
Printed in the United States of America

Library of Congress Cataloging-in-Publication Data

The diffusion of military technology and ideas / edited by Emily O.
Goldman and Leslie C. Eliason.
    p.   cm.
Includes bibliographical references and index.
    ISBN 0-8047-4535-8 (alk. paper)
1. Military art and science—Effect of technological innovations on—
History.   2. Military history, Modern.   I. Goldman, Emily O.
II. Eliason, Leslie C.

U39.D54   2003
355'.07—dc21   2003004549

This book is printed on acid-free, archival-quality paper.

Original printing 2003

Last figure below indicates year of this printing:
12   11   10   09   08   07   06   05   04   03

Typeset at Stanford University Press in 10/13 Minion

Contents

Tables, Map, and Figures

Tables

Map

Figures

Preface

The "revolution in military affairs" or "military transformation," as it is currently characterized in defense policy circles, has been the subject of a great deal of debate and speculation over the past decade. The idea that we are moving through a period when new technologies, particularly dramatic improvements in C⁴ISR—command, control, communications, computers, information, surveillance, and reconnaissance—harnessed to new operational concepts and organizational structures hold out the prospect of altering the nature of warfare is attractive, even if elusive. In policy circles and the academy, debates have raged at times between those who contend that the United States is on the verge of a new way of war and those who believe that warfare possesses innate characteristics, particularly "fog" and uncertainty that no amount of technology will ever dispel. The stakes have been raised far beyond mere academic debate because transformation has been seized upon by the Bush administration as one of its highest defense priorities and the heart of a new strategic approach committed to preventing any nation from surpassing or equaling the United States militarily.

The idea for this volume originated with the following question: "What will the international consequences be if the United States succeeds in realizing the dramatic increase in military effectiveness hinted at during the Gulf War?" It seemed logical to assume that the only way to get a handle on the broader international consequences of a U.S. military transformation was to understand how other militaries were likely to respond. To many, it is not self-evident that transformation will preserve America's military advantage and political influence in the long run. It may undermine them. Whether or not transformation preserves, augments, or undermines American military power and international influence depends among other factors on how likely other states are to try to alter their militaries when faced with similar opportunities presented by the information technology revolution.

Diffusion is a topic of tremendous importance to the current transformation debate because it tells us about the paths others might take and the speed at which any U.S. lead might be diminished. Even if we were not in the midst of a revolution in military affairs of sorts, the diffusion of military innovations would still be a key issue for the United States, because much of American foreign and security policy counts on a large and long-lasting U.S. conventional superiority over most possible challengers in most types of warfare. Proponents of aggressive transformation seem to believe that if the United States could somehow achieve this "new way of war," then America's current military superiority could be sustained for a very long time. Implicit assumptions about diffusion underlie this larger belief. These assumptions need to be subjected to intensive scrutiny, particularly because they cut against prevailing beliefs in scholarly circles that the technologies and practices associated with the information revolution are going to spread widely and rapidly, even more so than those associated with past military revolutions.

Historically, as innovations associated with revolutions in military affairs have spread, they have invariably altered the balance of military and political power globally. Ever since Michael Robert's seminal 1956 lecture on "The Military Revolution, 1560–1660," historians have been debating the nature and consequences of military revolutions. The final outcome of revolutionary change is necessarily unpredictable. How many of the scientists who urged President Roosevelt to build the atomic bomb anticipated its secondary and tertiary impact on international relations? What we can say is that the global impact of military revolution hinges on whether or not others can assimilate and exploit the innovations associated with it. In other words, it depends upon how the revolution diffuses. One need only reflect upon recent events—the nuclear tests conducted by India and Pakistan, the slow-motion breakdown of the international restraints on Iraq's ability to rearm, the war in the Balkans, North Korea's ballistic missile tests, China's saber rattling over Taiwan, the alarming reports for the Congress completed by the Rumsfeld and Cox committees—to see the consequences of the diffusion of advanced weapons and the capability to employ them.

In September 1997, the Joint Center for International and Security Studies, a collaboration between the University of California, Davis, and the Naval Postgraduate School in Monterey, launched a two-year study funded by the Smith Richardson Foundation to examine the international consequences of military innovation. This volume is the product of extended deliberations among political scientists, military historians, and defense practitioners about the international implications of revolutionary military change. The volume focuses on how innovative military ideas and practices spread, based on a recognition that discussion needed to move beyond a focus on the specific consequences of transformation for U.S. doctrine, strategy, and policy. We believe that dynamics outside of the United States will affect the future balance of military power as much as, if not more than, developments inside the United States. From a defense policy perspective, the United States must be able to respond to other states that we should assume are attempting to leverage information technologies for their own purposes. A vast RMA literature focuses on the new technologies, organization, and doctrine required for transforming the U.S. military. Too little focuses on how the innovations currently being leveraged most effectively by the United States are being adopted and adapted elsewhere.

For our academic audience, the book reflects the recent renaissance of interest in the relationship between culture and security. We examine different processes of diffusion in a way that is sensitive to the organizational, social, political, and cultural contexts that influence how innovations are absorbed by different national militaries. The analytical framework emphasizes sociological perspectives on organizational change that focus on the roles of local culture and worldwide norms, and mainstream political science approaches like realism and bureaucratic politics, all of which are relevant to the study of diffusion. Some chapters privilege the social conditions that structure military change, both norms that are peculiar to the state or organization, and norms institutionalized in the global environment. Other chapters privilege the material conditions that structure change. The book synthesizes these various theoretical perspectives in order to present the complex and contingent nature of diffusion.

The volume was written with both academic and practitioner audiences in mind. From the outset, we consulted closely with policy practitioners to ensure the relevance of our research for the current policy environment. We are indebted to Capt. Ed Smith, USN (Ret.) and Col. John Nelson, USA, who attended workshops and helped us to draw out the policy implications embedded in our work. Briefings based on the research delivered to members of the military, defense, and intelligence communities were very well received. We owe a particular debt of gratitude to Mr. Andrew Marshall, Director of Net Assessment (Office of the Secretary of Defense). He attended the workshop held in Washington, D.C., and his enthusiasm for our ideas and for the insights from the case studies sustained our belief that we were making a valuable and much needed contribution to the current dialogue on transformation. We are honored that he agreed to write a foreword for the volume. The final product, we hope, has steered a middle path between the policy world and the academic world, and resulted in a scholarly work on military diffusion that has clearly delineated implications for practitioners making decisions involving the diffusion of new forms of military knowledge and technology.

A number of people helped make this volume possible. This project could not have been carried out without the generous support of the Smith Richardson Foundation and the guidance of Nadia Schadlow. The project involved three workshops, the first at the University of California, Davis, in February 1998; the second at the Monterey Institute of International Studies in August 1998; and the final at the University of California Center in Washington, D.C., in February 1999. Alan Olmstead, Tom Mezzanares, and Martha Rehrman at UC Davis provided invaluable administrative support. Bruce Jentleson, then director of the UC Davis in Washington Center, helped to make the Washington briefings and workshop a success. At the workshops, we benefited from the comments and criticisms of David Ralston, Jan Breemer, and Colin Elman. George Quester and Scott Sagan read the manuscript and provided valuable feedback in the final stages. We have been helped by talented research assistants at UC Davis: Richard Andres, Leo Blanken, and Curtis Simon. Muriel Bell and John Feneron at Stanford University Press were very supportive during the editing and production process.

Emily owes a personal debt of gratitude to Barry Forrest, whose tolerance and support made this lengthy writing and editorial enterprise possible. For all of their support and love over the years, we dedicate this labor of love to our parents, Carolyn and Elmer Goldman, and Carol and Bill Eliason.

Davis, Calif., and Monterey, Calif., April 2003

Foreword

ANDREW W. MARSHALL

Director of Net Assessment, Office of the Secretary of Defense

This is an important and instructive book. Its original historical case studies, and its integrative analyses of those cases and of existing literature, develop a very good picture of the complex processes by which innovative military capabilities—including new technology, knowledge, and skills—diffuse from their originators to the military establishments of other nations. Among the book's many strengths is its grasp of how multidimensional the underlying sources of military capabilities are and, in particular, how important the organizational and cultural sources are. This collection of studies and analyses is especially important because it raises issues that U.S. policy-makers will have to address in developing a strategy to guide our actions in the revolution in military affairs that is currently unfolding.

The current revolution in military affairs (RMA) has been underway since perhaps the late 1960s with the emergence of the first generation of long-range precision strike systems. There is now a large literature about revolutions in military affairs, some of it reviewed in this volume. The only important point that needs to be made here is that the term refers to large, significant changes in warfare. Over the course of the last five centuries there have been a number of these periods of significant change in warfare, each taking place over a period of several decades. What is new is that—because of the work of Western military historians since the 1950s, and the use of this concept of revolutions in military affairs by Soviet military theorists beginning in their 1960's discussion of the impact on warfare of nuclear weapons and ballistic missiles—we are more self-aware about the process than were most previous generations who had this experience.

The reason that large changes in warfare take several decades is that it takes a good deal of time to develop new concepts of operation, to create the new military organizations that are required to execute these new concepts, for new skills to be acquired, and perhaps for new military careers and specialties to be created.

All of these things take time, and in addition, as Stephen P. Rosen's analyses of military innovation several years ago suggested, it may require generational change within the military establishment for the new ideas and new ways of fighting to establish themselves fully.[1] We may well be near the beginning of such a lengthy period of change, which makes this volume so interesting and potentially useful.

Although a number of people were aware in the late 1970s that the Soviet military theorists were beginning to write about a new military revolution that was underway, it was only in the late 1980s that a major assessment focused on the core question: Are the Soviets right? Are we really in another period of major change in warfare? If we are, important strategic management issues face the top-level managers of the Department of Defense. Among these issues are how to foster innovation, and how to change the weapons acquisition process to make field experimentation easier. A strategy would be needed, one part of which would be focused internally on how to make innovation and change easier so that the United States would more likely find the right ways of exploiting what technology would make possible. Another part of such a strategy would be focused externally on obtaining competitive advantages with respect to potential opponents and on the management of relations with allies. It is with respect to this latter, external aspect of a U.S. strategy that this volume is especially valuable. It provides information that may be useful in making judgments about the behavior and capabilities of other countries' militaries and the level of success that may obtain in their efforts to exploit available technologies. In the introductory chapter the editors say: "Our study takes up the question of how others are likely to respond to U.S. innovations and how this will affect America's position. The answer depends on whether and how others assimilate and exploit innovations. Anticipating the diffusion trajectories that are likely to accompany military innovation and transformation, and developing strategic responses, are core aspects of the RMA challenge."[2]

The volume is also of special value because of its discussion of the issues that surround the treatment of allies in a situation where the United States is ahead—indeed, currently by a large margin—in exploiting the newer technologies for military purposes. This is a matter of importance, but one that poses great difficulties.

As mentioned above, it is a particular strength of this book's various case studies and analyses that they describe and explore the complexities of the process of diffusion and change in military organizations. As the editors note,

[1]Stephen Peter Rosen, *Winning the Next War: Innovation and the Modern Military* (Ithaca, NY: Cornell University Press, 1991).
[2]Introduction, p. 2.

"Despite the vital concerns at stake, academics are just beginning to investigate the process of diffusion: how military knowledge, broadly defined to include hardware (e.g., technology) and software (e.g., doctrine, tactics, organizational form, etc.), diffuses throughout the international system, or what factors enhance or inhibit the ability of states to incorporate innovations into their defense structures."[3] This volume goes a long way toward filling that void. Moreover, it observes that how "[a] look at the historical record reveals far more variation in adoption and emulation across states and cultures than conventional international relations theory assumes. The process of diffusion appears far less deterministic and much more vulnerable to local conditions than the systemic view suggests."[4] In particular, this book shows that culture and local institutions have a major effect in shaping how military organizations incorporate and adapt to innovations by the militaries of their neighbor or potential opponent.

To repeat, this is an exceptionally valuable collection of case studies and analyses both for interested academics and policy-makers.

[3]Ibid., p. 7.
[4]Ibid., p. 8.

Contributors

JOHN ARQUILLA is Professor of Defense Analysis at the U.S. Naval Postgraduate School, and author of several books and many articles on military and security affairs. He is best known for his collaborative RAND studies with David Ronfeldt, most notably *In Athena's Camp* (1997) and *Networks & Netwars: The Future of Terror, Crime, and Militancy* (2001).

CHRIS C. DEMCHAK is Associate Professor at the University of Arizona. Her research focuses on how networks and associated complex systems affect the availability of technology and other aspects of the information revolution to various organizations around the globe. Her publications include a book and several articles comparing military modernizations. Demchak is a U.S. Army Reserve officer whose regional expertise includes Europe, Africa, and the Middle East. She is currently working on a comparative study of future paths of military change given new knowledge management tools, widening global integration (including effects of the internet), and growing U.S. involvement in coordinating antiterror operations.

MICHAEL J. EISENSTADT is a senior fellow at the Washington Institute for Near East Policy. A specialist on Arab-Israeli and Persian Gulf security affairs, his publications include several articles and monographs on the armed forces of Iraq, Iran, Syria, and Israel, and regional arms control. He has worked as a military analyst with the U.S. Army, where he is a reserve officer. He has served in Turkey and Iraq as part of Operation Provide Comfort. He holds an M.A. in Arab Studies from Georgetown University and has traveled widely in the Middle East.

LESLIE C. ELIASON is Associate Professor and Program Head at the Monterey Institute of International Studies. She is the coeditor of *Forward to the Past: Continuity and Change in Political Development in Hungary, Austria, and the Czech and Slovak Republics* (Aarhus University Press, 1997) and *Fascism, Liberalism and*

Social Democracy in Central Europe: Past and Present (Aarhus University Press, 2002). Her research focuses on organizational and bureaucratic change, including administrative capacity building in transitional states. She has also published several articles dealing with the process of policy learning and diffusion of innovation in the public sector in European welfare states. She received her Ph.D. from Stanford University and taught at the University of Washington from 1988 until 1997.

EMILY O. GOLDMAN is Associate Professor of Political Science at the University of California, Davis, and Codirector of the Joint Center for International and Security Studies. She is author of *Sunken Treaties: Naval Arms Control between the War* (Penn State Press, 1994) and articles on U.S. strategic, military, and arms control policy; strategic adaptation in peacetime; military innovation; organizational change; and defense resource allocation. Her current research focuses on the causes and consequences of military transformation in the information age. She has directed projects on the role of information in past revolutions in military affairs, and on the diffusion of the current revolution in military affairs in the Asia-Pacific region. She holds a Ph.D. from Stanford University and was a Secretary of the Navy Senior Research Fellow at the U.S. Naval War College.

GEOFFREY L. HERRERA is Assistant Professor of Political Science at Temple University. He holds a Ph.D. in Politics from Princeton University and has taught at The Johns Hopkins University, Swarthmore College, and the University of Pennsylvania. His scholarly work has appeared in the journal *Review of International Studies*.

TIMOTHY D. HOYT is Associate Professor of Strategy at the U.S. Naval War College. He received a Ph.D. in International Relations and Strategic Studies from The Johns Hopkins University in 1997. He has lectured, taught, and consulted for Georgetown University, the University of Maryland, the Naval War College, the Army War College, the National War College, the National Security Agency, and the Department of Defense. His research interests include technology and international security, strategy, and national security policies in the developing world. His recent publications include "Pakistani Nuclear Doctrine and the Dangers of Strategic Myopia," *Asian Survey* (Nov.–Dec. 2001), "Indian Views of the Emerging RMA," coauthored with Dr. Thomas G. Mahnken, *National Security Studies Quarterly* (summer 2000), and "Indian Military Modernization, 1990–2000," *Joint Force Quarterly* (summer 2000). Dr. Hoyt is the author of *Military Industry and Regional Power: Israel, Iraq and India* (forthcoming, 2003, Frank Cass).

CHRISTOPHER JONES is Associate Professor at the Henry M. Jackson School of International Studies at the University of Washington. He is also Director of the

UW's Institute for Global and Regional Security Studies. During 1999–2000, he was a fellow at the Woodrow Wilson International Center for Scholars. Among his most recent publications are "The Logic of NATO Enlargement: De-Nationalization, Democratization, Defense of Human Rights and De-Nuclearization," in Oles Smolensky, ed., *The Lost Equilibrium: International Relations in the Post-Soviet Era* (Lehigh University Press, 2001); "The Case of the Disappearing Army," a review article for *Contemporary Security Policy* (June 2001), and "Limits to Europe," a review article in *International Politics* (Dec. 2000).

JOHN A. LYNN is Professor of History at the University of Illinois and Adjunct Professor at the Ohio State University. He concentrates on military practice and warfare in early modern Europe, on eras of military transition, and on war and culture. His most recent works include *Giant of the Grand Siècle: The French Army, 1610–1715* (Cambridge, 1997), *The Wars of Louis XIV* (Longman, 1999), and *The French Wars, 1667–1714: The Sun King at War* (Osprey, 2002). He is currently revising his manuscript *Battle: A History of Combat and Culture*, which will appear in 2003.

THOMAS G. MAHNKEN is Professor of Strategy at the U.S. Naval War College. He is a graduate of the University of Southern California and holds a Ph.D. from the Paul H. Nitze School of Advanced International Studies of The Johns Hopkins University. He previously served in the Pentagon's Office of Net Assessment and as a member of the Secretary of the Air Force's Gulf War Air Power Survey. He is the author of *Uncovering Ways of War: U.S. Intelligence and Foreign Military Innovation, 1918–1941* (Cornell University Press, 2002).

KENNETH M. POLLACK is currently Director of Research and Senior Fellow at the Saban Center for Middle East Policy of the Brookings Institution. He was previously Director of National Security Studies at the Council on Foreign Relations. He has served twice as Director for Near East and South Asian Affairs at the National Security Council, in 1995–96 and 1999–2001. From 1988 until 1995 he was a Persian Gulf Military Analyst at the Central Intelligence Agency, where he wrote a classified history of Iraqi strategy and operations during the Persian Gulf War. Dr. Pollack has also been a Senior Research Professor at the Institute for National Strategic Studies of the National Defense University, and a Research Fellow at the Washington Institute for Near East Policy. He is the author of *Arabs at War: Military Effectiveness, 1948–1991* (University of Nebraska Press, 2002), and *The Threatening Storm: The Case for Invading Iraq* (Random House, 2002), in addition to numerous journal articles on Middle Eastern military and security issues. Dr. Pollack received his Ph.D. from the Massachusetts Institute of Technology.

WILLIAM C. POTTER is Institute Professor and Director of the Center for Nonproliferation Studies and the Center for Russian and Eurasian Studies at the Monterey Institute of International Studies. The author or editor of thirteen books dealing with nonproliferation and national security issues, his present research focuses on nuclear nonproliferation issues involving the post-Soviet states. A member of the Council on Foreign Relations and the International Institute for Strategic Studies, Dr. Potter serves on the UN Secretary-General's Advisory Board on Disarmament Matters, the Board of Trustees of the UN Institute for Disarmament Research, and the International Advisory Board of the Center for Policy Studies in Moscow. He was an advisor to the delegation of Kyrgyzstan to the 1995 NPT Review and Extension Conference and to the 1997, 1998, 1999, and 2002 sessions of the NPT Preparatory Committee, as well as to the 2000 NPT Review Conference.

ANDREW L. ROSS is Professor of Strategic Studies and Director of Studies in the Strategic Research Department of the U.S. Naval War College's Center for Naval Warfare Studies. He is also Director of the College's project on "Military Transformation and the Defense Industry after Next." His work on grand strategy, defense planning, regional security, weapons proliferation, the international arms market, defense industrialization, and security and development has appeared in numerous journals and books. Professor Ross is the editor of *The Political Economy of Defense: Issues and Perspectives* (Greenwood Press, 1991) and coeditor of three editions of *Strategy and Force Planning* (Naval War College Press, 1995; 2d ed., 1997; 3d ed., 2000). He has held research fellowships at Cornell, Princeton, Harvard, the University of Illinois, and the Naval War College, and previously taught at the University of Illinois and the University of Kentucky.

THOMAS-DURELL YOUNG is European Program Manager of the Center for Civil-Military Relations at the U.S. Naval Postgraduate School in Monterey, California, where he is responsible for developing and managing technical assistance projects throughout Europe. Previously he was a Research Professor at the Strategic Studies Institute of the U.S. Army War College. Dr Young received his Ph.D. and Certificat d'études supérieures from the Institut univérsitaire de Hautes Etudes internationales, Université de Genève, and is a 1990 graduate of the U.S. Army War College. His latest book, coauthored with the late John Borawski, *NATO after 2000: The Future of the Euro-Atlantic Alliance*, was published by Praeger in 2001. In 1999 he was the inaugural Eisenhower Fellow at the Royal Netherlands Military Academy in Breda. He is coeditor of *Small Wars and Insurgencies* and serves on the editorial board of *Defense and Security Analysis*. Dr. Young is also a consultant for the RAND Corporation.

The Diffusion of Military Technology and Ideas

CHAPTER ONE

Introduction

Theoretical and Comparative Perspectives on Innovation and Diffusion

LESLIE C. ELIASON AND EMILY O. GOLDMAN

Pursuing the Revolution in Military Affairs: The Policy Challenge

The idea of a revolution in military affairs, or RMA, has been discussed since the late 1970s, when Soviet General Nikolai Ogarkov argued that a range of recent innovations would become as important to waging war as nuclear weapons.[1] U.S. high-tech weapons in the 1991 Persian Gulf War—particularly the unprecedented integration of precision-guided munitions, C^4I (command, control, communications, computers, and information) and RSTA (reconnaissance, surveillance, targeting, and acquisition)—seemed to confirm that this transformation in warfare was well underway. The current RMA promises to link advances in precision weapons, surveillance satellites, and computer-based information processing to organizational changes that "network" military units to support a new way of war. Proponents of this information technology RMA (IT-RMA) argue that the United States must embrace emerging technologies and rapidly transform its armed forces to guarantee its military superiority for the foreseeable future. The current technological lead, if preserved, would increase our military strength while cutting costs (weapon systems and overseas deployments) and reducing the risk to U.S. troops. Advocates promote the RMA as the solution to the post–Cold

---

[1] These included new kinds of explosives, precision-guided weaponry, advances in C^3I (command, control, communications and information), sensor technology and automated control systems, and weapons based on new physical principles (e.g., particle beams and lasers). Leon Goure and Michael Deane, "The Soviet Strategic View," *Strategic Review* 12, no. 3 (summer 1984): 80–94.

War paradox of ill-defined threats, growing operational demands, and limited resources.

The RMA poses a set of challenges for decision-makers. At what point does it make sense to invest in new innovations, given that the costs (financial as well as organizational, doctrinal, and political) may be high and the payoff uncertain? Leader states need to know whether, when, and how to attempt to control or encourage the spread of innovations. Follower states must decide whether, when, and how to respond. Innovation and transformation are inextricably linked with diffusion. All three processes shape the strategic environment.

Decisions about pursuing the RMA depend on accurately assessing its potential. If revolutionary transformation is politically desirable, can the United States preserve its lead with competitors striving to emulate or counter the U.S. model? Are RMA technologies vulnerable to skillful hackers and terrorists? How long will the United States enjoy the benefits of the RMA?

So far, the RMA debate has focused on whether (1) a revolution is really underway, (2) its technological underpinnings are feasible, and (3) a presumed military advantage can be translated into political influence. Our study takes up the question of how others are likely to respond to U.S. innovations and how this will affect America's position. The answer depends on whether and how others assimilate and exploit innovations. Anticipating the diffusion trajectories that are likely to accompany military innovation and transformation, and developing strategic responses, are core aspects of the RMA challenge. This requires understanding the process by which innovations diffuse to other states and contexts.

Consequences of Pursuing the IT-RMA

Our central analytical concern in this volume is not how RMAs begin, but how they spread, to whom, how quickly, and with what consequences for U.S. national security and the global balance of military power. Thus, at the heart of our comparison of historical and contemporary cases is a concern with understanding the dynamics of the diffusion process. This has been a key aspect of the RMA debate ever since Michael Robert's seminal 1956 lecture on "The Military Revolution, 1560–1660."[2] Military historians and planners disagree about the nature and consequences of military revolutions, both in general and in specific cases. The Persian Gulf War reinvigorated this debate.

Not surprisingly, scholars and practitioners also diverge in their estimates of the likely consequences of the current RMA. The most dramatic version of twenty-first-century conflict envisions improvements in existing core technolo-

[2]Michael Roberts, "The Military Revolution, 1560–1660," reprinted in Clifford J. Rogers, ed., *The Military Revolution Debate* (Boulder, CO: Westview, 1995), pp. 13–35.

gies constituting the foundation for a fundamentally new way of war. These technologies include precision-guided munitions, surveillance satellites, and remote sensing (many of which were available in the 1970s) combined with advances in the speed, memory capacity, and networking capabilities of computers.[3] The significant U.S. edge in information technology, proponents of the RMA argue, can help us retain our military superiority for a significant period, allowing us to promote one of our key national objectives: "shaping" the international system.[4]

For supporters of further investment in the RMA, the revolution is an inevitable outgrowth of fundamental societal, economic, and political changes marking the information age.[5] Technically feasible, the RMA is the best way for the United States to maintain its leadership in international politics. Champions believe that the policy implications of an impending RMA are clear: the United States must take the lead or suffer the consequences. The technological "building blocks" of an RMA exist; this is not a U.S. choice. If the United States continues to invest in "legacy systems," others will leap ahead. The most benign consequence will be the need to "catch up" in circumstances of someone else's choosing. The United States should therefore take advantage of "a time of relative peace and reduced threats by radically changing the US military to capitalize on revolutionary technological advances and thereby be better prepared for the conflicts of the future—and within current spending levels."[6] Among the advantages of this approach are greater efficiency and flexibility achieved with fewer but more highly skilled troops, and the use of "smart" and automated technologies that will place fewer lives at risk.

Proponents marshal a variety of arguments to support their view that pursuing the RMA will sustain America's military lead as well as its political influence. Other states will tend to join with, rather than balance against, the United

[3] Equally important advances in genetic engineering and the biological sciences may revolutionize biological warfare, an area that had been a low priority until the anthrax incidents during the fall of 2001.

[4] Joseph S. Nye and William A. Owens, "America's Information Edge," *Foreign Affairs* 75, no. 2 (Mar.–Apr. 1996): 20–36.

[5] Alvin Toffler and Heidi Toffler, *War and Anti-War: Survival at the Dawn of the 21st Century* (Boston: Little, Brown, 1993); Andrew F. Krepinevich, "Cavalry to Computer: The Pattern of Military Revolutions," *National Interest* (fall 1994): 30–42; Eliot A. Cohen, "A Revolution in Warfare," *Foreign Affairs* 75, no. 2 (Mar.–Apr. 1996): 37–54; Nye and Owens, "America's Information Edge"; John Arquilla and David Ronfeldt, "Cyberwar Is Coming!" *Comparative Strategy* 12, no. 2 (1993): 141–65; Dan Goure, "Is There a Military-Technical Revolution in America's Future?" *Washington Quarterly* 16, no. 4 (autumn 1993): 175–91; *National Defense Panel, Transforming Defense Report: National Security in the 21st Century*, Report of the National Defense Panel (Dec. 1997), pp. 57–86.

[6] Council on Foreign Relations, *Future Visions for U.S. Defense Policy: Four Alternatives Presented as Presidential Speeches*, A Council Policy Initiative, John Hillen, Project Director (New York: Council on Foreign Relations, 1998), p. 35.

States—a benign hegemon. The spread of democratic institutions and norms has created a growing zone of peace that dampens international competition. Furthermore, organizational research has shown that assimilating new models from abroad is not easy, so it may be difficult for other countries to adopt IT-RMA innovations.[7]

Skeptics of the RMA challenge some of the proponents' key assumptions. They find little empirical evidence of a stable "third wave" economy. For them, the promise of technology assumed by RMA champions is exaggerated,[8] and translating uncontested military superiority into political influence is not automatic.[9] Aggressive transformation may in fact undermine U.S. power and influence. The U.S. policy of "enlargement" pursued via "shaping" is not viewed as a benign policy by more than a few states, to say nothing of nonstate actors and terrorist organizations. The fact that it is the U.S. Department of Defense seeking to "shape" the international environment leads some to claim that U.S. intentions are not benign.

Opponents of transforming the U.S. military are also quick to point out that U.S. technological superiority has its drawbacks. The further we pull ahead, the more difficult it becomes for us to coordinate our military efforts with those of our allies, because our equipment is not compatible with theirs. The post–Cold War situation increasingly has required U.S. forces to engage in peacemaking and peacekeeping operations with our allies and other friendly nations in multinational formations.

Operations in the Kosovo conflict revealed a significant gap between U.S. and European allied military capabilities. Interoperability problems have led to friction over sharing defense burdens. In more than a dozen interviews, a *Washington Post* reporter found evidence that the success of the air campaign "was tempered at NATO headquarters by the stark realization that Europe has fallen so far behind the United States in the use of precision-guided weapons, satellite recon-

---

[7]D. Eleanor Westney, *Imitation and Innovation: The Transfer of Western Organizational Patterns to Meiji Japan* (Cambridge: Harvard University Press, 1987); Everett M. Rogers, *Diffusion of Innovations*, 4th ed. (New York: Free Press, 1995).

[8]Michael O'Hanlon, "Can High Technology Bring U.S. Troops Home?" *Foreign Policy* 113 (winter 1998–99): 72–86.

[9]A. J. Bacevich, "Preserving the Well-Bred Horse," *National Interest* (fall 1994): 43–49; Stephen Biddle, "Assessing Theories of Future Warfare," *Security Studies* 8, no. 1 (autumn 1998): 1–74; Brian R. Sullivan, "What Distinguishes a Revolution in Military Affairs from a Military-Technical Revolution?" Paper presented at the Joint Center for International and Security Studies-*Security Studies* Conference on the Revolution in Military Affairs, Monterey, CA (26–29 Aug. 1996); Alex Roland, "Comparing Military Revolutions," Paper presented at the Joint Center for International and Security Studies-*Security Studies* Conference on the Revolution in Military Affairs, Monterey, CA (26–29 Aug. 1996); Colin S. Gray, "The Changing Nature of Warfare?" *Naval War College Review* 49, no. 2 (spring 1996): 7–22; Paul F. Herman, Jr., "The Military-Technical Revolution," *Defense Analysis* 10, no. 1 (Apr. 1994): 91–95.

naissance and other modern technologies that the allies are no longer equipped to fight the same way."[10] To date, the United States has been less than successful in transferring military innovations to allies and other countries undertaking combined missions. European and Canadian NATO members have little political will to adopt military innovations when defense spending is not only inadequate but declining. While ministries of defense are eager to obtain these capabilities, national governments refuse to fund them, as illustrated by the stalled "Defense Capabilities Initiative."

Lack of interoperability can create serious problems on the battlefield—with repercussions in the political realm. The difficulties of forging political consensus about how to execute military operations were all too evident during the war in Kosovo.[11] The linkages between allied technology investments and political divisiveness are not new. In the 1980s, West European governments viewed U.S. proposals for new conventional technologies as an unwelcome harbinger of U.S. disengagement,[12] while their publics marched in the streets to oppose new missiles. A persistent criticism was the U.S. tendency to make decisions first and consult later. More recently in Bosnia, the United States was perceived as "advising, and even pressuring, allies to take certain risks with their forces that it is not prepared to take with its own."[13] Freedman continues, "Given the centralizing nature of the 'system of systems,' there could be concern that Washington was in effective charge, even when its own liabilities in a situation were strictly limited. Nor would the allies relish the role of 'spear-carriers,' helping to create the appearance of a coalition to demonstrate that the US is acting on behalf of more than unilateral interest, yet deemed inadequate when it comes to participating in the most technologically demanding roles."[14]

Skeptics doubt that any decisive advantage can be realized. They contend that military leaders rarely sustain their lead without a challenge; a leader's very existence invites challengers.[15] An underlying assumption of some skeptics is that challengers arise because innovative military practices spread easily. Competition creates a powerful incentive for states to emulate the military practices of the most successful states in the system. States, like firms, "emulate successful inno-

---

[10]William Drozdiak, "In the Balkans, a Lopsided Division of Labor," *Washington Post National Weekly Edition* (5 July 1999): 16.

[11]Dana Priest, "A Decisive Battle that Never Was," *Washington Post* (19 Sept. 1999): A01; Dana Priest, "Bombing by Committee," *Washington Post* (20 Sept. 1999): A01; Dana Priest, "The Battle inside Headquarters," *Washington Post* (21 Sept. 1999): A01.

[12]Lawrence Freedman, *The Revolution in Strategic Affairs*. Adelphi Paper 318 (New York: Oxford University Press, 1998), p. 24.

[13]Ibid., p. 72.

[14]Ibid.

[15]Christopher Layne, "The Unipolar Illusion: Why New Great Powers Will Rise," *International Security* 17, no. 4 (spring 1993): 5–51.

vations of others out of fear of the disadvantages that arise from being less competitively organized and equipped. These disadvantages are particularly dangerous where military capabilities are concerned, and so improvements in military organizations and technology are quickly imitated."[16] In this respect, skeptics share the neorealist perspective in which the process of diffusion is seen as uncomplicated, and new technologies are readily acquired and unproblematically integrated into existing structures and practices.

RMA proponents and skeptics alike assume that innovations will spread. They differ in their assessment of the ease and speed with which this will occur. Proponents doubt that others can easily emulate the U.S. information technology–based military model in the foreseeable future, and so the United States will be able to enjoy a considerable advantage. Skeptics counter that military superiority has never remained uncontested for long. The United States is only encouraging competitors to challenge our lead, accelerating the pace at which the technological gap will close.

To understand the extent to which the United States is likely to be able to maintain its military lead, practitioners need to examine how the military knowledge and practices associated with the current RMA—technologies, doctrine, organizational forms, and behavioral practices—are likely to spread. Are all military organizations equal in their desire and ability to assimilate new technologies and forms? Will the United States have adequate time and resources to anticipate and adapt to these changes? Moreover, while U.S. leaders no doubt want to prevent or slow diffusion to potential adversaries, the need to ensure interoperability requires finding ways to encourage diffusion of innovations to our allies. How leading-edge military technologies, forms, and practices do—and do not—spread are questions of central importance to U.S. defense policy at the turn of the millennium.

The Policy Impact of Diffusion Research

By all accounts, the RMA debate is more than strictly academic: its outcome will have lasting consequences for U.S. budgets and military readiness, for R&D investment and U.S. leadership in international affairs. Recent events—both before and after September 11—emphasize how important controlling the diffusion of innovation is to national security. The Cox Report details a wide spectrum of Chinese government–sponsored activities to acquire U.S. technology with military applications in high-performance computing, missile and satellite technology, and

---

[16]Joao Resende-Santos, "Anarchy and Emulation of Military Systems: Military Organizations and Technology in South America, 1870–1930," *Security Studies* 5, no. 3 (spring 1996): 196.

thermonuclear warhead design.[17] Neither internal security at U.S. weapons laboratories nor export-licensing regulations have prevented the Chinese from acquiring sensitive security information and technologies. Advocates of "dual-use" technologies once favored their potential to spur economic growth via commercial applications that reach beyond their original government-contracted purpose. But demonstrated difficulties in controlling their spread to foreign competitors, both commercial and military, have dampened their enthusiasm. A better understanding of the dynamics of the diffusion process will inform policy choices in a wide range of domains, including not only defense and security but also trade, technology transfer, and cross-national collaboration in research and development.

Despite the vital concerns at stake, academics are just beginning to investigate the process of diffusion: how military knowledge, broadly defined to include hardware (e.g., technology) and software (e.g., doctrine, tactics, organizational form, etc.), diffuses throughout the international system, or what factors enhance or inhibit incorporating innovations into defense structures. The history of warfare has been marked by periods defined by certain innovations. Ross, Bracken, and others emphasize the importance of military innovations. How they diffuse can restructure power relations in the international system.[18] Despite the large body of scholarship on military innovation,[19] remarkably few studies explore either historical or contemporary processes of diffusion of military innovations.[20]

Part of the answer to the mystery about why international relations specialists have failed to take up the study of the process of diffusion of military innovation

[17]HR 105–851, Report of the Select Committee on U.S. National Security and Military/Commercial Concerns with the People's Republic of China.

[18]Andrew L. Ross, "The Dynamics of Military Technology," in David Dewitt, David Haglund, and John Kirton, eds., *Building a New Global Order: Emerging Trends in International Security* (New York: Oxford University Press, 1993); Paul Bracken, "Non-Standard Models of the Diffusion of Military Technologies," *Defense Analysis* 14, no. 2 (1998): 101–14.

[19]Barry R. Posen, *The Sources of Military Doctrine: France, Britain, and Germany between the World Wars* (Ithaca: Cornell University Press, 1984); Stephen Peter Rosen, *Winning the Next War: Innovation and the Modern Military* (Ithaca: Cornell University Press, 1991); Kimberly Martin Zisk, *Engaging the Enemy: Organization Theory and Soviet Military Innovation, 1955–1991* (Princeton: Princeton University Press, 1993); Leonard A. Humphreys, *The Way of the Heavenly Sword: The Japanese Army in the 1920s* (Stanford: Stanford University Press, 1995); Williamson Murray and Allan R. Millett, eds., *Military Innovation in the Interwar Period* (New York: Cambridge University Press, 1996); Emily O. Goldman, "The U.S. Military in Uncertain Times: Organizations, Ambiguity, and Strategic Adjustment," *Journal of Strategic Studies* 20, no. 2 (spring 1997): 31–74.

[20]Interestingly, Rogers's extensive bibliography drawing on the previous three editions of this work (dating back to 1962) includes very few of the contributions from political science or policy studies on the diffusion of innovations. Nevertheless, he makes the claim that the "trend toward a more unified cross-disciplinary viewpoint in diffusion research continues today; every diffusion scholar is fully aware of the parallel methodologies and results in other traditions" (Rogers, *Diffusion of Innovation*, p. 39).

lies in the implicit realist and neorealist assumptions that dominate current research. Neorealist theory holds that competition among states inevitably causes
pioneering military methods to diffuse rapidly among states. When militaries
confront new weapons and practices on the battlefield, they emulate them. Com-
petitor nations observe the successes and failures of other states and act accordingly, drawing the hard-learned lessons of others. The inherently competitive
nature of international politics, therefore, leads to the rapid spread of the most
successful organizational forms, practices, and technologies. As Waltz puts it,
"The possibility that conflict will be conducted by force leads to competition in
the arts and instruments of force. Competition produces a tendency toward the
sameness of competitors."[21] In this view, diffusion is a uniform and efficient
process driven by the threat of defeat by a superior power.

A look at the historical record reveals far more variation in adoption and
emulation across states and cultures than conventional international relations
theory assumes. The process of diffusion appears far less deterministic and much
more vulnerable to local conditions than the systemic view suggests. For example, Gustavian tactical systems spread relatively quickly across Europe and into
Russia. But it took nearly a century of military disaster before the Ottomans
adopted modern European training methods. Asian regimes lagged well behind
their European counterparts in making the market-oriented transformation crucial to the industrial expansion that helped stimulate the development of highly
effective armed forces in Europe. Chinese commercial behavior and their particular approach to the pursuit of wealth operated within limits defined by political authorities educated in Confucian traditions hostile to the ethos of the
(Western) marketplace. Why did Manchu China and nineteenth-century Ottoman Turkey fail to emulate superior Western military practices, while Meiji Japan made the transition? Why did Mongol practices, used so successfully in the
thirteenth century to dominate the largest geographical area before or since, fail
to spread to European armies? These puzzles demonstrate the contingent nature
of the diffusion process and suggest the need to search for factors that explain the
remarkably wide range of responses to innovation across societies, organizations,
cultures, contexts, and historical epochs.

Technologies and innovations have two important facets: "hardware" and
"software." Hardware refers to the artifacts, or *techne*, involved, while software is
used to describe the organizational or human application component of an innovation or technology. New inventions can be put to use in various ways and often
lead to changes in human behavior as their advantages become clear through use.

---

[21]Kenneth N. Waltz, *Theory of International Politics* (Reading, MA: Addison-Wesley, 1979),
p. 128.

This vital distinction points to the fundamental issue of the organizational, cultural, and societal basis for the introduction, application, and institutionalization of new technologies and practices.

One clue in the search for explanations of the variation in responses to innovation lies in the fact that new technologies do not exist in a cultural or organizational vacuum. They are not neutral instruments utilized uniformly anywhere, anytime, by anyone. Many of the case studies presented in this volume demonstrate that military innovations requiring significant changes in sociocultural values and behavioral patterns spread more slowly, less uniformly, and with more unpredictable outcomes. The rate of adoption may depend on how compatible the innovation is with existing values and practices, as well as past experience and current needs of the adopting state, society, or organization.

Furthermore, states may use innovations in novel ways. The utilization aspect of diffusion—whether and how an innovation is integrated into an acquiring state's organizational structures—is of central importance to contemporary defense practitioners. A conference sponsored by the Office of the Secretary of Defense/Net Assessment took as its point of departure the observation that "[often] during these periods of RMAs there is a realignment of power based on which nation or nations can best adapt new technologies to military uses."[22] Conference participants sought to identify criteria that could be applied to nations to determine their ability to successfully assimilate and implement key technologies associated with the current RMA. In particular, conference participants repeatedly pointed to the importance of organizational and cultural factors in inhibiting and promoting exploitation of military technologies, while observing that relatively little research on the diffusion of military innovations has systematically explored these dimensions. Studying the diffusion of military innovations involves examining not only how states and nonstate actors interact to acquire new ideas, practices, and hardware but also how they adapt and utilize new knowledge.

A repeated theme in discussions of the diffusion of innovation is the need for more empirical studies using qualitative approaches that can capture the subtleties of the processes involved in the spread of innovations. This approach might reveal how different contexts shape—and in turn are reshaped by—the introduction of new practices and technologies adopted or adapted from other places. The qualitative case studies of the diffusion of military innovations presented in this volume extend our theoretical understanding of various aspects of the diffusion process and also are profoundly important for current defense and foreign policy decisions.

[22]Ron St. Martin and Linda McCabe, *Final Report: Implications of Culture and History on Military Development* (McLean, VA: SAIC, Prepared for OSD/Net Assessment, 1996), p. i.

Goals of the Study

This volume seeks to remedy gaps in diffusion research by bringing together scholarship from a variety of disciplines—military history, strategic studies, political science, sociology, public policy, and international relations. The chapters encompass historical as well as contemporary cases of the diffusion of important military innovations. Early in the collaboration, we recognized that such a comparative study would require attention not only to characteristics of the new technologies and practices but also to the organizational, cultural, societal, and political contexts required to leverage the new technologies. Given the crucial importance of this subject to current policy debates, we also wanted to make our research useful both to scholars and (perhaps more importantly) to the policy community. Our goals therefore are twofold: to generate a set of hypotheses to guide future research on the diffusion of military innovations, and to provide insights useful to policy-makers during a period of military transformation.

Military innovations have been studied before. Missiles and weapons of mass destruction (WMD)—particularly nuclear weapons—are the most significant technologies to have been the subject of intense analysis in order to manage the diffusion process. Nonproliferation efforts provide useful insights for understanding the dynamics of the spread of the cluster of technologies and practices that undergird the current RMA. Inhibiting proliferation of WMD has involved negotiations among states of widely varying economic and conventional military power. Nonproliferation regimes have also addressed concerns about the spread of these technologies to nonstate actors. No study of the diffusion of military innovations would be complete without an examination of the WMD case, but conversely the study of diffusion of innovation must go beyond the special case of WMD proliferation. Accordingly, several of our case studies examine important innovations that have already been extensively analyzed (e.g., Napoleonic and Prussian warfare, carrier power, armored warfare). Other chapters explore less celebrated innovations or the diffusion of certain practices to less studied places (e.g., new technologies developed in the periphery, or Soviet battle techniques adopted by Arab states). Additional chapters focus on what enhances or impedes diffusion to friends and allies (e.g., the sepoys of India, cooperation among the post–World War II Anglo-Saxon nations). We also include several case studies of direct relevance to the IT-RMA.

The remainder of this introductory chapter has three objectives: a literature review, a presentation of our methodology, and a brief chapter overview. First we review the wide-ranging research literature on diffusion. Examining the current state of research on diffusion of innovation in other fields provides working hypotheses and conceptual clarification. Many of the debates in the diffusion literature informed the way we structured this study. They also influenced the questions that case study authors were asked to answer. The case studies address key debates in the diffusion literature. The literature review that follows is intended for the academic audience interested in this volume's contribution to theory development. We have aimed at producing a volume that speaks to both practitioners and scholars. Those with more practical concerns may wish to proceed directly to the overview of the section on methodology and the overview of the themes and cases covered in the book's four main sections.

Diffusion of Innovation Research

Three key debates emerge from diffusion research. The first debate concerns how one defines the diffusion process, which is critical for identifying whether or not diffusion has occurred. The key question here is whether the communication of information is sufficient to conclude that diffusion has taken place. How do we rule out independent discovery of an innovation? The second debate concerns the causes of diffusion. What motivates states to adopt innovations from abroad, and what is the mechanism by which knowledge is transferred? While scholars advance various typologies, three distinct processes—competition, socialization, and coercion—drive the spread of policies across societies, with different implications for what is modeled. The third debate concerns the patterns and effects of diffusion. Existing research supports our view that diffusion is a contingent process shaped by historical, institutional, and cultural factors.

What Is Diffusion? Access to Information vs. Adoption and Utilization

Webster's dictionary defines *diffusion* as "the spread of cultural elements from one area or group of people to others by contact."[23] A common definition therefore implies that an idea, thing, or practice is transmitted from one social group to another and that some kind of interaction must occur between these groups in order to constitute a process of diffusion. Much of the early research on the diffusion of innovation emphasized the transmission of information about a new practice or technology. This research was then challenged by researchers focusing on how receiving states adopt and utilize the knowledge that was transmitted. Empirical research dating back to the 1960s analyzed diffusion as a process of transnational communication.[24] Gray defined diffusion as "the communication of

[23]*Merriam Webster's Collegiate Dictionary*, 10th ed. (Springfield, MA: Merriam Webster, 1994), p. 323.
[24]Karl W. Deutsch, *The Nerves of Government* (New York: The Free Press, 1963).

a new idea in a social system over time."[25] Rogers defined diffusion as "the proc-esses by which an innovation is communicated through certain channels over time among the members of a social system."[26] Many of the basic arguments in diffusion analysis were offsprings of Torsten Hägerstrand's spatial diffusion the-ory, which isolated a single aspect of the underlying process: the communication of information about an innovation.[27]

More recently, Bennett in a series of articles distinguishes communication of information from adoption of an innovation.[28] He argues that "the words 'learn-ing,' 'diffusion,' 'emulation,' and 'lesson-drawing' have all appeared in the litera-ture to describe virtually the same phenomenon."[29] But a pattern of successive adoptions of a policy innovation does not mean that later adopters are necessarily using information from early adopters, that policy adoption in one place is at-tributable to similar actions elsewhere. Empirical evidence must demonstrate conscious copying, lesson-drawing, or adaptation. Otherwise, "there is no way to distinguish genuine learning from convergence determined by shared macro-level social and economic characteristics."[30]

To confirm the emulation hypothesis, Bennett argues, "requires the satisfac-tion of a number of conditions: a clear exemplar (a state that has adopted an in-novative stance); evidence of awareness and utilization of policy evidence from that exemplar; and a similarity in the goals, content or instruments of public policy."[31] Bennett distinguishes between "*knowledge* of a foreign program, *utiliza-tion* of that knowledge, and the *adoption* of the same program."[32] Awareness, utilization, and adoption are conceptually distinct. Wilensky and Turner also emphasize that just observing that diffusion has occurred "does not specify the path of causality running from the appearance of an idea or policy proposal to its adoption and implementation. Much intervenes between awareness and action, and diffusion accounts only for awareness."[33]

[25]Virginia Gray, "Innovation in the States: A Diffusion Study," *American Political Science Review* 67, no. 4 (Dec. 1973): 1175.

[26]Rogers, *The Military Revolution Debate*, p. 5.

[27]James M. Blaut, "Two Views of Diffusion," *Annals of the Association of American Geog-raphers* 67, no. 3 (Sept. 1997): 343.

[28]Colin J. Bennett, "How States Utilize Foreign Evidence," *Journal of Public Policy* 11, no. 1 (Jan.–Mar. 1991): 31–54; Colin J. Bennett, "Review Article: What Is Policy Convergence and What Causes It?" *British Journal of Political Science* 21, no. 2 (1991): 215–33; Colin J. Bennett, "Understanding Ripple Effects: The Cross-National Adoption of Policy Instruments for Bureau-cratic Accountability," *Governance: An International Journal of Policy and Administration* 10, no. 3 (July 1997): 213–33.

[29]Bennett, "How States Utilize Foreign Evidence," p. 32.

[30]Ibid.

[31]Bennett, "Review Article," p. 223.

[32]Bennett, "How States Utilize Foreign Evidence," pp. 32–33.

[33]Harold L. Wilensky and Lowell Turner, *Democratic Corporatism and Policy Linkages: The*

In the early literature, diffusion analysis focused on awareness, with little attention to the forces intervening between awareness and action. Interaction among policy officials and professionals across jurisdictions—including nation states—has grown steadily over the last century, constituting one of the primary modes of communicating policy innovation and "best practices." Several studies have shown that the adoption of a policy in one country influenced policy leaders in other countries to consider similar policies or similar policy goals.[34] But just as the transmission of sound requires both that sound waves be sent and that they be received, for diffusion to occur the receiving state must actually use the knowledge in some way, even if it is with significant modifications.

Research in the fields of anthropology, sociology, and business product cycle focused on the rate of adoption of innovations. The early identification of the S-curve pattern across a wide range of contexts and innovations was widely confirmed: early adoption progressed slowly, accelerated rapidly, and then tapered off toward the end of a diffusion cycle. Diffusion studies in political science replicated these findings, drawing in particular on the spread of new policies defined as the adoption of legislation with similar attributes or goals—for example, state lotteries,[35] tax policies, social security legislation,[36] and so forth. Quantitative modeling allowed researchers to assess the rate of diffusion of one or more innovations throughout a system. This analytical trend was fueled by advances in econometrics and data processing.

Robertson criticizes the largely quantitative modeling of diffusion.[37] Statistical analyses "necessarily define away many critical but hard-to-measure political variables, and often conclude that economic wealth[38] or physical contiguity[39] are the most important determinants of when a lesson will be drawn. Moreover, these studies do not usually examine whether and under what conditions a lesson will be drawn, but when it will be drawn; their dependent variables usually as-

Interdependence of Industrial, Labor-Market, Incomes, and Social Policies in Eight Countries (Berkeley: Institute of International Studies, University of California, 1987), p. 391.

[34]George Hoberg, "Sleeping with an Elephant: The American Influence on Canadian Environmental Regulation," *Journal of Public Policy* 11, no. 1 (Jan.–Mar. 1991): 107–32; Steven Kelman, *Regulating America, Regulating Sweden: A Comparative Study of Occupational and Health Policy* (Cambridge, MA: MIT Press, 1981).

[35]Frances Stokes Berry and William D. Berry, "State Lottery Adoptions as Policy Innovations: An Event History Analysis," *American Political Science Review* 84, no. 2 (June): 394–415.

[36]David Collier and Richard E. Messick, "Prerequisites versus Diffusion: Testing Alternative Explanations of Social Security Adoption," *American Political Science Review* 69, no. 4 (Dec. 1975): 1299–1315.

[37]David Brian Robertson, "Political Conflict and Lesson-Drawing," *Journal of Public Policy* 11, no. 1 (Jan.–Mar. 1991): 63.

[38]Harold L. Wilensky et al., *Comparative Social Policy: Theories, Methods, Findings* (Berkeley: University of California, 1985), p. 14.

[39]Collier and Messick, "Prerequisites Versus Diffusion," pp. 1299–1315.

sume that all the polities will eventually adopt the policy."[40] Robertson emphasizes the importance of political constraints and opportunities in determining the outcomes of lesson-drawing: "Systemic variables such as political culture, government constitutions, socioeconomic circumstances and the policy inheritance potentially obstruct policy emulation. Policymakers are likely to reject out of hand a program that is perceived to be normatively unacceptable to their citizens' shared norms and customs even if the program performs effectively elsewhere."[41] Indeed, Robertson's observation points to the fundamental importance of culture.

State security and power will be affected not only by the ability to acquire information about innovations observed elsewhere or encountered on the battlefield but also by how successfully the innovation can be incorporated into existing organizations, institutions, and practices. Bracken has pointed to important changes in the channels through which states such as Iraq or India can acquire (and have acquired) new military technologies, including the role of multinational corporations and new acquisition strategies.[42] Thus the cultural contact can be mediated by other individuals or organizations. But once it is acquired, can these states integrate and utilize new technologies and innovations?

Causes and Media of Diffusion

What causes diffusion and by what means do ideas and practices move from one culture or society to another? As discussed above, diffusion can be distinguished from the initiation of a new practice that is the by-product of underlying technological or socioeconomic change. States or organizations facing similar environmental circumstances may independently adopt similar policy responses. It is also possible that some combination of these two sources of innovation are operating in the same case. Major structural forces, such as industrialization, bring change wherever they operate, but one society might adapt first or more effectively and thus become a model for emulation by others. A process of trial and error may involve both independently developed policies along with examples drawn from other societies. A qualitative case study approach to the innovation process can detail these various attempts and allow for more subtle distinctions over an extended time period.

Several studies have sought to specify the transmission process. The causes of diffusion (or the forces that drive the spread of new knowledge) are sometimes taken together with the mechanisms for that spread (transmission belts), imply-

---

[40]Robertson, "Political Conflict and Lesson-Drawing," p. 68.
[41]Ibid.
[42]Bracken, "Non-Standard Models of the Diffusion of Military Technologies," pp. 101–14.

ing that the existence of a medium of transmission itself causes diffusion to occur. The distinction is important. It is conceivable that sufficiently strong causal forces or drivers of diffusion can lead to the development of new transmission modalities, especially in an era of rapidly advancing global communication. By distinguishing means from causes, practitioners may be able to identify which states or nonstate actors may be motivated to actively scan their environment for new practices (e.g., those facing a highly competitive environment; those aspiring to gain international respect or stature) looking for innovations to adopt. Analysis of the transmission paths may aid practitioners in selecting efficient media for encouraging diffusion while anticipating the means of transmission that must be safeguarded to prevent undesired diffusion.

Ikenberry's study of the spread of privatization policies focuses on the same questions guiding this study: "What are the patterns of diffusion? Through what processes does the spread of innovation occur? What variables influence the likelihood that the innovation will be adopted in secondary countries?"[43] He identifies three "types" of diffusion: external inducement, emulation or "policy bandwagoning," and social learning. External inducement occurs when direct external pressure, as benign as the loose structuring of incentives and inducements or as severe as overt coercion, is applied. Here the innovation is brought to the receiving state by an external actor, rather than the receiving state's having identified the innovation on its own.

With emulation, or policy bandwagoning, "elites monitor policy change abroad and, seeking similar successes, import the appropriate policies."[44] The competitive nature of the international system promotes emulation "for the brutal reason that a state which did not advance its economy would not be able to pay for sufficient military might to survive in the competitive international arena."[45] The competitive nature of international politics promotes successful diffusion: "It appears that emulation will be most pervasive when international competition is most intense, inasmuch as competition provides a powerful incentive to monitor and respond to innovations developed abroad. In cases of intense competition, the costs of failing to innovate are highest. Moreover, the diffusion of policy practices is most likely to occur when states share underlying cultural or social settings (although a highly competitive environment might serve to transcend obstacles presented by divergent cultural or societal circum-

[43]G. John Ikenberry, "The International Spread of Privatization Policies: Inducements, Learning, and Policy Bandwagoning," in Ezra Suleiman and John Waterbury, eds., *The Political Economy of Public Sector Reform and Privatization* (Boulder, CO: Westview, 1990), p. 99.

[44]Ibid., p. 101.

[45]John A. Hall, *Liberalism: Politics, Ideology and the Market* (London: Paladin, 1988), p. 24, quoted in Ikenberry, "The International Spread of Privatization," p. 101.

stances)."[46] In this type of diffusion, the receiving state is actively engaged in scanning the environment to identify possible innovations to adopt.

Ikenberry reserves the term "social learning" for a particular kind of diffusion in which policy-relevant knowledge about the way the world works is transmitted as "consensual knowledge" and spreads among specialists located in "the policy-making apparatus of government."[47] Keynesian economic management ideas are a key demonstration of social learning. As professionals and experts in the field of economics became incorporated within the political decision-making system of government, new ideas emerging within the professional discourse were trans-mitted to a wide range of governments. "That is, politicians and bureaucrats are attempting to solve problems and therefore are open, at particular historical moments, to new information. Those moments may be provoked by crisis: old reigning ideas are discredited. Or those moments may come with a change in re-gime (itself provoked, perhaps, by crisis). In these various ways, a new conven-tional wisdom emerges," giving rise to an international policy culture.[48] A defin-ing characteristic of the knowledge elite is their professional attentiveness to new developments in their field. Indeed, professional reputations are built on the ability to discover new approaches and to enlist the support and approval of other experts in the field.

These various types of diffusion may be very difficult to distinguish in prac-tice. Ikenberry concludes that the spread of Keynesianism and privatization must be understood with reference to all three models of diffusion: "Power matters. . . . Likewise, the adoption of the preferred policies of the dominant nation will be greatly facilitated when those policies come to be associated with economic and political success. Finally, the presence of a coherent body of theoretical knowl-edge . . . will help propel the spread of policy."[49] As he concedes, his analysis has offered some analytical tools but is no substitute for detailed case studies, pre-sumably because only richer, more contextualized empirical research can accu-rately determine which types of diffusion are most influential.

Bennett identifies four causes of "policy convergence": emulation, elite net-working, harmonization, and penetration. Emulation requires only the utilization of policies from abroad that "might serve as a blueprint that pushes a general idea on the political agenda."[50] Elite networking and policy communities result from interaction and consensus among elites. Policy communities, issue networks,[51]

---

[46]Ikenberry, "The International Spread of Privatization," p. 103.
[47]Ibid., p. 104.
[48]Ibid., p. 105.
[49]Ibid., p. 106.
[50]Bennett, "Review Article," p. 221.
[51]Hugh Heclo, "Issue Networks and the Executive Establishment," in Anthony King, ed., *The New American Political System* (Washington, DC: American Enterprise Institute, 1978).

epistemic communities,[52] and transnational networks[53] all capture this basic medium for the transmission of innovations. Research in the area of environmental protection has identified this modality as particularly important in the diffusion of new practices throughout the international system, leading to the adoption of new international agreements and treaties to protect natural resources. Through professional organizations, journals, meetings, and other forms of communication, experts and practitioners develop a common set of understandings and an established body of knowledge about how to solve particular problems and even what should be defined as an appropriate problem to be solved. New technologies introduced into these communities can contribute to redefining the problem, as well as changing the way the problem is addressed.[54]

When consensual knowledge becomes embedded in intergovernmental organizations and supranational institutions, convergence results from explicit attempts to harmonize policies by "a coherent group of transnational actors" who have "a broad consonance of motivation and concern and regular opportunities for interaction."[55] Harmonization also requires "authoritative action by responsible intergovernmental organizations. Under this process, convergence is driven by a recognition of *interdependence*, a vague notion signifying a reliance on others for the performance of specific tasks to ensure complete and successful implementation or to avoid troubling inconsistencies."[56]

The final convergence cause identified by Bennett is penetration. This occurs when one state is forced to adopt an innovation by external actors, either a colonial power or, more recently, international organizations like the IMF or international actors such as multinational corporations. The risk of an external actor's withholding investment capital to aid a troubled economy or loans to support a weak currency usually induces a weak government of a poor state to adopt new policies. In states transitioning to a market economy, the models and advice of external actors (international organizations, aid agencies, and others) can be a complex mixture of enticement, inducement, and social learning.

DiMaggio and Powell examine policy convergence, specifically convergence in organizational form. They call the process of convergence "isomorphism" and

[52]Peter M. Haas, "Introduction: Epistemic Communities and International Policy Coordination," *International Organization* 46, no. 1 (1992): 1–35; Peter M. Haas, Robert O. Keohane, and Marc A. Levy, eds., *Institutions for the Earth: Sources of Effective International Environmental Protection* (Cambridge: MIT Press, 1993).

[53]Robert O. Keohane and Joseph S. Nye, *Power and Interdependence: World Politics in Transition* (Boston: Little, Brown, 1977).

[54]Karen Litfin, *Ozone Discourses: Science and Politics in Global Environmental Cooperation* (New York: Columbia University Press, 1994).

[55]Bennett, "Review Article," p. 225.

[56]Ibid.

identify two types: competitive and institutional. Competitive isomorphism occurs as organizations strive to achieve efficiency, and it explains the early spread of organizational models. The later spread of a model—that is, beyond the point at which the model can be attributed to efficiency demands—results from institutional pressures: coercive, mimetic, or normative. Wider pressures in the community—for example, common legal standards, centralization of capital—may coerce organizations to conform to one model. Organizations may opt to imitate or borrow practices from another organization to cope with uncertainty. Convergence may also result from the adoption of professional standards. Normative pressures to adopt new models and policies are transmitted by educational and professional networks.[57]

The typologies advanced by DiMaggio and Powell, Bennett, and Ikenberry to identify the causal mechanisms driving diffusion are similar. Bennett's emulation resembles DiMaggio and Powell's mimetic processes and Ikenberry's policy bandwagoning. Elite networking and harmonization resemble normative pressures and social learning; penetration is similar to coercive pressures and external inducement. All concur that competition, socialization, and coercion are three social mechanisms that help spread new policy practices across societal boundaries.

Patterns and Effects of Diffusion

Is it possible to discern patterns emerging from the diffusion of technological and organizational innovations that occur over time and space? This analytical approach would include patterns in the way an innovation spreads (e.g., spatially, hierarchically) and the extent to which innovations are faithfully adopted (fully emulated, selectively emulated, or adapted).

The conventional wisdom in international relations research supports a hierarchical view of the diffusion process in which "innovations appear in the most advanced or largest centers and are then adopted by successively less advanced or smaller units."[58] Collier and Messick distinguish this from spatial diffusion in which "diffusion [occurs] along lines of spatial proximity or, alternatively, along major lines of communication."[59] This is, however, an empirical question, and the record of the diffusion of innovations in military affairs should allow us to test the validity and conditionality of the hierarchical model. For policy-makers, working with a set of assumptions based on the hierarchical model might be particularly dangerous, especially if it turns out that innovations emanating from

[57]Paul J. DiMaggio and Walter W. Powell, "The Iron Cage Revisited: Institutional Isomorphism and Collective Rationality in Organizational Fields," *American Sociological Review* 48 (Apr. 1983): 147–60.
[58]Collier and Messick, "Prerequisites Versus Diffusion," p. 1306.
[59]Ibid.

places other than the core start to diffuse in ways that challenge peace and security but go undetected too long because they did not start in the core. Failing to look for innovations that originate and spread from places other than the richest or most powerful nations creates a serious blind spot in defense planning.

Spatial theories of diffusion patterns predict that innovations will diffuse relatively quickly among geographically proximate states or states with an established sense of "regional" identification created by similar political cultures. As Foster puts it: "Adjoining states may have similar political traditions and climates, but quite different levels of industrialization and urbanization."[60] This might account for why states lacking "prerequisite" levels of other factors nevertheless adopt an innovation found in a nearby (or culturally similar) state. Foster finds that "a portion of the variance in American state innovation adoption rates appears to be associated with regional proximity, even after the impact of similar economic and political structures is controlled for."[61]

In American policy studies, proximity plays a key role in who adopts from whom and how quickly, but "proximity" may be something other than strict geographical closeness. States may tend to copy or adapt policies similar to those of states with which the leadership identifies. Thus "southern" states may be more likely to adopt policies that have taken hold in other southern states than they are to adopt policies in perhaps geographically closer "northern" or "border" states.[62]

Identity or cultural proximity as factors influencing the rate of diffusion echo Blaut's findings. He argues that "diffusion is an incident of culture change, and culture change is an exceedingly complex process."[63] Blaut focuses on cultural factors as barriers to diffusion, though his analysis implies that similarity in culture can contribute to diffusion (or speed up the rate of diffusion). He concludes that "technological innovations tend to be rejected, not through ignorance, but through incompatibility with the existing cultural system as a whole."[64] Hall and Ikenberry also note that the common cultural heritage of the European states may explain the ease of diffusion of policy innovations among them.[65]

[60]John L. Foster, "Regionalism and Innovation in the American States," *Journal of Politics* 40, no. 1 (Feb. 1978): 181.

[61]Ibid., p. 186.

[62]Jack L. Walker, "The Diffusion of Innovations among the American States," *American Political Science Review* 63, no. 3 (Sept. 1969): 880–99; Jack L. Walker, "Comment: Problems in Research on the Diffusion of Policy Innovations," *American Political Science Review* 67, no. 4 (Dec. 1973): 1186–91; Gray, "Innovation in the States"; Virginia Gray, "Rejoinder to 'Comment' by Jack L. Walker," *American Political Science Review* 67, no. 4 (Dec. 1973): 1192–93.

[63]Blaut, "Two Views of Diffusion," p. 343.

[64]Ibid., p. 346.

[65]John A. Hall and G. John Ikenberry, *The State* (Milton Keynes, UK: Open University Press, 1989).

In her analysis of the massive transfer of Western organizational models to Meiji Japan, Westney observes that the models under consideration came from only a few societies: Britain, France, the United States, and Germany.[66] International prestige does not entirely account for this list. Rather, adoption was linked to accessibility of information, which in turn was tied to the scale of each model nation's presence in East Asia. "The nations which were the most important sources of organizational models were those which were dominant in Japan's immediate environment. Selection tended to be a cumulative process; that is, the selection of a model from one society increased the likelihood that the same society would serve as a source for further models."[67] Westney believes that this "contagion" effect applies to other nations and other time periods. "The pressure on developing nations to emulate the patterns of the nations that are most powerful in their immediate environment is significant today. . . . Information is skewed toward those countries with which the developing nation has the closest interactions."[68] The implication is that patterns of adoption may not be based on optimal compatibility with the adopting state's environment or needs, but rather on the powerful presence of key model states.

Rose distinguishes "lesson-drawing" from innovation, emulation, or diffusion. A lesson is "an action-oriented conclusion about a programme or programmes in operation elsewhere."[69] The emphasis in Rose's analysis is on knowledge acquisition and adaptation to local circumstances. "Lessons must identify circumstances that are different as well as those that are the same."[70] The object of study in lesson-drawing is quite different from the traditional emphasis in diffusion studies: "Nearly all studies of the diffusion of measures, whether public programmes or agricultural or pharmaceutical products, assume that not only are there common problems but also a common response, regardless of partisan values or political cultures. The emphasis is upon the sequence of diffusion, rather than concentrating upon what is transferred."[71] According to Rose, "[D]iffusion studies seek to identify states or countries that are leaders and laggards in adopting programmes, and to account for the difference."[72] The focus on diffusion rates, however, did not pay adequate attention to adaptation, a process that almost always accompanies the implementation of an innovation observed elsewhere. In the same special issue of the *Journal of Public Policy* devoted to lesson-drawing, Ma-

---

[66]Westney, *Imitation and Innovation*, p. 22.
[67]Ibid., p. 21.
[68]Ibid., p. 23.
[69]Richard Rose, "What Is Lesson-Drawing?" *Journal of Public Policy* 11, no. 1 (Jan.–Mar. 1991): 7.
[70]Ibid., p. 9.
[71]Ibid.
[72]Ibid.

jone's contribution emphasizes the importance of adaptation in the diffusion of regulatory policies throughout Europe.[73]

Recent editions of Rogers's classic volume *Diffusion of Innovations* offer many examples of cultural and social barriers to the diffusion of innovation and the limits on the kind of technocratic determinism that Rose associates with the diffusion literature.[74] However, the common shortcomings of earlier, especially quantitative, models of diffusion are two assumptions: that (1) the object of diffusion is essentially similar across cases, or at least that the differences in policies, practices, or legislation (for example) are uninteresting or unimportant in understanding the rate and pattern of diffusion; and (2) the failure to adopt a particular innovation means that the state (country, society, or organization) that lags is deficient in some key factor possessed by the "leaders" or early adopters. In fact, adaptation may be a more appropriate or sophisticated response than emulation.

These biases are perhaps even more present although less recognized in the international relations literature precisely because of the Darwinesque assumption that competition leads either to efficient, rapid diffusion with faithful emulation or to defeat. To borrow Rose's phraseology, competition breeds diffusion without any "consideration of institutional constraints, normative preferences, and the inertia of established programmes."[75] Evaluation and the ability to project into the future are essential to the lesson-drawing process. "Since the applicability of a lesson is contingent, prospective evaluation is necessary to justify the conclusion that a programme that works elsewhere will or will not work here."[76] Somewhere between the view that an innovation that is technically feasible necessarily will be adopted and the opposite extreme in which historical, institutional, or cultural obstacles prevent diffusion lies the conditional and contingent process of lesson-drawing. "Lesson-drawing is not a mechanical set of deterministic procedures leading to unalterable conclusions. Obstacles to transferring programmes would be permanent only if the present differences in space could not be bridged by time. By thinking in terms of both time and space we can undertake a prospective evaluation that not only identifies the blockages that exist today, but also highlights steps that can be taken to make a programme effective in one country succeed elsewhere tomorrow."[77] Rose offers a perspective that emphasizes the importance of the comparative case study approach of this volume in which comparisons can be drawn across cultures, locations, and time.

[73]Giandomenico Majone, "Cross-National Sources of Regulatory Policymaking in Europe and the United States," *Journal of Public Policy* 11, no. 1 (Jan.–Mar. 1991): 79–106.

[74]Rogers, *Diffusion of Innovations*.

[75]Rose, "What Is Lesson-Drawing," p. 8.

[76]Ibid., p. 23. It is worth noting that this observation also applies to lesson learning *about* diffusion processes, including the cases presented in this volume.

[77]Ibid., p. 24.

We examine both historical and contemporary cases not to draw ironclad laws or simple analogies. Rather we want to understand a wide range of the possible paths that the diffusion process can take. In some instances, the attempt to emulate or copy goes badly awry, with disastrous consequences; in other places and times, the innovation is adapted—sometimes intentionally, sometimes not—to local conditions. The result may not entirely resemble the model innovation, and it may or may not yield better outcomes for the adopting state. Adaptation to local circumstances including important cultural considerations, economic and political development, and resources, in our view, combined with the pressures of not only competition but also cooperation in the international system, have generated a rich historical record of the diffusion of various military innovations.

While we can never fully anticipate or predict the innovations that may be developed in the future, or how, when, under what circumstances, or with what consequences they will diffuse, our case studies provide ample material for understanding the importance of the "software" dimension of the diffusion process. We are convinced that the primary exploratory research in this vital area requires detailed qualitative case studies that allow us to identify important dimensions of the process of diffusion as well as the key factors that affect the pace, timing, and success of the adoption of an innovation from one place to another.

Methodology

The analytical approach informing this volume is the comparative case study method.[78] Case studies are used inductively to develop and refine typological theory through a "building block" approach. This approach is particularly useful in new or emerging research programs to generate theory.[79] The authors in this volume employ methods of within-case analysis, particularly process tracing, in order to trace the causal pathways for diffusion. Most chapters also function as mini–comparative studies by examining several cases of a particular innovation, and in two instances the global universe of relevant cases. George advocates the comparative case study approach as an important link in bridging the gap between the re-

[78]Alexander L. George, "Case Studies and Theory Development: The Method of Structured, Focused Comparison," in Paul Gordon Lauren, ed., *Diplomacy: New Approaches in History, Theory, and Policy* (New York: Free Press, 1979), pp. 43–68; Alexander L. George and Timothy J. McKeown, "Case Studies and Theories of Organizational Decision Making," in Robert F. Coulam and Richard A. Smith, eds., *Advances in Information Processing in Organizations*, vol. 2 (Stamford, CT: JAI, 1985), pp. 21–58.

[79]Charles C. Ragin, *Constructing Social Research: The Unity and Diversity of Method* (Thousand Oaks, CA: Pine Forge Press, 1994).

search and policy communities in international affairs.[80] An explicit and coherent analytical framework is crucial if we are to find ways to cumulate the knowledge gained from a wide variety of individual case studies, particularly if we are to draw policy-relevant insights. Each chapter adheres to a common understanding of explanatory variables and data requirements. This allows us to bring together the findings of a team of researchers pursuing different cases by developing a common approach to build and extend theory. The cases were selected to develop contingent generalizations about a variety of diffusion processes (competition, cooperation, and socialization). The cases also allow us to trace the impact of a variety of factors (cultural, economic, political, organizational, and technological) affecting how innovations are assimilated, and their consequences for strategic behavior.

To study the dynamics of the diffusion process across time, space, and innovation, the authors contributing to this volume have focused on the following questions to identify the conditions and paths that lead to the spread of military knowledge.

(1) What forces motivated the import and export of new technologies, ideas, and practices that were developed in settings and against competitors that may not resemble the particular strategic circumstances of the importing state?

(2) What was the process of transnational communication, influence, and learning?

(3) What factors facilitate and hinder the spread of military knowledge?

(4) What factors encourage or hinder retention of the new idea by the receiving state and/or organization (e.g., resource capacity, threat environment, social, political, cultural, organizational, or other)?

(5) What are the effects of diffusion? Effects are analyzed in terms of responses (how the receiving state responds to the innovation and incorporates it into their organizations and practices—emulation, adaptation, off-set, other) and trajectories, the systemic outcomes of diffusion. Systemic outcomes include how widely the innovation spreads, how rapidly, and in what pattern (e.g., who emulates? who off-sets? is there rising or declining convergence in military forms across space?).

In addition to rigorous adherence to a common set of research questions, the success of this approach also depends on careful case selection. We opted not to base case selection on historical patterns associated with actual military revolutions, but rather on different diffusion processes and transmission paths. Experts do not agree on what constitutes a revolution in military affairs. Any rigorous definition that would adequately serve our empirical analysis would undoubtedly

[80]Alexander L. George, *Bridging the Gap: Theory and Practice in Foreign Policy*. Washington, DC: United States Institute of Peace Press, 1993.

be challenged by military specialists.[81] Since we are more interested in under-standing the process of diffusion associated with innovation regardless of whether it constitutes a full-scale revolution in military affairs, we chose to select cases so that the study would not be sidetracked by definitional debates such as how to subdivide the history of warfare into distinct RMAs. Furthermore, revolutions in military affairs are generated when innovations spread throughout the interna-tional system, altering the conduct of war. Studying the spread of innovations contributes to understanding the trajectory of RMAs.

Cases were selected to allow for broad coverage and variations in the diffusion process. The cases cover several historical eras, spanning from the Napoleonic era to the information age, in both peace and war. Regional coverage includes Europe, Asia, and the Middle East, with some chapters providing global coverage. Many types of innovation are examined. Some innovations are purely military, such as tank warfare. Others such as railroads and information systems involve dual-use technologies. Some are highly technological in character, such as nu-clear weapons, while others involve large macro-social changes such as the Na-poleonic military system. We also examine diffusion among friends and foes to assess the effect of the relationship between states (competitive or cooperative). Finally, some cases examine diffusion exclusively among core great powers, oth-ers from core to periphery, and one chapter from periphery to core.

This broad range of cases sheds light on some of the most important policy questions of the day: whether and how the United States can preserve its current lead; whether the United States should try to promote or hinder diffusion; which practices the United States should seek to disseminate and which to monopolize; how we might prevent the diffusion of what we want to monopolize, and the ob-stacles we might encounter when transferring military practices to our allies; who might attempt to emulate us, and whether there are countries other than the United States whose military practices some will seek to emulate or counter; and to what extent diffusion concerns should shape how the United States pursues the current transformation.

The goals of this study are to offer new insights into the following questions:

(1) What affects the scope and pace of diffusion? Are there consistent patterns of initiators and adopters? Early adopters and late adopters?

(2) What transmission paths are the most powerful conduits for diffusion (e.g., competition, collaboration, coercion or imposition, socialization, others)?

(3) What forces drive diffusion (e.g., strategic necessity, technology push, commercialization, normative pressures, others)?

---

[81]Indeed, contributors to this volume debated at length which of the innovations covered in this volume qualified as RMAs.

(4) Is diffusion a uniform and even process? Or do military innovations penetrate various regions of the world differently? Are there variations within regions?

(5) What factors are most important in influencing rates of adoption (nature of the innovation, local threat environment, societal and cultural factors, political and organizational contexts, others)?

(6) What are the consequences of diffusion? Is diffusion more likely to produce emulation (full or selective), adaptation, or off-sets? Under what conditions should we expect these outcomes?

Themes and Chapter Contributions

The volume is organized around four major themes. The first section examines a recurring motif in many of the case studies—the way local culture shapes and redirects even the most assiduous attempts at emulation. Imported practices rarely fit a new environment without modification. In some cases, the importing society fails to achieve the objective of integrating a new way of war; in others, indigenous culture enhances the effectiveness of the particular innovation. The essential elements of effectiveness in one cultural environment may be irrelevant or counterproductive in another.

John Lynn examines the diffusion of key Western military innovations to South Asia during the second half of the eighteenth century. The sepoy case demonstrates the ways in which cultural translation can multiply rather than erode the effectiveness of imported practices. The introduction of British regimental culture into South Asia was highly effective for two reasons. First, it mobilized South Asian values. The regiment turned out to be a well-crafted repository for indigenous cultural mores, a better repository than the native military organizational forms that preceded it. The regiment tapped into native codes of personal and community honor in ways that temporary or irregular home-grown military units could not. Second, the sepoys (native soldiers trained according to European military fashion) were highly effective because the British allowed adaptation of regimental culture to South Asian values and norms. Within a generation, the sepoy regiment, while maintaining key advantages of European regimental structure, had become a decidedly Indian institution.

The adoption of Soviet doctrine and organizational forms by the most powerful Arab states (Egypt, Syria, and Iraq) during the Cold War—the subject of Michael Eisenstadt and Kenneth Pollack's study—further demonstrates the ways in which indigenous culture shapes diffusion. Although the Soviet Union emerged as the principal military patron of these states, Soviet doctrine and organization were not adopted wholesale, contrary to the assumptions of many

military historians and analysts. Arab military doctrine and practice often diverged from Soviet doctrinal norms. In some cases, Arab armies preferred to pick
and choose, combining Soviet, British, and/or French doctrinal influences. In
other cases, they innovated according to their specific operational requirements.
Arab armies could not faithfully emulate Soviet doctrine, which called for high
degrees of tactical initiative, innovation, and flexibility. They lacked the tactical
leadership to wage fast-paced conventional maneuver warfare. Arab culture required conformity to group norms over initiative and independent thinking.
Arab armies functioned most effectively when they adapted Soviet (or British or
French) doctrine to their own needs, norms, and styles of war. These arguments
are demonstrated with examples drawn from the armed forces of Egypt, Syria,
and Iraq during the 1967 and 1973 Arab-Israeli wars, and the 1991 Persian Gulf
War.

Defense cooperation among America, Britain, Canada, Australia, and New
Zealand demonstrates that cultural affinity allows transmission of military expertise far exceeding identifiable security requirements. Thomas-Durell Young looks
at the largely unexamined programs that promote defense cooperation among
these five countries and argues that they surpass that which Washington enjoys
with its other NATO allies. For example, since the end of the Cold War, the armies of the ABCA countries (minus "active" participation by the New Zealand
Army) have cooperated in field training exercises and divisional command-post
exercises employing RMA technologies and techniques—all despite the lack of a
common, overriding security agreement, or any identifiable common threat. The
Australian Army, in particular, has been closely associated with the U.S. Army's
Force XXI modernization program and has initiated reforms that introduced
RMA organizational concepts and technologies. This case suggests that sharing
modern, sophisticated, and sensitive information and techniques to enhance
military capability tends to be extensive among nations sharing not only common interests and objectives but also common cultural values.

The second section examines whether and to what extent it is possible to
shape, direct, and manage the diffusion process. Attempts to promote diffusion
of some technologies and knowledge and impede the diffusion of others have
met with mixed success. The Soviets attempted to restructure the military organizations of the Warsaw Pact states by controlling the diffusion of technology.
Global efforts to control the spread of nuclear weapons have been extensive, but
management has been partial and temporary. The notion that technologies and
ideas can be controlled by a set of core states underestimates the potential for innovation in the periphery, which is often driven by local strategic circumstances
rather than by developments in the core.

Christopher Jones studies Soviet efforts to manage diffusion in the Warsaw

Pact, where dissemination of military knowledge was used as a political tool. The Soviets were virtually the sole supplier of East European military technology, either by export or licensing. Soviet technology was distributed within the Warsaw Pact for political, rather than military, reasons. Diffusion of ground force technology—the offensive weapons at the heart of mounting an offensive campaign against the West—is the focus of the case study. The Soviets provided their allies with a token number of state-of-the-art tanks, and substantial quantities of second-hand tanks. The East Germans received more state-of-the-art tanks than other states, even though they had the smallest of the non-Soviet Warsaw Pact armies. The East Germans were the only national units trained and equipped at the level of Soviet forces. The goal was to provide a rival to the *Bundeswehr*, the German national armed forces created by NATO in 1955, as heir to the German national military tradition.

In no other area have efforts to control and manage the spread of military technology and knowledge been as extensive as in nuclear weapons. Efforts to shape and manage diffusion have often failed because they were based on mistaken assumptions about why nations seek or refrain from acquiring nuclear weapons. William Potter's analysis of nuclear weapons decision-making in two dozen countries reveals that there is no typical proliferator profile. He explores a variety of propositions regarding technology diffusion and policy innovation, including the relative importance of domestic and external factors, variation in the explanatory power of alternative proliferation incentives and disincentives, and the relationship between technological developments and political decisions to acquire nuclear weapons. The comparison yields insights about the utility of alternative nonproliferation strategies.

The challenge of controlling diffusion is often treated as a problem of preventing the spread of sophisticated military knowledge from more to less industrialized or informatized states. Timothy Hoyt's contribution to this volume serves as a useful reminder that innovation can also originate in the periphery, raising the question of when and how diffusion travels from the periphery to the core. His chapter examines four innovations directly derived from peripheral conflicts: (1) the Fast Missile Attack Craft in Israel; (2) Remotely Piloted Vehicles/Unmanned Aerial Vehicles in Israel; (3) the use of chemical weapons by Iraq against Iran; and (4) Iraq's use of ballistic missiles in the Iran-Iraq and Persian Gulf wars. In each instance, peripheral powers developed capabilities that surprised observers, achieved significant strategic results within the region, and had important effects on core military preparations for future conflicts. Innovations in the periphery usually take the form of "niches": subsets in a broader revolution that, nevertheless, may have striking effects on both local and global power balances.

The third section focuses on diffusion during periods of rapid military transformation, similar to the current period. Several insights emerge, sharpening our understanding of how revolutionary military innovations spread. First, diffusion does not always originate from a single source and spread outward; rather, different players may simultaneously experiment with new technologies, hit upon different solutions at the same time, and learn from each other. Second, demonstration effects are particularly important in periods of rapid transformation, though the lessons learned are often contested. Finally, the patterns of diffusion are influenced by broader economic and macro-social changes. Societal change creates strong pressures for military change, while the desire to preserve existing social and political structures may limit the scope, speed, and extent of diffusion.

Geoffrey Herrera and Thomas Mahnken take on the two waves of military innovation that swept through Europe in the nineteenth century, each of which triggered attempts to emulate or counter the successful practices. By the beginning of the century, the Napoleonic military system had repeatedly defeated Old Regime armies. Napoleon's adversaries then tried to emulate and counter this style of warfare. By midcentury, Prussia's unique response to Napoleonic innovations transformed warfare once again and spawned attempts at emulation. Both sets of innovations depended on broader social and economic changes. Napoleonic warfare was the result of nationalism, while the Prussian military system drew on the industrial revolution to transform the military sphere. The ability to emulate these new military practices depended on the fit between the new techniques and access to the broader social, political, and economic resources that made these innovations possible.

Mahnken then takes on the much-debated case of Germany's successful use of combined-arms armored warfare in Poland and France during World War II, which forced Britain, the United States, and the Soviet Union to establish large armored forces to remain competitive on the battlefield. Each state attempted to emulate German armored warfare practices. All three greatly expanded the size of their armored forces and adopted some variety of combined-arms warfare. The result was a dramatic change in the character and conduct of warfare. This chapter traces the process of diffusion in wartime and qualifies the argument that a competitive security environment promotes the spread of military innovations. Uncertainty about how best to employ the innovation, as well as organizational culture, can prevent, delay, or shape the diffusion of particular innovations, even during times of military engagement.

Emily Goldman examines the application of air power at sea, which transformed naval warfare. Over a twenty-five-year period from the introduction of aircraft at sea to the Battle of Midway, naval establishments made the conceptual leap that brought changes in doctrine, tactics, and organization to exploit the potential of carrier air power. This chapter analyzes why navies integrated air power into their naval operations, organizations, and doctrines differently. The Americans and Japanese made offensive air power their navies' centerpiece. The British grafted air power onto existing doctrine, keeping the carrier in a defensive role. The Germans and Italians came to appreciate the value of the aircraft carrier, but only after it was too late in World War II to shift resources. The carrier air power case shows that diffusion is a path-dependent process and that choices made early on often foreclose later options. Moreover, during periods of rapid military transformation, several leader states may simultaneously try to apply the innovation while learning from one another. Even in the most competitive of environments, diffusion of highly effective military practices may face political and organizational obstacles.

The final section examines diffusion of the information revolution in military affairs, including the global spread of the U.S. information technology–based military model and commercial and dual-use technologies.

Chris Demchak's contribution examines today's unprecedentedly rapid global diffusion of the U.S. model of military modernization. Her analysis examines modernizing and nonmodernizing militaries around the world. Assessing the influence of resources, likely threats, alliance obligations, industrial pressures, and acquisition of advanced technologies on the direction of military modernization, Demchak finds that none of these factors account for observed trends. The most plausible explanation for the extraordinary spread and depth of interest in this particular model of military modernization is a process of growing institutional convergence across militaries. States are seeking to acquire the characteristics of a "modern" military as defined by U.S. practices. As this process matures, it is likely to produce a new, widespread community of highly lethal, deployable, long-range military forces with far-reaching implications for international security and global stability.

High-performance computing challenges U.S. policy-makers to come up with workable policies that serve the needs and demands of both commercial interests and security priorities. As John Arquilla shows, supercomputing and encryption are key examples of dual-use advanced information technologies, and there are unique problems associated with their diffusion. For the United States, the global market leader, the commercial benefits of exploiting existing advantages by means of aggressive marketing of supercomputers are offset by: (1) the risks that the spread of these machines will increase the opacity of the testing activities related to nuclear proliferation; and (2) the chances that other militaries will be able to use high-performance computing to achieve the command and control capabilities associated with a successful RMA. When a product has security as well as commercial applications, the perceived need to avoid "arming one's adversaries" runs headlong into private firms' profit motive. The stark choice in this case is between commercial gains and security threats.

The conclusion synthesizes the book's findings for scholarly and practitioner audiences. It discusses the motivations that stimulate diffusion, key transmission paths, and their relative efficacy as conduits for diffusion. We identify the range of responses to new military practices, and provide some general insights into the scope, rate, and patterns of diffusion of military knowledge over time and across space. We summarize the ways our analysis might aid practitioners in anticipating the course of diffusion of the current RMA.

The information revolution and its associated technological innovations continue to transform daily life as well as warfare. As these technologies spread around the globe, penetrating deeper into an ever-widening circle of societies, they are also transforming how we interact. Few organizations will be left untouched by the impact of these innovations—and the military is no exception. The recent conflicts in Afghanistan and Iraq provided ample demonstration of the impressive U.S. capabilities. But the United States can not expect to remain the sole power capable of mustering such superior power. Potential opponents are also likely to learn from each encounter with superior military capacity, both in technology and in organization. NATO allies are preparing for their own "military transformation," partly as a result of U.S. policy intended to secure our continued leadership role. How these changes will affect the international system and the global distribution of power may be impossible to predict with precision. However, we remain convinced that a greater comprehension of the range of variation in how innovations with military applications have spread in the past—and the factors that have helped or hindered their successful integration in a variety of societies—will better prepare us to meet the challenges of security and conflict prevention in the twenty-first century.

PART I

Culture and Diffusion

CHAPTER TWO

Heart of the Sepoy

The Adoption and Adaptation of European Military Practice in South Asia, 1740–1805

JOHN A. LYNN

A map of the South Asian subcontinent circa 1740 pictures a land still ruled by native principalities with names that strike a Western ear as exotic; European trading outposts cling only to the periphery of that triangle, splitting the Arabian Sea from the Bay of Bengal. But with succeeding decades, the red identifying domains of the British East India Company spreads across the page, as the course of conquest turns India crimson. The story told here concerns British triumph, but it does not chronicle the march of conquest itself; rather it tells of those native soldiers, the sepoys, who fought and won for the Company, staining the dust and the map with their blood.

Diffusion and synergy constitute the plot of this essay: the diffusion of military innovations pioneered in Europe and transferred to India, and the synergy between those imported military practices and indigenous culture. The East India Company's most important invention was the sepoy, a native soldier armed and trained according to eighteenth-century European military fashion. (The term "sepoy" was an English form of the Persian word for soldier, *sipahi*.)[1] At first glance, the sepoy may have seemed little more than a brown imitation of his British comrades in arms; however, the concepts of honor, duty, and loyalty that inspired the sepoy had their roots deep in South Asian soil, and it was those, not simply his fusil, uniform, and drill that made "Jack Sepoy" such a fine soldier.

[1]Persian was the official language of the Mughal Empire, thus the use of this language imported from West Asia for an institution imported from Europe.

While the analysis presented here generally fits all the Company's native forces, it is particularly apt concerning the army of the Bengal presidency. By the eighteenth century, the East India Company had three loci, Bombay, Madras, and Calcutta; each served as the center of an administrative area, known as a presidency, and each had its own army. Because the three armies operated independently and drew their recruits from different populations, each displayed somewhat contrasting characters. The army of the Bombay presidency prided itself on traveling light and marching fast. The army of the Madras presidency enjoyed a reputation as that most willing to deploy by sea to meet imperial needs. The army of the Bengal presidency, with its capital at Calcutta, ranked as the most important of the three armies by the end of the eighteenth century. It also differed in composition; unlike the other two armies, high-caste Hindus made up the majority of its rank and file. As a consequence, the army of the Bengal presidency obeyed the practices and honored the restrictions imposed by the Hindu religion to a greater degree than did the other forces. This army has also been the most studied of the Company's forces, less for its victories than the crisis it brought, for it alone mutinied in 1857, shaking British India at its foundations.

An example from the eighteenth century—and from the periphery at that—may seem remote to a present-minded audience, but it touches on matters of current relevance. Diffusion brings unexpected complexities, and military practices rarely, if ever, transplant whole. Furthermore, while diffusion from great power to great power seems to worry most analysts today, the real future problem is more likely to be diffusion from a great power to a lesser power, as occurred two hundred years ago in South Asia.

Contrasting Military Systems

Evolving Indian Military Tradition

During the millennia before the importation of European military practice, the subcontinent followed a path that contrasted sharply with the evolving European art of war. By the epic period of Indian history, ca. 800 B.C., great armies of the subcontinent were composed of four elements: infantry, cavalry, chariots, and elephants.[2] Within this *caturanga-bala*, or four-part army, infantry could claim only the lowest status; foot soldiers served as garrisons, guards, and laborers. Neither was cavalry rated as supreme. Never the core of Hindu armies before the arrival of Islamic invaders, Hindu cavalry lacked stirrups and did not employ the bow from horseback. In contrast, the chariot arm garnered more prestige; the

[2]Concerning military practice in ancient India, see P. C. Chakravarti, *The Art of War in Ancient India* (Delhi: Oriental Publishers, 1972); and V. R. Ramchandra Dikshitar, *War in Ancient India* (Delhi: Motilal Banarsidass, 1987).

warrior-hero of this age was typically a charioteer, wielding, along with other weapons, a bow.

Of the four parts of the army, elephants inspired the greatest fear and wonder. On the battlefield, war elephants were formidable, trampling and crushing anything in their paths, while on their backs they carried *howdahs* with sides tall enough to protect archers and spearmen within. Moreover, an elephant provided a high and impressive perch from which a commander might survey the battlefield as the fighting surged around him. And in this case it was as important to be seen as to see, for Indian hosts were held together by the general or prince, and the troops needed to see him like a banner. In siege warfare, elephants could stave in fortress gates with their massive heads; so real was this threat, that large spikes often studded gates in order to deter elephants. Yet beyond their obvious utility, one is left with the impression that elephants enjoyed renown because of their symbolic value. They graced religious processions for the same reason that they accompanied armies—their awesome presence. They seemed to promise invincibility, and contemporaries ascribed almost miraculous powers to them; one authority proclaimed, "One elephant is capable of slaying 6,000 horses."[3] The mystique of elephants seduced all Indian states and conquerors before the British.

Ancient texts, particularly the *Arthashastra* (ca. 250 B.C.), provide a window into Indian diplomacy and warfare. The *Arthashastra* portrays a state system typified by rivalry and animosity. War was the normal condition.[4] This classic work assumes a natural enmity between neighbors, and then extends this elaborate form of "the enemy of my enemy is my friend" in the *mandala*, or circle, theory of foreign policy. "That which encircles him on all sides and prevails in the territory immediately adjacent to his is the constituent of the circle of states known as the enemy. Similarly, that which prevails in the territory that is separated from the conqueror's territory by one [namely, by the enemy's lands] is the constituent known as friend."[5] This series of concentric circles of enemies and friends extends out to the twelfth ring. The *Arthashastra* also councils that deceit, subversion, and intrigue are superior ways to defeat an enemy. This theme of intrigue as an essential aspect of warfare marks Indian conflicts right through the eighteenth century and would, it will be seen, provide great possibilities for the British East India Company.[6]

[3]Chakravari, *The Art of War*, p. 48.
[4]This classic work, often compared to Machiavelli's *The Prince*, is ascribed to Kautilya, the brahmin prime minister of the emperor Chandragupta Maurya (322–298 B.C.). However, linguistic analysis indicates that it was not the work of a single author, and the work probably did not assume its final form until about 250 B.C.
[5]The *Arthashastra*, 6.2. This is available in the handy and accessible collection edited by Ainslie T. Embree, *Sources of the Indian Tradition*, vol. 1, 2d ed. (New York: Columbia University Press, 1988), p. 247.
[6]Steven Peter Rosen, in his recent book *Societies and Military Power: India and Its Armies* (Ithaca,

This weakness facilitated the arrival of Islam by the edge of the sword.[7] The first Moslem conquerors were Arabs who took Sind in 711–15, but they stopped there. During the late tenth and eleventh centuries, Turks raided into the Ganges valley from the principality of Ghazni that they had established in Afghanistan. The Ghaznis were eventually defeated by other Turks, the Ghorids, who captured Delhi in 1193 and made it their base in India. In 1206, Qutb-ud-din proclaimed himself sultan, giving birth to the Delhi Sultanate, which for the next three centuries would be the major Moslem state in India. The Mongol chieftain Tamerlane led a devastating, though passing, raid into India in 1398–99, sacking even Delhi. The Delhi Sultanate survived, but it never fully recovered. A century later, Babur, who claimed descent from Tamerlane, led Chatatai Turks who first occupied Kabul and then, in 1526 defeated the last sultan of Delhi at the Battle of Panipat, to establish the great Mughal Empire that ruled over north India and parts of the Deccan.

Moslem invaders brought with them new military forms and practices along with their new proselytizing religion. Turkish Moslems—the Ghaznids, Ghurids, and finally the Mughals—came from a military tradition that contrasted sharply with that of South Asia. They were horse peoples from the steppes of Central Asia who fought in classic horse-archer fashion, engaging as much as possible at bow range and closing in only for the final kill. This form of combat relied on rapid mobility in fluid battle, thus differing from the massive, lumbering armies of Hindu India and from the steady and powerful lines of European infantry. Armed with a superior bow and equipped with stirrups, Turkish cavalry swept all before it. By as early as the eighth century, chariots had disappeared from the Indian arsenal, giving indigenous cavalry greater importance, but the Indians failed to develop horse archers of their own before the Moslem invasions; and, of course, the lack of stirrups limited the effectiveness of Indian horseman. The Mughals never forgot their origins as raiding horse-people; the court itself was known as "the exalted stirrup," and custom demanded that emperors' sons be born on a horse blanket placed on the floor of a tent. Their mounted armies contained relatively little infantry and restricted its roles.

Yet it did not take the Mughals long to succumb to the power of elephants. Abul Fazl, a major commentator during the reign of the great emperor Akbar (1556–1605), wrote, "This wonderful animal is in bulk and strength like a moun-

---

NY: Cornell University Press, 1996), argues that the key constant in the various forms of the Indian state system remained the division of Indian society by caste and later religion, inspiring internecine warfare and limiting the ability of Indian political entities to unite.

[7]On medieval India, which is to say India after the arrival of the Moslems, see Jagadish Narayan Sarkar, *The Art of War in Medieval India* (New Delhi: Munshiram Manoharlal, 1984); and William Irvine, *The Army of the Mughals* (New Delhi: Eurasia Publishing House, 1962).

tain, and in courage and ferocity like a lion. It adds materially to the pomp of a King and to the success of a conqueror; and is of the greatest use for the army."[8] Akbar put musketeers in the howdahs, and by the eighteenth century, elephants even carried small cannon on their backs.

The Mughals did appreciate the power of more modern weaponry and adapted over time. The center of Babur's line at Panipat was occupied by thundering cannon and musketeers, and the next year at Khanwa, his artillery stopped the charging Rajput cavalry. Mughals favored the very large cannon in vogue among the Ottoman Turks. The Mughal term for artillery, *top*, was an Ottoman word, and the early Mughals employed Turks from Istanbul as artillerymen. Matchlock muskets, an early form of shoulder-fired small arm, used by Babur's infantry were essentially the same as those employed in contemporary Europe, although the Mughals failed to develop European close-order infantry tactics to go along with these weapons. Akbar even improved that matchlock's design, and he also lightened Mughal artillery pieces.

Eventually, encumbered by large artillery pieces and elephants, the swollen Mughal armies became as unwieldy as those of earlier Hindu states, although the Mughals never lost their overwhelming preference for cavalry. The last of the great conquering Mughals, the emperor Aurangzeb (1658–1707), campaigned with an army that numbered in the hundreds of thousands. He literally took his capital into the field, forming a moving tent city thirty miles in circumference with thirty thousand elephants and a half-million camp followers.[9] By the time that Aurangzeb turned back from his twenty-five-year-long campaign to conquer the Deccan in 1705, he had exhausted the empire; thus, Mughal excess prepared the way for Persian and Afghan invasions during the mid-eighteenth century and, finally, complete conquest of the subcontinent by the East India Company.

The Mughals raised their military forces in a variant of the feudal system, known as the *mansabdari* system.[10] The emperor had only a small number of paid household troops; the rest were supplied by men to whom the emperor granted wealth and rank in exchange for raising a prescribed number of troops. These powerful individuals held office, a *mansab*, and so were known as *mansabdars*. Akbar, who instituted the mansabdari system, paid his mansabdars in money, but soon they received land grants, *jagirs*, as compensation. This age-old exchange of land for military service received a twist under the Mughals: jagirs could not be inherited, and at the death of a mansabdar his jagir reverted to the emperor, who then reassigned it. While this concept of meritocracy would seem to be more en-

[8]Fazl in Sarkar, *The Art of War in Medieval India*, pp. 105–6.
[9]Stanley Wolpert, *A New History of India*, 5th ed. (Oxford: Oxford University Press, 1997), p. 167.
[10]See Abdul Aziz, *The Mansabdari System and the Mughul Army* (Delhi: Idarah-i Adabiyat-i Delli, 1972).

lightened than European hereditary feudalism, it suffered a serious drawback. Because a mansabdar could not retain his jagir within his family, there was little incentive to husband its resources, and mansabdars lived in wasteful luxury at the expense of the inhabitants of their jagirs. Local welfare suffered. By requiring that jagirs revert to the emperor, the Mughal emperors hoped to avoid the natural tendency for great landholders to become powers unto themselves. However, when the empire became weak, that is exactly what happened, and in the eighteenth century, powerful local rulers became increasingly independent, although still expressing loyalty to the emperor.

While the official tally of troops to be furnished by a mansabdar was set, from ten up to seven thousand (only princes held mansabs larger than seven thousand), the actual contingent supplied would be much smaller and of no regular size. The Mughal army, then, lacked standardized or permanent units. Moreover, the system grew more inefficient with time, and attempts to regulate and reform the mansabdari system did not arrest its decline.

Individual Mughal warriors could be superbly capable with their personal weapons, of fine physique, and excellently mounted, but there was little drill or discipline. Units neither deployed in a regular manner nor maintained set formations; they fought as collections of skilled individuals, not as cohesive units obedient to central command. The code of the warrior ruled, that of the soldier was unknown. This explains why European observers thought little of the Mughal army; Sir Thomas Roe at the emperor's court in the early seventeenth century wrote that he saw "what they call an army; but I see no soldiers, though multitudes entertained in that quality."[11] These armies with only a rudimentary organizational structure remained very centered on their leaders, whose deaths could determine the fate of battles. R. O. Cambridge would later comment that the British could "end the battle by one discharge of a six-pounder at the Raja's elephant."[12]

European Military Innovation, 1650–1750

Western ways of war overwhelmed the traditional military forms of Mughal India in the second half of the eighteenth century. In the hundred years before the Company's conquest of the subcontinent, Europeans forged the weaponry, practices, and institutions they later transported to South Asia. The flintlock-bayonet combination and new more mobile artillery proved particularly effective

---

[11]Roe in Bruce Lenman, "The Transition to European Military Ascendancy in India, 1600–1800," in John A. Lynn, ed., *Tools of War: Instruments, Ideas, and Institutions of Warfare, 1445–1871* (Urbana: University of Illinois Press, 1990), p. 10. The original spelling has been modernized.

[12]Cambridge, *Account of the War in India* (first pub. 1772), in Philip Mason, ed., *A Matter of Honor: An Account of the Indian Army, Its Officers and Men* (London: Papermac, 1986), p. 55.

on the subcontinent. But nontechnological military practices—the battle culture of forbearance, the Western reliance upon drill, and the foundation of the regimental community—were even more important.

The transition from musket and pike to the fusil and bayonet constituted a prerequisite of European conquest in South Asia. The new weaponry maximized the power of Company forces especially when they were at a numerical disadvantage, as was usually, and sometimes dramatically, the case.

At the start of the 1600s, the musket/pike combination determined European tactics. The musket employed a matchlock, which ignited the powder charge by means of a lighted "match," a cord of flax or hemp. The complicated loading procedure limited the rate of fire to one shot per minute, and the rate of misfire could rise to 50 percent.[13] The lighted match was dangerous to the musketeer and those around him, since it might accidentally ignite nearby gunpowder. Musketeers, lacking offensive shock potential and unable to defend themselves effectively from cavalry charges, were joined by pikemen. The pikemen grouped together in the center files of a battalion, flanked on either side by files of musketeers. Over time, the proportion of firearms to pikes in the battalion increased, from 1:1 at the start of the 1600s, to 2:1 at midcentury, to 4:1 by 1690.

After the mid-seventeenth century, the more easily loaded, more reliable, and safer flintlock fusil began to replace the older musket, and the pike gave way to the bayonet. Instead of using a lit match to set off gunpowder, the fusil used flint striking steel to generate a spark. A fusilier could load in a third fewer movements, since he need not worry about the match.[14] He could also prime and load his fusil, set it on half-cock, sling or stack the weapon, and still be instantly ready to fire the moment he seized the firearm. As early as 1660, General Monck rearmed his regiment, the Coldstream Guards, with fusils, which became common in the English army in the 1690s. The French were a bit slower to adopt the new weapon because the cost of conversion was substantial, but an ordinance of 1699 finally prescribed the universal adoption of the fusil.[15] About the same time, the socket bayonet eliminated the need for the pike. A desire to give muskets some shock value for defense and offense by turning them into short but stout pikes had produced earlier and more primitive bayonets about 1640.[16] But these "plug" bayonets fit down the barrel of the musket, meaning that with bayonet in place,

---

[13]David Chandler, *The Art of Warfare in the Age of Marlborough* (New York: Hippocrene, 1976), pp. 76–77.

[14]Corvisier says that the number of movements fell from thirty-six to twenty-three. Philippe Contamine, ed., *Histoire militaire de la France*, vol. 1, series editor André Corvisier (Paris: Presses Universitaires de France, 1992), vol. 1, p. 409.

[15]Jean Colin, *L'infanterie au XVIIIe siècle: La tactique* (Paris: Berger-Levrault, 1907), p. 26.

[16]Louis André, *Michel Le Tellier et l'organisation de l'armée monarchique* (Geneva: Slatkine Reprints, 1980 [1906]), p. 344.

the weapon could not be loaded or fired.[17] In 1687 the famous French military engineer Vauban created the socket bayonet that attached via a collar that slipped around the barrel, leaving the muzzle free for loading and firing. The socket bayonet soon became standard issue in European armies during the 1690s, and early in the War of the Spanish Succession (1701–14) the pike disappeared from European battlefields.[18]

Adopting the fusil/bayonet combination improved the effectiveness of European infantry. Firepower multiplied, as every infantryman carried a firearm, and shock capacity increased because every soldier now had an effective edged weapon. The implications of this improved weaponry for Europe were significant enough, but in a South Asian context they would be even greater. There, infantry would often face massive cavalry forces, so the need for entire units to form impenetrable hedgehogs of bayonets was critical. With limited numbers of Europeans and sepoys, every man needed an effective firearm.

The logic of maximizing firepower to overcome larger native armies with smaller numbers of Europeans and sepoys applies even more to artillery pieces. Artillery allowed one adversary to use the chemical energy of gunpowder to counter the human energy of his enemy's multitude. Shipboard cannon were essential to the initial expansion of European domains in South Asia long before the Company pushed inland, where lighter field artillery proved vital.[19] The chief physical factor that limited the utility of cannon on the early modern battlefield was weight. In the 1620s, the barrel alone of a thirty-four-pounder weighed fifty-six hundred pounds, and the cannon on its carriage required twenty horses to pull it and a crew of thirty-five to serve it.[20] Once in place, cumbrous cannon could not be shifted to keep pace with the movement of the action. Major European states conducted quiet but essential reforms that lightened and improved cannon in the period from 1740 to 1780. Frederick the Great improved and expanded the artillery branch of his army, making such technical improvements as the addition of ammunition boxes to gun limbers and employing screw mechanisms instead of clumsy wedges to elevate guns in 1747.[21] Under the direction of Liechtenstein, the Austrians instituted an impressive program of experimentation

[17]Puységur in Chandler, *The Art of Warfare*, p. 83.

[18]In 1703 the French abandoned the pike altogether. Jacques-François de Chastenet de Puységur, *Art de la guerre par règles et principes*, 2 vols. (Paris: Jombert, 1748), vol. 1, pp. 51, 57; Louis XIV, *Oeuvres de Louis XIV*, edited by Philippe Grimoard and Philippe Grouvelle, 6 vols. (Paris: Treuttel et Würtz, 1806), vol. 4, pp. 396–97fn.

[19]Lenman, "The Transition to European Military Ascendancy in India," argues that until Buxar, naval gunfire played a key role in supporting the army.

[20]These figures come from the very interesting and very detailed tables supplied by Du Praissac, *Les discours militaires* (Paris: Guillemont et S. Thiboust, 1622), pp. 112–30. I have called his 33-1/3-pounder a 34-pounder for convenience.

[21]Lenman, "The Transition to European Military Ascendancy in India," 119.

and innovation. The most renowned artillery reform came in France, where Gribeauval brought in a new system of field pieces in the 1770s. The artillery pieces of the Gribeauval system were half the weight of earlier guns; for example, the standard eight-pounder cannon was cut from 2,260 pounds to 1,280.[22] In addition, these new pieces boasted more precise manufacture. Thus by the late eighteenth century, the artillery environment had changed significantly.

Other, nontechnological, military innovations were important. The first was a new and counterinstinctual attitude toward losses in battle—dubbed here the "battle culture of forbearance."[23] The great French monarch Louis XIV (1643–1715) wrote, "Good order makes us look assured, and it seems enough to look brave, because most often our enemies do not wait for us to approach near enough for us to have to show if we are in fact brave."[24] For Louis, good order equated with victory, and he was not alone in this opinion. Seventeenth-century generals came to regard infantry combat as a test of wills, with victory going to the force that could absorb casualties and still maintain order, rather than to the side that inflicted the greatest physical casualties on its foe. Marshal Catinat, a practical soldier, in describing an assault, insisted, "One prepares the soldier to not fire and to realize that it is necessary to suffer the enemy's fire, expecting that the enemy who fires is assuredly beaten when one receives his entire fire."[25] At the Battle of Bunker Hill in 1775, American troops received the standard instruction: "Don't fire till you see the whites of their eyes." Europe developed a battle culture based less on fury than on forbearance.

Performing with forbearance in the face of danger and chaos was far from natural; it had to be learned. European infantry training emphasized obedience and restraint. Not only must the soldier master the tools of war—he himself must be mastered. In later ages, officers would come to trust in the initiative of their troops, but during the seventeenth and eighteenth centuries, aristocratic commanders assumed a low level of honor and motivation among the rank and file.

[22]These are weights for the barrel of the gun only. Gribeauval also developed improved carriages. See comparative weights in Denis Diderot et al., *Encyclopédie*, 17 vols. (Paris: Briasson, 1751–65), vol. 2, p. 608; Louis Jouan and Ernest Picard, *L'artillerie française au XVIIIe siècle* (Paris: Berger-Levrault, 1906), pp. 44–47; Howard Rosen, "The Système Gribeauval: A Study of Technological Development and Institutional Change in Eighteenth-Century France," Ph.D. dissertation, University of Chicago, 1981, p. 130; Matti Lauerma, *L'artillerie de campagne française pendant les guerres de la Révolution* (Helsinki: Akateeminen Kirjakauppa, 1956), pp. 10, 16; Chandler, *The Art of Warfare*, p. 178. Weights are translated from *livres* to kilograms with the *livre* figured at 489.50585 grams. Marcel Marion, *Dictionnaire des institutions de la France aux XVIIe et XVIIIe siècles* (Paris: 1972), p. 375.

[23]For a discussion of the battle of forbearance and its implications for the French, see chapter 15 of John A. Lynn, *Giant of the Grand Siècle: The French Army, 1610–1715* (New York: Cambridge University Press, 1997).

[24]Louis XIV, *Mémoires de Louis XIV pour l'instruction du dauphin*, edited by Charles Dreyss, 2 vols. (Paris: Didier, 1860), vol. 2, pp. 112–13.

[25]Catinat in Colin, *L'infanterie au XVIIIe siècle*, p. 25.

42 JOHN A. LYNN

So rather than rely on the common soldier's self-control, military practice put faith only in supervision by officers. The key to teaching both skill and obedience was drill, the repetitive practice of rigidly prescribed movements in marching and the manual of arms.[26]

witnessed the emergence of a new form of military community epitomized in the regiment. This military community, typical of European armies by the late 1600s, possessed several innovative characteristics; it was standardized, permanent, and male. While it was not the first military formation or unit to serve as the focus of life and devotion of its members, the European regiment gave new intensity and direction to loyalty. The regiment was a regular, standardized unit of a defined size that included one or more battalions, each containing a set number of companies of an ordained size. An officially prescribed cadre of officers with standardized ranks and functions commanded. This carefully defined chain of command established and enforced the concept and reality of a hierarchy defined by military rank rather than by social prestige or personal bonds. The regiment persisted beyond a particular campaign, war, or commander, because this new military community was in theory a permanent one. The long life of these formations created a regimental culture typified by a strong identity and intense devotion.

A coincident and integral development was the exclusion of the vast majority of women and children from the military community. Armies of the sixteenth and early seventeenth centuries included huge numbers of camp followers. Herbert Langer mentions a forty-thousand-man imperial force in Germany during the Thirty Years' War accompanied by a hundred thousand camp followers, mostly women.[31] This logistical nightmare also undermined discipline, so authorities drove noncombatants from armies during the late seventeenth century.[32] Morality was also a concern; thus in the 1680s Louis XIV banned prostitutes from his armies, as the elector of Brandenburg had a generation before. In addition, authorities discouraged or forbade common soldiers to marry. The regiment became an essentially male community isolated from other bonds and focused loyalty on the military unit, profoundly affecting motivation, morale, and professionalization.

Removing soldiers from women helped cut them off from civil society. The common soldier chose an employment of last resort, and the impoverished man in the ranks retained few ties with home. Even when a French soldier received a rare permission, he could not marry a woman from his garrison town for fear of developing local ties. In fact, all French regiments, except the guards, changed post every year to keep soldiers from identifying with the civilians around them. This form of regimental community would be severely altered to suit the South Asian environment.

[31]Langer in Barton Hacker, "Women and Military Institutions in Early Modern Europe: A Reconnaissance," *Signs* 6, no. 4 (summer 1981): 648.
[32]For the French case, see Lynn, *Giant of the Grand Siècle*, pp. 337–43,

Diffusion of European Weaponry and Military Practice

Diffusion of Technology

Imported European military practice of the mid-eighteenth century shaped the era of conflict that followed; however, the East India Company did not enjoy a monopoly of Western weapons, tactics, or organization. No informed author ascribes the conquest of South Asia by the East India Company simply to a weapons gap. Indians had used cannon and small arms for many generations before they had to fight the British and their sepoys. Indians lagged somewhat in technology during the seventeenth century, but when the critical clashes came after 1740, Indian military hardware soon equaled that of the Europeans.[33]

Indians possessed the requisites for manufacturing modern weapons: a generally sophisticated technological base, prodigious wealth, and good raw materials. Not only did South Asia provide superior iron, it also exported high-quality saltpeter for gunpowder to Europe.[34] Moreover, the agents of technological diffusion abounded on the subcontinent. At first the Mughals employed Ottoman Turkish technological experts, but over the sixteenth century they turned more and more to Europeans; one contemporary estimated that five thousand Portuguese renegades served Asian potentates from Bengal to Makassar during the early seventeenth century.[35] This tradition continued: "Frenchmen, Germans, Portuguese, Armenians, and Topasses" also served Mir Kasim's artillery at Buxar in 1764.[36]

Native rulers secured European-style weaponry one way or another. Hector Munro reported to Parliament in 1772: "There is hardly a ship that comes to India that does not sell . . . cannon and small arms."[37] But the most important source of weapons was Indian arsenals. Mir Kasim, the *nawab* of Bengal, supplied his Western-style battalions with excellent flintlock muskets of Indian manufacture that tests showed to be superior to British weapons, owing to superior Indian Birbhum iron and Rajmahal flints.[38] In the late 1760s, Shuja-ud-daula, nawab of Oudh, in an effort to create Western-style forces, founded an arsenal with a workforce of more than five hundred under the direction of a native Bengali and a French officer; it produced light cannon, muskets, and about 150 to 200 flint-

[33]Rosen, *Societies and Military Power*, p. 162, judges that there existed "rough equality of British and Indian military technology."

[34]See Rosen, ibid., p. 165, for comments on superior Indian iron, which was exported worldwide. Hoover, "Origins of the Sepoy Military System," p. 44.

[35]João Rubeiro in Lenman, "The Transition to European Military Ascendancy in India," p. 103. See as well Jean Baptiste Tavernier in ibid., p. 102.

[36]Second Report of the Select Committee of the House of Commons, 1772, p. 8, in Hoover, "Origins of the Sepoy Military System," p. 128.

[37]Munro from Select Committee, H.C., 1772, in ibid., p. 126.

[38]Mason, *A Matter of Honor*, p. 40.

locks per month.[39] The Indians retained a preference for heavy artillery pieces too long; however, when confronted with the example of successful Company campaigns, Indian rulers secured up-to-date lighter artillery pieces of their own, complete with the latest European refinements in artillery design, such as elevating screws.[40] Toward the end of the century the Marathas employed a Scottish expert, Sangster, to establish the most successful Indian arms production at Agra, where five factories turned out flintlock fusils, gunpowder, and cannon at the close of the eighteenth century.[41] No less an authority than the Duke of Wellington, who faced these guns at Assaye, found the "ordnance so good and so well equipped that it answers for our service," and he incorporated Maratha guns into his own artillery train.[42]

Diffusion of European Military Practice and Expansion of Sepoy Forces

The superficial and fleeting nature of the European advantage in weaponry suggests that Western tactics and regimental culture were the more important military innovations on the subcontinent. Their primary diffusion came via efforts of French and British trading companies, but soon native rulers pursued a secondary diffusion of Western practice to gain advantage over their neighbors and to survive the East India Company's onslaught.

From the establishment of European trading stations on the subcontinent during the sixteenth and seventeenth centuries, foreign interlopers employed local manpower to help defend their coastal bastions. However, these forces were limited in number and purpose. Not until the mid-eighteenth century did the French Compagnie des Indes set off the race to create Western-style infantry units manned by native recruits. At Pondicherry, some ninety miles south of Madras, French governor Dumas mustered four to five thousand sepoys in 1740 in response to raids launched by the Marathas, who demanded tribute throughout much of India.[43]

Meanwhile, war clouds gathered in Europe. The War of the Austrian Succession (1740–48) once again pitted Great Britain against France. Truces had spared the British and French from importing their wars to India before, but in 1744, Westminster forbade any new truce. The aggressive governor Dupleix, who had replaced Dumas, used his new troops to capture Madras in September 1746. This

---

[39]Seema Alavi, *The Sepoys and the Company: Tradition and Transition in Northern India, 1770–1830* (Delhi: Oxford University Press, 1995), pp. 23–24.

[40]Lenman, "The Transition to European Military Ascendancy in India," p. 119.

[41]Pradeep Barua, "Military Developments in India, 1750–1850," *Journal of Military History* 58, no. 4 (Oct. 1994): 607.

[42]Wellington to his brother Henry, Mason, *A Matter of Honor*, p. 161.

[43]G. B. Malleson, *History of the French in India*, 2d ed. (Delhi: Renaissance Publishing House, 1984 [1909]), p. 87.

seizure embroiled the French in a struggle with the nawab of the Carnatic, who had forbidden the Europeans to attack one another. The nawab dispatched an army of ten thousand men with artillery to punish the French. Paradis, in command of only three hundred French and seven hundred "*çypayes*," devoid of any cannon, defeated the Indian host at the Battle of St. Thomé in November. Paradis's triumph was the first in what became a pattern of victories by surprisingly small European forces with sepoy support over far larger indigenous armies.

The East India Company responded to the loss of Madras by raising sepoys of its own. Major Stringer Lawrence created a body of three thousand natives trained to fight in European fashion and led them at the Battle of Cuddalore in 1748.[44] Robert Clive, a clerk turned soldier, won the first dramatic victories with East India Company sepoys. Although the war in Europe ended in 1748, the fighting in India continued as the two rival trading companies backed different local candidates to rule the Carnatic. Clive led a small force of two hundred British plus three hundred sepoys to seize the French-sponsored nawab's capital at Arcot. Soon Clive found himself besieged there, but he outlasted the three thousand troops sent against him. By the 1750s the Compagnie des Indes and the East India Company each maintained roughly ten thousand sepoys; however, from early on, the British seem to have done better with their native troops, paying them with greater regularity and demanding and achieving greater discipline.[45]

The East India Company's sepoy army continually evolved during its first half-century. Initially, sepoys formed only in companies, which, though trained by Europeans, were commanded by Indian officers. It took some time before sepoys donned military uniforms, although by 1756 the authorities at Madras reported that its sepoys were attired in a "uniform of Europe cloth."[46] Alongside the sepoys, the Company always fielded its own battalions of Europeans independent of the royal army. The Company mounted little native cavalry in regular units but employed many irregular native horsemen. Artillerymen were generally European, the Company being reluctant at first to share the secrets of gunnery with the native population. In the initial clashes between French and British, sepoys played a secondary role in support of European troops, but with the years, the number of sepoys and their effectiveness increased.

The next crisis came when the nawab of Bengal took Calcutta in June 1756. Responding to the fall of Calcutta and the atrocities following the capitulation—

---

[44]Lawrence insisted that there, "the military, both black and white . . . behaved extremely well." Lawrence in Hoover, "Origins of the Sepoy Military System," p. 91.

[45]As Dupleix's Indian assistant, Ananda Pillai, reported, native troops in French service tended to degenerate into lawlessness in cantonments. Ibid., p. 77.

[46]Apr. 1756 report by the president of Fort St. George to the Company in Mason, *A Matter of Honor*, p. 62.

including the infamous Black Hole—Clive landed in Bengal. With an army of just over three thousand troops, two-thirds of them sepoys, he defeated an enemy host of fifty thousand at the Battle of Plassey, 23 June 1757. In fact, this battle was something of a put-up job, because Clive had split the enemy alliance before the battle, and most of the troops arrayed against him did not actually fight.

Soon after his arrival in Bengal, Clive recruited the first full battalion of native infantry, the Lal Paltan, or red battalion, because of its British-style uniforms. In 1759 an ordinance completed the conversion to battalions by grouping all sepoy companies into battalions led by a small European cadre, in addition to the native officers. At this time, the army of the Bengal presidency also expanded; in a letter of April 1760, the directors of the Company in London stipulated to Calcutta that its European forces should number no more than fifteen hundred, but allowed "blacks at your discretion."[47] From five native battalions in 1760, the Bengal army doubled to ten in 1763.

Harder fighting against a variety of enemies followed the easy victory at Plassey. By now the Seven Years' War (1756–63) had engulfed Europe, and the French dispatched reinforcements to Pondicherry and besieged Madras, but failed. A bitter struggle between the French and British also raged in the Northern Circars, a rich area held by the French, and a Company army dispatched by Clive performed marvels there. Finally, in 1761, Pondicherry fell, and the French were finished as a major colonial power in India. Even though the Treaty of Paris in 1763 restored Pondicherry to them, it so restricted their activities that the Compagnie des Indes soon dissolved. The French could and did play the role of spoilers, but they never again constituted the main threat to Company ambitions.

Indigenous rulers observed the success of the Company's new military units and copied them; as the decades progressed, the native powers became more and more capable of fielding European-style armies of their own. The *nizam* of Hyderabad did so in the 1750s; next the *peshwa*, leader of the Maratha confederation, lured away the nizam's commander in 1758 and employed him to fashion a body of eight thousand infantry. After 1760, Mir Kasim, made nawab of Bengal after Plassey, created Western-style infantry led by two European adventurers. When war broke out between Mir Kasim and the Company in 1763, he allied with other regional rulers, including Shuja-ud-daula, nawab of Oudh, but the native coalition met defeat at the Battle of Buxar in 1764. This battle was probably the single most important for the Company, for by winning there, the Company secured its hold over Bengal, Bihar, and Orissa—rich territory that provided the wealth and manpower necessary for further British conquest. Shuja-ud-daula, though de-

---

[47]Amiya Barat, *The Bengal Native Infantry: Its Organization and Discipline, 1796–1852* (Calcutta: Firma K. L. Mukhopadhyay, 1962), p. 11.

feated at Buxar, had learned the value of the Company's troops, and he created
his own Western-style infantry soon after the battle; however, his army disappeared with Shuja's death in 1774. The raja of Banaras also created such forces,
which the Company defeated and dissolved in 1782.

From 1780 to 1799 the Company engaged in a struggle with Mysore, led by the
gifted military adventurer Haider Ali until his death in 1782, and then by his son
Tippoo Sultan. Haider Ali showed no great partiality for European-style infantry,
although he maintained a body of European mercenaries. However, Tippoo Sultan raised a sizable force of native infantry armed and trained in European fashion with the guidance of European, mainly French, drillmasters. Tippoo staffed
his new regiments (*cushoons*) and brigades (*chucheris*) with salaried native officers, rather than Europeans. Tippoo gave his "sepoys" uniforms, insignia, and
drill manuals, but the language of command was Persian.

After the final defeat and death of Tippoo Sultan, the East India Company
faced its most severe challenge in the Marathas. Maratha power dated back to the
close of the seventeenth century when their legendary leader, Shivaji, fought Aurangzeb to establish Maratha independence. After Shivaji's death in 1680, the
Marathas fractured, with a peshwa enjoying theoretical leadership, but with several Maratha rulers, or *sardars*, exercising their own independence. One sardar,
Madhaji Sindia, seemed likely to reunite the Maratha confederacy in the late
eighteenth century. A cornerstone of his authority was a new European-style infantry army forged with the aid of a Savoyard mercenary, Benoit de Boigne, who
had once led native troops with the Company army. Sindia hired de Boigne to
form two battalions in 1784, and later expanded this force into the four brigades
of the Army of Hindustan, totaling twenty-seven thousand troops drilled according to British manuals and armed with weapons manufactured by his arsenal
at Agra. De Boigne recruited a polyglot band of European officers, many of them
British, to train and command this impressive force. In 1794, Sindia's death cut
short efforts to reunite the Marathas. The far less able Daulatrao Sindia succeeded
his great uncle, and de Boigne returned to Europe, leaving his subordinate, Pierre
Cuillier Perron, in command.

A moment of Maratha cooperation preceded their political demise. Late in
1794, the young peshwa summoned the Maratha sardars to join in a war against
the nizam of Hyderabad. Daulatrao Sindia committed his Army of Hindustan. At
the Battle of Kharda, the Marathas triumphed over the nizam, whose army of
130,000 boasted 12,000 sepoys under Raymond.[48] The key clash occurred between
Perron's infantry and Raymond's troops, as the two opposing forces fired volley
after volley in good eighteenth-century fashion.

[48]Rosen, *Societies and Military Power*, p. 183.

TABLE 2.1

Size of the Armies of the East India Company Presidencies
at the Close of Major Wars

	Bengal	Madras	Bombay
1763	6,680	9,000	2,550
1782	52,400	48,000	15,000
1805	64,000	64,000	26,500

SOURCE: Callahan, *The East India Company and Army Reform*, p. 6.

The Company army that would soon confront the Marathas was larger and more highly evolved than that of Clive and Munro. Further regularization and reorganization arrived via the army reform of 1796, replacing the older brigade organization of sepoy forces with regiments of two battalions each.[49] Forty-two European officers commanded each sepoy regiment. Each sepoy company would also be led by a *subadar* aided by a *jemadar*, resulting in a dual chain of command, one British and one South Asian. The reformed army expanded as well; Table 2.1 details the permanent sepoy and European troops.

The Second Maratha War (1803–5) pitted the Company against the disunited Marathas.[50] In 1802 the British exploited a civil war among the Marathas by offering protection to the threatened peshwa. The *sardars* entered the fray piecemeal, so at first only Sindia buttressed by some allies took the field against the Company. The British undermined the Army of Hindustan before the real fighting started by offering substantial bounties to its European officers who would come over to the Company. Perron mishandled the situation by not extending a counterbid; instead, he dismissed all British and Anglo-Indian officers in his brigades, regarding them as untrustworthy. This stripped his units of the majority of their veteran officers at the moment of crisis.

Marathas did not concentrate their forces, putting them at further disadvantage. In September 1803, at the Battle of Assaye, Arthur Wellesley, the future Duke of Wellington, commanded an army of six thousand troops in a hard-fought victory against a Maratha army of sixty thousand, including Sindia's First Brigade, numbering about six thousand. Wellington would later call Assaye "the bloodiest [battle] for the numbers that I ever saw."[51] In the north, General Gerald Lake de-

[49]On the process of achieving army reform in 1796, see Raymond Callahan, *The East India Company and Army Reform, 1783–1798* (Cambridge, MA: Harvard University Press, 1972). Barat, *The Bengal Native Infantry*, may overstate the case, but he makes clear the importance of the 1796 reform: "[O]nly in 1796 did the rapid, haphazard growth of Clive's original forces give way to some regularity of organization."

[50]The First Maratha War (1779–82) had been a minor and inconclusive affair fought in North India.

[51]Wellesley in Mason, *A Matter of Honor*, p. 161.

feated the other Maratha brigades in separate fights, with the climax coming at Laswari in November. The hostile Sardar, Holkar, entered the war after this battle, but the destruction of Sindia's army determined the fate of the Marathas.

After defeating the Marathas, the Company faced only one more great rival on the subcontinent, the Sikhs, but victory in the Second Maratha War, in addition to the earlier conquests of Bengal and the south India, gave the Company such overwhelming wealth and manpower that the Sikhs were doomed to defeat in the Sikh wars of the 1840s. Once again, the key leader, Ranjit Singh, who had created a large force of Western-style infantry and forged Sikh unity, died before the war broke out with the Company, and the Sikhs fell out among themselves.

The narrative of the conquest provides no simple formula that prescribed victory for the East India Company. It enjoyed no monopoly over modern weaponry, and the diffusion of European military practice, though begun by the French and British, later spread to native powers who enthusiastically copied Western ways of war. Learning was a two-way street, and the Company discovered much through its South Asian experience. The British became adept at South Asian politics, characterized by divisive intrigue as well as sheer power. Company commanders eventually mastered difficult aspects of military campaigning in India, notably logistics.[52] Fundamental to the success of its sepoy armies was the way in which the Company preserved and focused South Asian values to buttress military performance. To understand the sepoy's effectiveness, one must not simply catalog the formation of Company units but also comprehend South Asian culture as well.

Building upon South Asian Tradition

Indian Military Values

The ability to create a highly effective South Asian army that appeared so European depended not on the similarities between the British soldier and the sepoy, but on their profound differences. Weaponry and uniforms, organization and command may have all had their origins in Europe, but the passion, resolve, and loyalty of the sepoy had roots in South Asian society and culture. Thus in the key springs of martial action—motivation and morale—the sepoy units of the Company army were essentially South Asian, not mere imitations of the British.

The Company benefited from an indigenous culture that produced a social and religious climate optimal for the professional soldier of the prenationalist epoch. Before the onset of the French Revolution, European officers had devel-

---

[52]See Barua, "Military Developments in India, 1750–1850," p. 609. These conclusions concerning British political and military adaptation are borrowed from Barua.

oped a tradition of military service that was its own rationale, but the rank and file remained outside this matrix of motivation and reward.[53] In contrast, South Asian culture explained the necessity of, and bestowed honor upon, military service by the common soldier. To grasp the sepoy's mentality requires some understanding of social and religious tradition on the subcontinent. For the army of the Bengal presidency, that means the Hindu tradition.

In its ancient form, the religious and social caste system in India elevated the warrior to high status that was preserved into the twentieth century. According to the traditional *varna* system of caste dating back to ancient Vedic times, the highest caste, the *brahmins*, were priests, but the second caste, the *kshatriyas*, were warriors and rulers, followed by the *vaishyas* (commoners, merchants, and landholders) and the *shudras* (menial workers, peasants).

In daily life, the fairly simple caste schema of varna was far more varied and complicated by the principle of *jati*, literally "birth." Jati defined occupation and status by lineage. Moreover, one's jati determined not only one's profession but also whom one could marry; thus, jati were limited regional endogamous groups throughout India. Each jati claimed to belong to one of the classical varna. On a ritual level, the Hindu religion is much concerned with pollution and purification, and to eat, smoke, or drink with an individual of a lower ranked jati or caste would pollute the higher-caste Hindu. Even accepting water drawn by the wrong caste endangered the individual. The definitions, status, and restrictions of jati varied from one area of India to another, but whatever the variants, within Hindu society jati was the most important reality of caste.

Hindu society saw humanity divided into groups set by birth and occupation, sanctioned by religion, governed by distinct codes of conduct, and guided by principles suited to one's condition. Overlaid on this caste society came the new profession of sepoy, soldier in service to the East India Company. Effectively it became another jati, if not initially defined by birth, then by calling. The sepoy identity fit well into a world defined by function and community, and now, added to his civilian identity came the new military collectivities he served: company, battalion, and regiment. Sepoys easily assumed an approach that regarded the individual as a representative of a community, which required the individual to be faithful to its standards. Of course, group identification and loyalty seem to be omnipresent aspects of military motivation, whether in the form of primary group cohesion or esprit de corps. However, the difference in India was the impressive level of religious and social sanction attached to group and occupational identity.

In a jati, right conduct was defined in terms of living by the dictates of that particular jati: the highest virtue for the sepoy was to fulfill the role of soldier. His

[53]See Lynn, *Giant of the Grand Siècle*, chap. 8.

most fundamental loyalty would not, then, be to the Company or to the British Crown, but to his military unit and to his own duty, or *dharma*, conceived of in a very Indian way.[54]

Hindu literature discusses dharma at length, and for the soldier, the *Bhagavad Gita* ("Song of God") offers guidance. The *Gita*, described by one author as "the most typical expression of Hinduism as a whole," dates from the first or second century A.D.[55] It is a dialogue between the hero Arjuna and the great god Krishna, who appears as the driver of Arjuna's chariot. The interchange comes before the climactic battle of Kurukshetra between the related but rival Pandavas and Kauravas. Arjuna, the leading prince and champion of the Pandavas, sees his kin in the opposing army, shows humane compassion for them, and expresses his desire to throw down his arms rather than fight. But Krishna insists that this course of action would be sinful, for by it Arjuna would fail to fulfill his dharma as a kshatriya.

Krishna first argues that because of the cycle of birth, death, and rebirth, Arjuna should not think that he is killing a foe by slaying a body; bodily death is inevitable and irrelevant. "Never, indeed, was there a time when I was not, nor when you were not, nor these lords of men. Never, too, will there be a time when we shall not be." And thus, "He who regards him [i.e., the soul] as a slayer, and he who regards him as slain—both of them do not know the truth; for this one neither slays nor is slain" (2.19) (2.27).[56] Krishna follows with an appeal that Arjuna's dharma requires him to fight, and he must accept this because of his kshatriya caste. "For a kshatriya there does not exist another greater good than war enjoined by dharma" (2.31). "But if you do not fight this battle which is enjoined by dharma, then you will have given up your own dharma as well as glory, and you will incur sin" (2.33). So Arjuna has a moral duty to fight because of his caste, and the consequences of battle will be positive whatever his fate: "Either, being slain, you will attain heaven; or being victorious, you will enjoy the earth. Therefore arise . . . intent on battle" (2.37). Krishna next explains that morality lies in the action itself, not its fruits. By extension, the proper standard by which to judge personal morality in combat is not the violence and destruction of war, but the fulfillment of the warrior's dharma.

The Hindu community's great reverence for the *Gita* derives from its discus-

---

[54]Dharma is subject to a number of definitions, including duty, responsibility, law, and truth.

[55]Embree, *Sources of the Indian Tradition*, p. 276. The *Gita* is interpreted in many ways that have nothing to do with soldierly conduct. The analysis here is not meant to question more religious interpretations of the *Gita*, but simply to see it in a literal sense as having a great deal to do with the ethics of the warrior.

[56]The quotations from the *Gita* are found in Embree, *Sources of the Indian Tradition*, pp. 281–86. I have inserted chapter and line notations. Conveniently, the entire text of the Gita is available on the internet. *Bhagavad Gita*, trans. Ramanand Prasad (American Gita Society, 1988) at ht.://eawc.evansville. edu/anthology/gita.htm.

sion of dharma, rather than its specific reference to conduct in war; nonetheless, the *Bhagavad Gita* preaches a powerful warrior's code. The combatant has a duty to fight because it is the duty of the warrior. Since death is both inevitable and followed by rebirth, the warrior need not feel guilt for the death of his enemies. To die in battle is blessed; to be victorious wins earthly rewards. The morality of personal martial action is best understood as separate from the consequences of that action, a proposition that could be interpreted as measuring a soldier's right conduct in terms of warrior values—prowess, courage, self-sacrifice—rather than in terms of the interests and aims of those for whom he fights.

While the notions of caste, action, and duty examined above are definably Hindu, these conceptions influenced the Moslem and Sikh communities of South Asia as well; Indian notions of honor are even more completely shared. In martial morality and motivation, there is no more complex concept than honor. It can refer both to an individual's highly personal, internalized code of behavior and to the reputation that an individual enjoys within a group or community.[57] Modern soldiers prefer to speak of honor as if it were brave and independent, but in fact, the measure of honor is primarily in official and unofficial recognition by the military community—in other words, by reputation.

To this inherent complexity, Indian society adds another layer of meaning, distinguishing between personal honor, *izzat*, and community honor, *rasuq*.[58] Izzat corresponds to Western notions that an individual's actions earn praise or condemnation for himself. The definitions of honorable action may differ, but the consequence is still focused on the individual. Beyond this, rasuq is community honor, meaning that the community, be it a village or a jati, holds to a standard it expects of its members. Their actions bring honor or disgrace to the community as a whole. In accord with rasuq, the individual represents his community and therefore must live up to community standards.

Another distinctly South Asian attitude with great implications for the code of the warrior was the concept of being "true to one's salt." Salt, *namak*, implied sustenance and a compact that an individual must feel great loyalty to whoever provided that livelihood. More than simply a quid pro quo between employee and employer, it involved a covenant of strong moral force on an individual

---

[57]On honor, see Bertram Wyatt-Brown, *Honor and Violence in the Old South* (New York: Oxford University Press, 1986); John A. Lynn, "Towards an Army of Honor: The Moral Evolution of the French Army 1789–1815," *French Historical Studies* 16, no. 1 (spring 1989): 152–73; and Lynn, *Giant of the Grand Siècle*, chapters 8 and 13.

[58]See the stress placed on izzat and rasuq in Pradeep Dhillon, *Multiple Identities: A Phenomenological of Multicultural Communication* (Frankfurt: Peter Lang, 1994). In contrast, McNeill, *Keeping Together in Time*, p. 135, sees sepoys as developing European esprit de corps when exposed to Western drill, but this conclusion misses the point. While native regiments certainly developed esprit de corps, this took a very South Asian flavor for very South Asian reasons.

compelling him to give not only service but also fidelity to an institution or authority that paid and fed him. Perhaps this code of loyalty grew out of the frequent change of masters, particularly political masters, in South Asia; perhaps it was an extension of the values expressed in the *Bhagavad Gita*. In any case, it gave leverage to any institution that possessed the resources necessary to hire troops to further its interests. Such an institution could expect more of its servants than a hireling's sense of giving a day's work for a day's wage. Not surprisingly, the Company made reference to salt in the oath administered to new recruits in 1766: "I . . . do swear to serve the Honorable Company faithfully and truly against all their enemies, while I continue to receive their pay and eat their salt."[59]

The East India Company's Exploitation of Tradition

The Company learned to create an environment for its sepoys that fostered native concepts of jati, duty, honor, and loyalty. Because tradition demanded high standards, the Company had a stake in allowing the sepoy to remain integrated into his original community as much as the demands of military service allowed. To be sure, battalions and later regiments became communities themselves, but these benefited from association with their sepoys' civilian identities and bonds.

The Hindu texts and jati apply to all three Company armies, but they were crucial to the Bengal army, because it was overwhelmingly composed of high-caste Hindus. The other presidency armies also contained a large percentage of Hindus, because that was the majority religion of South Asia. However, the high-caste nature of the Bengal army was unique. Madras, for example, had no kshatriyas, and its brahmins were so adamant about their orthodoxy and special position in society that they did not serve in the army. While complete records are not available, existing evidence stresses the high-caste nature of the Bengal army, which was recruited predominately in Bihar and Oudh. A Benares regiment raised in 1814–15, for which a detailed account exists, lists 696 brahmins and Rajputs (kshatriyas) and only 108 Hindus of lower castes, along with 92 Muslims.[60] This seems to have been the pattern as early as the administration of Charles Cornwallis, governor-general and commander-in-chief, 1786–93.[61] High-caste brahmins and kshatriyas were most likely to take to heart the classical outlook of the *Bhagavad Gita*, although all Hindus lived in a world defined by jati.

The Company provided a menu of incentives to the sepoy that would seem effective in any army, but in a South Asian context these carried added dimensions. Pay, promotion, and pensions mattered most. The rate of pay offered by the Com-

---

[59]Mason, *A Matter of Honor*, p. 66.
[60]Barat, *The Bengal Native Infantry*, p. 122. See as well the general observations of the court of inquiry, Barrackpore, 2 Jan. 1825, in ibid., p. 121.
[61]Callahan, *The East India Company and Army Reform*, pp. 3–4.

pany to a sepoy was not particularly handsome, but it was enough that men sought Company service.[62] The basic pay of a sepoy was seven rupees per month, while a jemadar earned seventeen rupees and a subadar fifty-two rupees. In addition to base pay, the Company added allowances, known as *batta*, when a soldier was on the march, in the field, or posted at a distance from his home base. While pay was sustenance, it was also a bond among a soldiery pledged to be true to its salt.

Promotion provided added incentive. Common soldiers within a European regiment, Company or royal, had little or no chance of attaining an officer's commission; however, promotion to Indian rank in a native battalion depended only upon seniority. The recruit could advance from sepoy to *naik*, to *havildar*, to jemadar, and finally to subadar, the highest ranking Indian officer in a company. The climb up the ladder took a very long time; a young man entering the service at age sixteen could probably not expect to become a naik until he reached age thirty-six, and he would have to survive to sixty before rising to subadar.[63] Since not only authority but also pay and pensions increased with promotion, a jump in grade was much desired.

The Company also instituted regular pensions for men who had served a minimum of twenty years, a period later reduced to fifteen. A pensioner received half his pay rate, and even the smallest pension, when supplemented by a minimal income in the village, allowed a man to live in a tolerable condition. In 1852, Viscount Gough declared, "The pension is our great hold on India."[64] The pension was a considerable draw for recruits. From 1782 the Company in Bengal offered a small land grant, a jagir, as an alternative to a straightforward monetary pension in order to ease the strain on the Company treasury.[65]

Beyond these obvious compensations, the life of the sepoy brought to bear rewards and sanctions specific to Indian society. The creation of a high-caste army in the Bengal presidency may have been an accident of regional politics and the operation of the manpower market in northeast India, but the British quickly came to terms with this reality and encouraged it.[66] Before the Mutiny of 1857, the Company catered to caste by creating single-class battalions—that is, battalions composed of men of the same ethnicity, religion, and caste.[67]

[62]See Barat, *The Bengal Native Infantry*, pp. 132–43, on pay in the company armies.

[63]Ibid., p. 154.

[64]Gough in ibid., p. 142.

[65]Ibid., p. 51.

[66]On the Hindu nature of the Bengal army and on recruitment patterns in North India, see ibid.; and Alavi, *The Sepoys and the Company*. See Hastings on trying to maintain caste to divide Hindus. Add. 29234, Warren Hastings, Copies of essays, etc., Warren Hastings Papers, in Alavi, *The Sepoys and the Company*, pp. 44–45.

[67]Incidentally, the preference for high-caste recruits in the army of the Bengal presidency also played into the hands of the brahmin peasantry, particularly in Bihar, where that peasantry used this leverage to strengthen their claims for elevated status. Alavi, *The Sepoys and the Company*, p. 51.

The Company accommodated the religious scruples of the sepoys and facilitated maintaining ties with their home communities by allowing native battalions
to live in a manner very different from that imposed on European Company
troops and those royal regiments posted to India. European regimental culture
strove to isolate the soldier from his personal background and from the population of the garrison towns. The soldier himself tended to be an economic castoff,
and he was held in low repute by society and his officers. Cornwallis wrote in
dismay about European recruits for the Company armies: "I did not think Britain
could have furnished such a set of wretched objects," and he went on to condemn them as "contemptible trash."[68] Two particular eighteenth-century trends
increased the isolation of soldiers in Europe: they were more and more consigned
to barracks, and only a minority were allowed to marry.

While European soldiers in India continued to live in barracks, troops of native battalions inhabited small huts. An individual sepoy built this simple accommodation on a spot of ground assigned him. Soldiers living without families
might share their space with fellow sepoys, but married men were likely to have
wife and children with them and perhaps other relatives. Europeans tended to regard the huts as dirty and "unmilitary," but they were a sensible adjustment to
the needs of South Asian troops.

Family obligations figured prominently in sepoy life. Even if wives were not
physically present in the cantonment, married sepoys could hardly forget their
duties to them. A sepoy with wife and family was expected to remit part of his
salary back home to his wife, and should he fail to do so, sepoys from his own
village would condemn him.[69] Sepoys living with their families regularly brought
wives and children with them when they changed stations.[70] The presence of
women and children, as well as the bazaar peddlers necessary for supply in India,
meant that armies marched with great numbers of noncombatants. When troops
marched from Bengal to Bombay in 1778–79, the 103 Europeans and 6,624 sepoys
were accompanied by 19,777 camp followers plus an additional 12,000 bazaar
people.[71] Such crowds had once been common in European armies but had disappeared in the male world of eighteenth-century professional armies.

Whereas European troops messed together, Hindu practice insisted that a man
of one caste could not eat food prepared by one of a lower caste or even drink

---

[68]*The Correspondence of Charles, First Marquis of Cornwallis*, ed. Charles Ross (London, 1859), vol. 1,
pp. 268–69, 299. See as well the comments of a Major Scott, who reported in 1760 that "among the 900
raw highlanders" sent to India, "not one could speak English or even use a firelock. . . . [T]he men's inducement to enlist was a promise that they might carry ten women in the company." Letter from Scott,
16 Mar. 1760, Home Misc., vol. 95, p. 373, in Barat, *The Bengal Native Infantry*, p. 17.

[69]Barat, *The Bengal Native Infantry*, p. 177.

[70]Rosen, *Societies and Military Power*, p. 179.

[71]Callahan, *The East India Company and Army Reform*, p. 9.

water given him by an inferior. In practice, Hindu restrictions on high-caste individuals required each sepoy to prepare his own food, rather than eating in a common mess. By European standards this was hopelessly confused, but it was a religious and social necessity in India, and authorities respected it.

Such regard for religious practice was hardly accidental; it was a matter of policy intended to avoid unnecessary friction between Company and sepoy. But the by-product was to reinforce religion and traditional martial values within the Company army. Cornwallis wrote in 1789: "We cannot too forcibly impress on you the important light in which we view an attention to objects connected with the means of indulging these religious prejudices."[72] In 1793 a general order specifically sanctioned religious festivals: "The Commander-in-Chief has no objection to the native troops amusing themselves at the celebration of their festivals according to their respective rites and customs."[73] In the late eighteenth and early nineteenth centuries British officers even took part in Hindu ceremonies, notably in the annual blessing of the battalion's weapons and flags. Custom commanded each jati to celebrate the tools of its appointed trade, and the sepoys grafted this practice into battalion life. In this effort to tolerate, even foster, native religion so as to avoid alienating its sepoys, the government and army even deflected well-meaning Christian attempts to convert the troops. When the bishop of Salisbury approached Cornwallis in an effort to foster missionary activities, Cornwallis rebuffed him. "It is likewise a matter for serious considerations how far the impudence of intemperate zeal of one teacher might endanger a Government which owes its principal support to a native army composed of men of high caste whose fidelity and affections we have hitherto secured by an unremitted attention not to offend their religious scruples and superstition."[74] While parrying missionary efforts, so careful was the Company to respect the caste of its Bengal soldiers that it supplied them only with brahmin and kshatriya prostitutes.[75]

In order to ensure that its officers respected the men's language, customs, and religion, the Company tried to make its officers become knowledgeable about them. As early as 1768, the Company directors in London declared that "no Officer should rise to the Command of a Battalion until he has made himself sufficiently Master of the Language to acquit himself in his Duty without the Assis-

[72]Cornwallis to Deputy governor and council at Fort Malbro, 1789, PRO, Cornwallis papers, PRO/30/11/184. in Alavi, *The Sepoys and the Company*, p. 46. See as well Pennington P.P., 1831–32, XIII, (735-v), H.C., Ap. B, in Barat, *The Bengal Native Infantry*, pp. 174–75.

[73]General Order of 17 Mar. 1793, in Alavi, *The Sepoys and the Company*, p. 79. On Ramlila, a celebration of Rama's conquest over the demon Ravana with the aid of the money-king and his general, Hanumant, see ibid., p. 81.

[74]Cornwallis to the Bishop of Salisbury, 1788, Cornwallis papers, PRO/30/11/187 in ibid., p. 47. Cornwallis's brother was a bishop.

[75]Ibid., p. 83. This citation deals with a report dated 1850.

tance of an Interpreter."[76] A generation later, Cornwallis insisted that "a perfect
knowledge of the Language, and a minute attention to the prejudices of the Sepoys" were absolutely essential for all officers of native troops.[77] Sita Ram testifies
that in his early days as a sepoy, ca. 1815, officers spoke Indian languages well,
lived with Indian women, and took part in the entertainments of the sepoys.[78]
The dual chain of command within battalions, one European and the other native, ensured that even if a British captain fell short in understanding his men's
values and customs, the subadar and jemadar could inform the officer and serve
as a buffer between him and the sepoys.

Maintaining a soldiers' religion meant maintaining his relationship with his
home community. Amiya Barat concludes that "by enlistment he gained status in
his society to which he continued to retain his allegiance. He therefore remained
a civilian at heart through becoming a soldier by profession."[79] Recruitment conspired to reinforce family and village ties. When the Company first formed native
armies, the men who presented themselves for service were mercenaries similar
to those who had offered their swords to native rulers, but the Company soon
evolved its own system of recruitment. In Bihar and Oudh this produced respect-
able young sepoys who arrived with their families' blessings. Pensioners who returned to their villages with a red coat and a comfortable income demonstrated
the generosity of the Company, and their stories of adventure whetted the appe-
tites of the young. The army introduced a furlough system that sent soldiers
home to their villages periodically. One contemporary officer endorsed this practice as a means by which the "good feelings of their families continue through the
whole period of their service to exercise a salutary influence over their conduct as
men and as soldiers."[80] It was hoped that men on furlough would bring back recruits.[81] The autobiography of Sita Ram tells how he was recruited by his uncle, a
jemadar in the infantry, on furlough to visit his family.[82] While returning to
camp, they were joined by two sepoys from the uncle's regiment, one of whom

[76]11 Nov. 1768 letter from Court of Directors to the president and council at Fort William in Lenman, "The Transition to European Military Ascendancy in India," 122.

[77]Cornwallis to Dundas, 4 Apr. 1790, PRO 30/11/150, in Callahan, *The East India Company and Army
Reform,* p. 107.

[78]Sita Ram, *From Sepoy to Subedar,* ed. James Lunt (Delhi: 1970), pp. 24–25. Controversy surrounds
this work, which is supposed to be the autobiography of an old *subadar* who served in the Bengal army
from 1812 to 1860. Some believe that Norgate, who published the work first in 1873, invented the story.
In any case, if this is not the true story of a particular sepoy, it is a distillation of the experience of
many, and is regularly used as a source by Indian historians, as it is the only detailed account of a sepoy's life.

[79]Barat, *The Bengal Native Infantry,* p. 126.

[80]Sleeman in Mason, *A Matter of Honor,* p. 166.

[81]See Barat, *The Bengal Native Infantry,* p. 126, on furloughs and recruitment.

[82]Sita Ram, *From Sepoy to Subedar,* chapter 1.

brought his brother to enlist. Such new recruits were given no housing allowance at first, because it was expected that they would share the hut of a relative or friend during the first year of their life in the ranks. Amiya Barat interprets this last practice as "another example of the way in which the army authorities encouraged the growth of family or village ties within the army."[83]

Over time, regiments became not only extensions of peasant villages but also communities in their own right, taking on some of the characteristics of a village. Braithwaite, commander-in-chief of the Madras army, wrote in 1793: "In native corps of any standing, the ties of caste and consanguinity are . . . strong and numerous, from frequent intermarriages."[84]

Families had good reason to encourage a son to enlist. Of course, salary and pension promised lifelong income in an honored profession, but added incentives also lured the high-caste Hindus who so dominated the ranks in the army of the Bengal presidency. Such families were often landowners, and from 1796 to 1815 official policy gave preference to the pleadings of sepoys; thus a family could gain real advantage if a son served in the ranks.[85] These benefits could extend to other dealings with civil authority. Sita Ram's mother lamented his wish to become a sepoy, but his father approved because he had a case pending concerning a disputed mango grove; he knew that he might reap a benefit in court once his son was in uniform.[86] The fact that a family might expect to enjoy advantages from the military performance of a son put all the more pressure on sepoys to conduct themselves with honor.

The emphasis here on the strong bonds between sepoy and community runs counter to Steven Rosen's thesis in *Societies and Military Power: India and Its Armies.* He insists that the success of the British sepoy army was due to "a form of military organization that increased the cohesion of the army by divorcing it from society."[87] The record presented here demonstrates the opposite claim: European regimental culture succeeded in South Asia only by being profoundly altered so as to link the military unit with family, village, jati, and religion. Cohesion and motivation within native battalions depended on focusing the force of indigenous community identity and honor. In the absence of patriotism, the army exploited identification with the battalion or regiment, but it also took advantage of the wider set of societal and religious norms that sepoys brought with them from their civil communities.

[83]Barat, *The Bengal Native Infantry*, p. 130.
[84]Mason, *A Matter of Honor*, p. 123.
[85]Barat, *The Bengal Native Infantry*, p. 150.
[86]Sita Ram, *From Sepoy to Subedar*, pp. 5, 49.
[87]Rosen, *Societies and Military Power*, p. 196.

Conclusion: Diffusion, Stability, and Motivation

Many of the chapters in this volume demonstrate that diffusion has not occurred whole; rather, diffusion requires selection and modification, as Herrera and Mahnken argue. For good reason, alterations are most profound when technologies and practices travel across cultural barriers, as illustrated here and in the contribution by Eisenstadt and Pollack. Even Young's consideration of cooperation between the Anglo-Saxon militaries provides an exception that proves the rule, in the sense that cultural similarity allows integration and coordination even when military forces do not adopt a common technology. This present chapter contrasts with other contributions by detailing a case in which cultural translation multiplied the effectiveness of imported practices, rather than eroding it.

The sepoy experience suggests that for diffusion to succeed, imported practice must ultimately be compatible with indigenous culture. Such a conclusion would seem straightforward. However, imported practice can rarely if ever be expected to fit a new environment without modification. Therefore, successful diffusion requires (1) the realization that elements considered essential to effectiveness in one cultural environment can be irrelevant or counterproductive in another, and (2) the wisdom to recognize which elements can be altered, and to what degree, without detracting from effectiveness.

In the case of the East India Company's sepoy battalions, the match between European regimental forms, as modified, and South Asian culture far exceeded mere compatibility. A great affinity for the new regimental system produced a powerful synergy between innovation and tradition, creating native forces that were far better than they would have been as an awkward brown duplicate of the contemporary British army. The result was neither intended nor expected; it can only be explained as a fortunate convergence. History tells the fascinating tale of how European regimental culture metamorphosed into a form that optimized South Asian values in the service of a colonial power.

From the 1740s, European trading companies, first the French and then the British, expanded their military presence in South Asia by adding sepoy troops. This primary diffusion imported not only European fusils and artillery but also Western discipline, drill, and battle culture. Above all, the colonial British instituted standardized, permanent units in imitation of European regiments. The full flower of the East India Company's system could only be expected to bloom with time; successful diffusion was a process, not an event. In fact, the passing years allowed the system to mature. In the first decade of fighting, sepoys played a supporting role to European troops in the Anglo-French struggle for south India. Clive would not lay the foundations for the army of the Bengal presidency until the 1760s, and the greatest enemies of the Company, Mysore and the Marathas, would not really take the field against the British until the 1780s and later.

By this point the British appreciated the benefits of letting well enough alone. The East India Company had been on the subcontinent for well over a century before they created sepoy infantry, and the British had already learned to adapt in order to survive in India. The Company first raised sepoy units precisely because Europeans were so scarce on the subcontinent. The shortage of European officers and the experimental nature of the first sepoy companies dictated that they be commanded by Indians, and apart from the specifically tactical aspects of their duties, drill, and training, for example, sepoys enjoyed their own style of life under direct supervision of indigenous leaders. When the Company reformed and expanded sepoy units in later years, it imposed European officers as company, battalion, and regimental commanders, but it left religious and community aspects of sepoy life to continue in an Indian pattern. Military and political leaders, exemplified by Cornwallis, clearly appreciated the wisdom in doing so.

Agents of the East India Company in South Asia enjoyed considerable flexibility in dealing with native troops—more, in fact, than with European soldiers who, because of their higher cost, were of greater concern to the directors back in England. As a private, or at least quasi-private, trading company, the Company was insulated from royal government interference during the formative period of the sepoy army, although the Crown became much more directly involved with the subcontinent through the India Act of 1784. In any case, the greatest guarantee of Company flexibility regarding its sepoy army was probably distance and unconcern. London was a very long voyage away, and, at any rate, the handling of sepoy regiments was hardly a priority there.

After European trading company armies demonstrated the ability of their sepoy forces in battle, native states created Western-style armies of their own. This secondary diffusion copied the pattern established by the East India Company, and, in fact, hired Europeans to direct the transformation. Stewart N. Gordon points out how the involvement of indigenous elites with traditional forms of land-holding, wealth, and honor inhibited the adoption of Western-style infantry by native states; nonetheless, several important Indian rulers forged such armies.[88] Measured by the quality of their weaponry, the precision of their drill, and the character of their organization, the diffusion of European military technology and practice was a success. However, measured by victory and defeat, indigenous diffusion failed, for no native army could stand for long against the Company.

Thus arises an apparent paradox. If the East India Company sepoy's success

[88]Stewart N. Gordon, "An Analysis of the Limited Adoption of European Style Military Forces by Eighteenth Century Rulers in India," paper sent to me by Stewart Gordon.

depended more upon his South Asian values than upon his Western weaponry, why would not the European-trained soldiers of indigenous rulers, such as the Marathas, be far more effective on the battlefield, since it would seem that native powers would be able to draw upon the wells of South Asian culture better than could the British? The logic here is attractive, but it is also misleading. Mobilizing the military potential of the sepoy required an institutional stability that no South Asian power enjoyed during the eighteenth century.

Certainly, its political and territorial base provided the East India Company with an ability to tap money, materiel, and manpower over the long haul, and without this it could not have conquered the subcontinent. However, indigenous powers were not lacking for these assets. The richness of British coffers or the number of sepoy armies cannot alone explain the conquest. After all, Company armies were generally outnumbered on campaign, and against the Marathas, for example, they were matched in the number of Western-style troops and out-gunned by superior artillery.

In contrast to native powers, the Company's institutional stability freed it from the fates of individual leaders. For the sepoy army, this provided the time needed to develop the concert between military innovation and traditional culture. While South Asian states remained at the mercy of shifting feudal loyalties and the survival of particular rulers, the Company endured and prospered. After the East India Company initiated the use of Western military practices by native troops, this stability bought time for sepoy battalions and regiments to develop identities and for bonds to grow between the sepoys and their surrounding and supporting communities, for rasuq to multiply izzat. The Company's permanent European-style military units benefited from Indian military values in a way that the irregular and transitory forces of the older South Asian pattern could not. While some native states imitated European-style forces, these lacked the constancy and longevity required to maximize their effectiveness. In this unique manner, the political environment determined the success of military diffusion.

The triumph of the Company's sepoy armies involved a great deal more than the introduction of European weaponry to the subcontinent. In fact, the spread of European technology was the least interesting and least important aspect of the conquest of South Asia by the British East India Company in the second half of the eighteenth century. The diffusion of European military weapons and practices owed its great success to the synergy between innovation and tradition. The sepoy fought so well because even though he carried European weapons and wore a European uniform, he was inspired by South Asian ideals. Therefore be not misled by the British red coat of the sepoy; beneath it beat the heart of India.

CHAPTER THREE

Armies of Snow and Armies of Sand

The Impact of Soviet Military Doctrine on Arab Militaries

MICHAEL J. EISENSTADT AND

KENNETH M. POLLACK

On 12 October 1973, the Iraqi army joined the October War. That afternoon, newly arrived tanks and armored personnel carriers (APCs) of the 12th Armored Brigade, 3d Armored Division, lumbered into combat against the battle-weary Israeli 240th Armored Division, as it threatened to outflank Syrian defenses on the northern Damascus plain. The Iraqis drove Soviet T-55 tanks. They advanced in two long lines under a heavy but indiscriminate supporting artillery barrage. To some of the Israelis lying in wait on the tels (small volcanic mounds) of southwestern Syria, the Iraqis looked like the very model of the Soviet army on the attack.[1]

Nothing was further from the truth. Although using Soviet equipment, the Iraqi army employed a British-based system, with some U.S., French, Soviet, and indigenous Iraqi practices. Although difficult for the Israelis to discern at the time, the Iraqi tanks and APCs were not moving to contact in Soviet-style assault lines, but rather were attempting to use British-style overwatch techniques. This would not be the last time that observers would mistake appearance for reality.

During the Cold War, the Soviet Union was the principal military patron of Egypt, Syria, and Iraq. At the height of their military exertions—the 1973 Arab-Israeli war for Egypt and Syria, the 1991 Gulf War for Iraq—the armed forces of

The authors would like to thank David Isby, Amnon Sella, Dov Tamari, and Steve Zaloga for commenting on earlier drafts of this paper.
[1]Interviews with Israeli military officers, Tel Aviv and Jerusalem, Sept. 1996.

these states were equipped largely, if not exclusively, with Soviet arms. Many military analysts assumed that these states had also adopted Soviet doctrine[2] and organizational forms, factors that accounted for their lackluster performance on the battlefield. Egyptian authors Anwar Abdel-Malek and Mahmud Hussein have argued that from the late 1950s onward, "reliance on Russian weapons systems and the tactics which such systems necessarily implied," hindered Egyptian military operations.[3] One Western scholar writes that "Arab disadvantages were heightened by the application of Russian introduced tactical models that were ill-suited for the fluid situation created by Israeli deep penetration tactics."[4] These assertions are incorrect.

Egyptian, Syrian, and Iraqi military doctrine and practice during the Cold War often diverged—sometimes dramatically—from Soviet norms.[5] The Syrian army most closely resembled the Red Army in its organization, tactics, and operations. The Egyptian military employed Soviet organizational models, but borrowed from Soviet tactics, and to a lesser extent Soviet operational thought, based on lessons learned in past wars and specific operational needs. The Iraqis were eclectic. They employed Soviet arms but added Soviet, French, U.S., and indigenous Iraqi touches to their predominately British tactics.

Contrary to conventional wisdom, Soviet practices helped more than hindered these Arabs. The Soviets provided a carefully thought through and battle-tested doctrine. Soviet practices were often better suited to the dominant cultural and organizational patterns of Arab societies than alternative Western models. Nonetheless, Syrian, Egyptian, and Iraqi armed forces fared best in battle when they adopted, and adapted, those Soviet practices that were most compatible with Arab cultural and societal norms and the operational requirements of their armed forces. They were less successful when they failed to do so.

Algeria, Libya, and North Yemen also acquired Soviet weapons and developed close military ties with Moscow. We focus on Egypt, Syria, and Iraq, the most militarily active Arab states during the Cold War and those with the closest military ties to the Soviet Union, because their military experiences have been thor-

---

[2] The Soviets used the term "doctrine" in a very precise sense, to refer to the entire body of military thought based on "scientific" Marxist-Leninist principles, including small unit tactics, operational art, strategy, and national security policy. Anglo-American usage of the term is typically much less precise. "Doctrine" is often used interchangeably to refer to tactics, techniques, and procedures, operations, or strategy. This paper will employ the latter usage, which is probably more familiar to most of its readers.

[3] Quoted in Roger Owen, "The Role of the Army in Middle East Politics: A Critique of Existing Analyses," *Review of Middle East Studies* 3 (1978): 70.

[4] Jon D. Glassman, *Arms for the Arabs* (Baltimore, MD: Johns Hopkins University Press, 1975), p. 50.

[5] For a pioneering work on this subject, which provided some inspiration for this study, see Amnon Sella, "Soviet Military Doctrine and Arab War Aims," in Itamar Rabinovich and Haim Shaked, eds., *From June to October: The Middle East Between 1967 and 1973* (New Jersey: Transaction, 1978), pp. 77–92.

oughly documented. Moreover, Egyptian and Syrian performance in 1973 and Iraqi performance in 1991 provide a good basis for comparison, since they reflect the range of Soviet influence on Arab militaries. The Syrians tried to emulate the Soviets closely; the Iraqis emulated least, while the Egyptians fell somewhere in between.

The Soviet Role in the Middle East

After gaining independence after World War II, the armies of Egypt, Syria, and Iraq continued to reflect the organization, doctrine, and traditions of their former colonial masters: Britain (Egypt and Iraq) and France (Syria). After a series of military coups brought to power radical, secular-nationalist regimes in Syria in 1949, Egypt in 1952, and Iraq in 1958, the Soviet Union became their principal political and military patron.

All three states aspired to a leadership role in the Arab world and hoped that Moscow would help them achieve that goal. Each depended on Soviet hardware and know-how to build a modern, effective military. As an ally, Moscow offered critical advantages over the West. First, the Soviet Union had no colonial past, and countries in the region were struggling to overcome the legacy—if not the continuing reality (in Algeria)—of British and French colonial rule. Second, "Arab socialist" ideologies were compatible with Soviet ideology. The ideology of radical nationalist regimes in Egypt, Syria, and Iraq borrowed heavily from the lexicon of Marxism and Soviet communism, and shared with the Soviet Union a deep-seated antipathy toward Western "imperialism" and the forces of Arab "reaction." Third, the radical Arab states and the Soviet Union were committed to an anti–status quo political agenda. The Soviets supported the Arabs in their conflict with Israel (if not Arab aspirations to destroy the Jewish state), and efforts to overthrow conservative, pro-Western Arab monarchies. Finally, unlike the UK, France, and the United States, which imposed restrictions on arms transfers to the region beginning in 1950, the Soviets were willing to provide large quantities of hardware at low prices and on favorable credit terms. Sometimes they provided weapons not yet available to their Eastern European allies.

The Soviets were eager to exploit their opening in the Arab world. Moscow recognized the Middle East as a key arena in its competition with the West, because of its location and its oil resources. Soviet military involvement in the region essentially began with the famous "Czech arms deal" of 1955, in which the USSR used its Czech allies to funnel arms to Egypt.[6] Soviet arms sales to Syria and

---

[6]The Czech arms deal provided Egypt with 230 tanks (primarily T-34/85s), 200 APCs (mostly BTRs), 100 Su-100 self-propelled guns, 500 artillery pieces, 200 jet combat aircraft (120 MiG-15s, 50 Il-28s, and 20 Il-14s), as well as several destroyers, submarines, and motor torpedo boats. Prior to the deal, Egypt

Iraq commenced shortly thereafter, in 1958. This assistance, however, failed to transform the armed forces of these countries—still deeply mired in domestic political intrigue and coup-plotting—into modern, professional armies.

The stunning defeat of the Arabs by Israel in 1967 changed all this. The Arabs recognized that they would not only have to rearm but also have to develop a more professional approach to waging war. Only the Soviets were willing or able to provide the necessary assistance.

The 1967 defeat transformed the relationship between the Soviet Union and its Arab allies. Moscow undertook a major resupply effort to replace Arab war losses. Both sides agreed that the Soviets had to provide more extensive support if the Arabs were to achieve the professionalism and competence needed to regain lost territory. Between 1967 and 1973, the Soviet advisory presence in Egypt grew from a few hundred to nearly two thousand.[7] In Syria it topped fifteen hundred. The function of Soviet advisors in Egypt and Syria also changed, from providing instruction on Soviet-made weapons to training the Arab armed forces in all manner of planning and operations. Soviet advisors were placed in all branches of the Egyptian and Syrian armed forces.[8] They were assigned to every military training facility, air and naval base, and maintenance depot; attached to all ground units down to battalion level; air units down to squadron level; and on every major naval combatant.[9] The Soviet advisory presence in Iraq was less extensive; it grew from some four hundred men in 1970 to about one thousand by 1975, remaining at those levels through the 1991 Gulf War.[10] Baghdad desired to

---

and Israel had fewer than 200 tanks apiece: Egypt possessed 80 old British jet aircraft (mostly Vampires), while Israel boasted only 50 early-model French and British jets (Ouragons and Meteors). Moshe Dayan, *Diary of the Sinai Campaign* (New York: Shocken, 1967), pp. 4–5; and Nadav Safran, *From War to War* (New York: Pegasus, 1969), p. 209.

[7]The latter figure does not include fifteen to eighteen thousand Soviet air defense personnel in Egypt during and after the 1969–70 Egyptian-Israeli War of Attrition.

[8]In the words of one scholar, "[T]he magnitude of the Soviet commitment was unprecedented, surpassing in both quantity and quality the aid given to North Vietnam and exceeding the rate at which aid had hitherto been given to allied or friendly countries." Alvin Z. Rubinstein, *Red Star on the Nile: The Soviet-Egyptian Influence Relationship since the June War* (Princeton: Princeton University Press, 1977), p. 30.

[9]Ibid, p. 30; Efraim Karsh, "Soviet Arms for the Love of Allah," *U.S. Naval Institute Proceedings* 110, no. 4 (Apr. 1984): 48. Numbers provided by various Egyptian sources vary considerably. Former Egyptian chief of operations during the 1973 War, Field Marshal 'Abd al-Ghany al-Gamasy, puts the total Soviet presence at about half this number—about 850 advisors, and 100 technicians; former minister of war Lt. Gen. Muhammad Fawzi puts the number of Soviet advisors in 1970 at 1,200; while the authoritative Egyptian journalist Muhammad Heikal puts the number of Soviet advisors and technicians in Egypt after the 1967 war at 1,500. Mohamed Abdel Ghani El-Gamasy, *The October War* (Cairo: American University in Cairo Press, 1989), p. 146; Muhammad Fawzi, *Harb al-Thalath Sanawat 1967–1970: Mudhakkirat al-Fariq Awwal Muhammad Fawzi* [*The Three Years War: 1967–1970: Memoirs of General Muhammad Fawzi*] (Cairo: Dar al-Mustaqbal al-'Arabi, 1983), p. 357; Mohamed Heikal, *The Road to Ramadan* (New York: Quadrangle, 1975), p. 41.

[10]Efraim Karsh, *Soviet Arms Transfers to the Middle East in the 1970s*, Jaffee Center for Strategic

limit Soviet influence (based in part on fear of Soviet meddling in Iraq's domestic politics) and relied less on Soviet doctrine.

Large numbers of Arab officers went to the USSR for training. Between 1955 and 1979, twenty thousand Arab military personnel studied or trained in the Soviet Union, most being Egyptians, Syrians, and Iraqis. The advisory presence, and the training and education of Arab military personnel in the Soviet Union, were the principal means by which Soviet organization, tactics, and operational concepts were transmitted to their Arab allies.[11]

The effort did not always go well. In the Egyptian case, Russian cultural insensitivity, Egyptian distrust of Soviet political and military motivations, and Egyptian pride hindered the transmission of Soviet doctrine.[12] The Egyptians complained that the Soviets withheld weapons systems that they needed to go to war, and that the systems delivered were "two steps" behind those provided to Israel by the United States. This ensured continued Arab military inferiority.[13] Many Egyptian officers considered the Soviet advisors to be inexperienced and condescending. The decision to increase dramatically the Soviet advisory presence after the 1967 war created tension and resentment in the Egyptian military, and dissension between the Egyptians and their Soviet advisors. According to Egyptian president Nasser's confidant, journalist Mohamed Heikal, the presence of Soviet advisors created a dilemma for the Egyptians:

> Was the advice given by the experts indeed only advice, or had it to be acted on? When a ruling was given that the advice was, in fact binding on them, many officers felt humiliated. They did not feel that Egypt's defeat [in 1967] was due to any deficiencies on their part that foreign experts, again with their inevitable interpreters, were likely to be able to put right. After all, they had combat experience and their advisors had not. They had commanded tanks and flown MiGs against the enemy, which the Russians had not. They knew local conditions in a way that their advisors never could. Russian military thought was still conditioned by memories of World War Two's 2,000 mile front, whereas the Egyptian army was never likely to operate on a front of more than 100 miles (the length of the Suez Canal). Moreover, the quality of the experts varied considerably [and while some] commanded universal respect . . . others were less admirable and, as the proverb says, sickness is contagious but health is not.[14]

Studies Paper No. 22, Tel Aviv University, Israel, Dec. 1983, pp. 23–27; Michael R. Gordon, "Pentagon, Disputing Moscow, Says 500 to 1,000 Soviet Advisors Are in Iraq," *New York Times*, 26 Sept. 1990, p. A6.

[11]Karsh, *Soviet Arms Transfers*, p. 23.

[12]Colonel E. V. Badolato, "A Clash of Cultures: The Expulsion of Soviet Military Advisors from Egypt," *Naval War College Review* 37 (Mar.–Apr. 1984): 69–81.

[13]See, for instance, Anwar el-Sadat, *In Search of Identity: An Autobiography* (New York: Harper and Row, 1977), pp. 197–98, 219–21, 225–26, 228–31, 238.

[14]Heikal, *Road to Ramadan*, pp. 181–82. Similar frictions developed between the Soviets and the Syrians and Iraqis. In 1972, the Soviet ambassador to Damascus fumed: "These damned Syrians, they will take anything except advice." Insight Team of the *London Sunday Times*, *The Yom Kippur War* (New York: Doubleday, 1974), p. 72. Likewise, Soviet advisors often found that cultural factors frustrated their

Many Soviet personnel held a low opinion of their Egyptian counterparts. In the words of a former Soviet air defense officer who served in Egypt during the 1969–70 War of Attrition:

> The Egyptians . . . had no confidence in the Soviet hardware, which they often said was inferior. But it was by no means the Soviet equipment that was to blame for their defeats, it was rather the low training standard of their missile crews. For example, they would promptly vacate their work stations upon firing a missile, and it never occurred to them that a missile needed to be guided in flight. The Soviet advisors with the Egyptian battalions could not do much and they often perished with the poorly trained crews. For you to compare, our battalion took 32 minutes to take up a new position, and an Egyptian one required 3 to 4 hours.[15]

The decision by Egypt, Syria, and Iraq to import Soviet equipment and know-how was dictated by shared interests and the availability of large numbers of advanced arms on favorable terms, rather than on any natural affinities between the parties, or any objective assessment of the relative merits or suitability of the Soviet military system. Nonetheless, the Soviet Union was an appealing ally to many Third World countries during the Cold War. The Soviets boasted a large, modern, and sophisticated military that enjoyed a certain prestige as a result of its epic achievements during World War II. Soviet military thought during much of the Cold War was more advanced and developed (and therefore, in a certain sense, superior to and more appealing) than its Western counterparts. To many in the Arab world, it also appeared that the Soviet Union was a rising power with history on its side. Aligning with the Soviet Union was an investment in the future.

Soviet Military Doctrine

Soviet military experience during World War II and the development of tactical nuclear weapons in the 1950s heavily influenced Soviet military doctrine during the Cold War. The main principles of modern Soviet doctrine had crystallized by the early 1960s, with only relatively minor modifications thereafter. The salient characteristics of the Soviet military model are described below.[16]

---

ability to train their charges. Lt. Col. Sergey Ivanovich Belzyudnyy, "Former Soviet Advisor Describes Experiences in Iraq: I Taught Saddam's Aces to Fly," *Komsomolskaya Pravda*, 23 Feb. 1991, translated by the U.S. government's Joint Publication Research Service, in JPRS-UMA-91-014, 5 June 1991, pp. 62–63.

[15]Col. D. Povkh, "Courage for Export: Interview with Lt. Col. Nikolai Kutyntsev," *Soviet Soldier*, no. 11 (November 1991), p. 65.

[16]This section is based on the following sources: V. D. Sokolovskiy, *Soviet Military Strategy* (New York: Crane, Russak, 1968); William F. Scott, "Changes in Tactical Concepts within the Soviet Forces," and John Erickson, "Soviet Theater-Warfare Capability: Doctrines, Deployments, and Capabilities," in Lawrence L. Whetten, ed., *The Future of Soviet Military Power* (New York: Crane, Russak, 1976); Friedrich Wiener and William Lewis, *The Warsaw Pact Armies* (Vienna: Carl Ueberreuter, 1977); John Erickson, "The Soviet Military System: Doctrine, Technology and 'Style,'" in John Erickson and E. J.

Organization

To implement tactics that emphasized rapid movement based on superior reconnaissance and firepower and a high degree of mechanization, Soviet ground
combat units generally possessed a large organic reconnaissance capability, large
numbers of antitank and artillery systems, and were tank heavy. The Red Army's
order of battle was highly mechanized, with a relatively large number of tank
formations—reflecting the central role of the tank in Soviet doctrine. In 1973 the
Red Army consisted of 50 tank, 107 motorized rifle, and 7 airborne divisions, for a
total of 164 divisions.[17] Soviet forces in Eastern Europe had an even higher proportion of tank to motorized rifle divisions, reflecting their preferred ratio of tank
to motorized rifle units in a potential military theater of operations. In 1973, the
Red Army in Eastern Europe fielded 15 tank and 16 motorized rifle divisions—out
of a total of 31 divisions deployed there.[18]

Tactics and Operations

In Soviet military thought, tactics concerned the use of fire and maneuver to
win battles; operations concerned the waging of campaigns (usually consisting of
a series of battles); while military strategy concerned the use of a nation's political, economic, moral, and military instruments to achieve political aims. The Soviets firmly believed that wars were won at the operational, not the tactical, level.
They concentrated resources and authority at operational levels of command.
The Soviets purposely limited the freedom of tactical commanders, while maximizing that of operational level commanders. A key element of this differentiated
command structure was the Soviet reliance on "battle drills" up to battalion level,
to ensure that tactical commanders could not, through initiatives of their own,
upset the plans or foreclose the options of senior commanders. Battle drills were
well-defined sets of tactical maneuvers (very much like football plays) that tactical commanders were expected to employ to conduct their missions. For any
given situation, the Soviets had a certain number of battle drills available. Small-
unit commanders were expected to draw on and adapt them to accomplish their
tasks while tailoring the selected battle drill to the mission, enemy, terrain, and

---

Feuchtwanger, eds., *Soviet Military Power and Performance* (Hamden, CT: Archon, 1979); David Isby,
Weapons and Tactics of the Soviet Army (New York: Jane's, 1981); Christopher Donnelly, *Red Banner:
The Soviet Military System in Peace and War* (Alexandria, VA: Jane's, 1988); Col. Ghulam Dastagir War-
dak, *The Voroshilov Lectures: Materials from the Soviet General Staff Academy* (Washington, DC: NDU,
1989); Raymond L. Garthoff, "Continuity and Change in Soviet Military Doctrine," in Bruce Parrott,
ed., *The Dynamics of Soviet Defense Policy* (Washington, DC: Woodrow Wilson Center, 1990), pp. 143–
85; and Aleksandr A. Svechin, *Strategy* (Minneapolis: East View, 1992).

[17]International Institute for Strategic Studies, *The Military Balance, 1973–1974* (London: IISS, 1974),
cited in Jeffrey Record, *Sizing up the Soviet Army* (Washington, DC: Brookings Institution, 1975), p. 12.

[18]Erickson, "Soviet Theater-Warfare Capability," p. 125.

other relevant considerations. Likewise, Soviet air force pilots were vectored to their engagements by ground controllers, though it was up to the flight leaders and individual pilots to choose the appropriate tactics once they had engaged the enemy.

Red Army tactics also emphasized the importance of combined arms integration, firepower, and maneuver to success in small-unit engagements. Moscow taught that ground and air forces, and combat and combat support units, had to work closely together to achieve an effect that was greater than the sum of their parts. The Soviets were also famous for the massive use of artillery to blast holes in an enemy's lines. Nonetheless, tactical units were still expected to secure an advantage over their adversaries through combat maneuver, by enveloping or flanking enemy formations where possible, breaking through enemy positions when necessary, and driving into the enemy's rear in order to cut his communications, disrupt the deployment of his reserves, seize key objectives, and envelop enemy units.

Hallmarks of Soviet Operations

Soviet military thought placed heavy emphasis on achieving rapid victory through offensive operations intended to annihilate the enemy's forces. The defense was a temporary expedient to be abandoned as soon as sufficient force had been built up to move to the offensive. Soviet offensives stressed shock and surprise; attacking throughout the enemy's depth; achieving superiority in numbers and firepower at the point of decision; seeking multiple penetrations in order to minimize the vulnerability of their forces to tactical nuclear weapons;[19] and conducting continuous operations (achieving advance rates as high as 50 to 70 kilometers per day) to maintain the initiative and momentum of attack. Soviet offensives were conducted in echeloned assaults. Waves of fresh troops eventually would overpower a defender through weight of numbers, firepower, and relentless pressure. Motorized rifle divisions led in the first echelon of an attack, followed by tank divisions that exploited successful penetrations by first echelon forces. Operational maneuver groups consisting of armor and mechanized formations would drive deep into the enemy's rear, wreaking havoc along the way, while airmobile and airborne forces would seize key objectives in the enemy's rear, eventually linking up with advancing ground forces.

Flexibility was a critical element of Soviet operational thought. In the attack, Soviet operational commanders kept their most powerful forces in reserve to re-

[19]During World War II, Soviet tactics emphasized the need to mass forces at a small number of breakthrough points. With the development of tactical nuclear weapons, massing forces could be dangerous. Doctrine was modified, calling for rapid concentration of forces at multiple breakthrough points, followed by their rapid dispersal thereafter.

TABLE 3.1

The "Soviet Model" and Its Impact on Three Arab Armies

Elements of the "Soviet Model"	Syria 1973	Egypt 1973	Iraq 1988/1991
War serves political objectives; military operations and strategy should be subordinate to these. Force and diplomacy are linked.	War planning was divorced from diplomatic considerations. The sole objective: retake the Golan by force.	War planned in accordance with a well-defined political-military concept: seize a foothold on the east bank of the Suez Canal to lay the groundwork for the return of the rest of Sinai through negotiations.	1988: retake territory lost to Iran through short, set-piece offensives, to facilitate a cease-fire on favorable terms. 1990: occupy and annex Kuwait through a short, set-piece offensive, to resolve Iraq's financial and geostrategic problems. 1991: fight the coalition to a stand-still, and negotiate a diplomatic solution to the crisis.
Rapid victory through surprise, offensive action, and continuous operations to annihilate enemy forces.	Regain the Golan through a short, limited, surprise attack with geographic objectives.	Seize a foothold in the Sinai through a limited surprise attack, assume a defensive posture, and defeat Israeli counterattacks.	1988: victory through a series of short, set-piece offensives to seize and hold occupied territory. 1990: victory through a short, set-piece offensive to occupy Kuwait. 1991: allow coalition forces to attack, fight them to a stand-still, and negotiate a favorable diplomatic solution.
Echeloned attacks with multiple breakthrough points.	Echeloned attack with two or three breakthrough points.	No echelonment, attacked along a broad front, no real attempt at a breakthrough.	1988: no echelonment, attacks limited to narrow fronts with multiple breakthrough points. 1990: echeloned attack along two axes. 1991: not applicable.
Attack through the enemy's depth (through air and missile operations, and use of operational maneuver groups).	Limited, unsuccessful efforts to strike deep.	Limited, partially successful efforts to strike deep.	Limited, partially successful efforts to strike deep.
Preemptive strikes on enemy tactical nuclear forces.	Not applicable.	Not applicable.	Not applicable.
Fast-paced operations, high rates of advance (50–70km/day).	Not applicable or feasible.	Not applicable or feasible.	Not applicable or feasible.
Maximize freedom of operational commanders; tactical commanders choose from a small repertoire of battle drills.	Only the president has authority to make key operational-level decisions; rigid hierarchy at all levels below.	Make key operational-level decisions; rigid hierarchy at all levels below.	Only the president has authority to make key operational-level decisions; rigid hierarchy at all levels below.

TABLE 3.1—*cont.*

Elements of the "Soviet Model"	Syria 1973	Egypt 1973	Iraq 1988/1991
Emphasis on detailed planning of operations, but a plan is still the basis for change.	Emphasis on scripting of set-piece operations: the plan is sacrosanct.	Emphasis on scripting of set-piece operations: the plan is sacrosanct.	Emphasis on scripting of set-piece operations: the plan is sacrosanct.
Highly mechanized forces.	Partially mechanized forces.	Infantry-heavy forces.	Infantry-heavy forces, but key units mechanized.
Tank plays central role in force structure.	Tank plays central role.	Infantry armed with antitank weapons plays central role.	Tank plays central role.
Emphasis on combined arms.	Attempted, but largely unsuccessful.	Attempted, but largely unsuccessful.	Attempted, but largely unsuccessful.
Air superiority key to success.	Deny Israel air superiority with Soviet-style, ground-based integrated air defense system.	Deny Israel air superiority with Soviet-style, ground-based integrated air defense system.	1988/1990: air power contributes to success, which hinges on overwhelming numerical superiority on the ground. 1991: Deny coalition forces air superiority with Soviet-style, ground-based integrated air defense systems during the coalition air campaign, then commit air power during the decisive ground phase of the war.

inforce success. Commanders were taught to commit all available forces to exploit the most successful penetrations in order to achieve breakthroughs, and to deploy forces and issue orders so that plans could quickly be altered if an unexpected opportunity or problem arose. On the defensive, Soviet commanders deployed their forces in depth and held their most powerful units in reserve. Weaker forces would absorb the enemy's initial blows. Senior commanders could then identify the enemy's main effort and concentrate reserves against them in powerful counterattacks.

Similar principles guided Soviet air operations. Air and ground operations were closely coordinated. Soviet operational-level ground commanders could use aviation assets as circumstances required. Air superiority was considered crucial to success. Moscow preferred to secure air superiority by conducting large-scale offensive counter-air operations including relentless fighter sweeps over enemy territory coupled with air, rocket, and missile strikes against enemy air defenses, air bases, and command and control facilities (including assets such as AWACS—Airborne Warning and Control System aircraft). From the start of a

Soviet ground assault, air, rocket, and missile forces would strike enemy nuclear forces, assembly and staging areas, command and control (C^2) facilities, transportation choke points, operational reserves, and other key targets throughout the enemy's depth. Heavy emphasis was placed on neutralizing enemy nuclear forces before they could be used, and defeating the enemy's forces before his reserves and military-industrial potential could be mobilized.

To facilitate implementation of their operational concept, the Soviets maximized the forces controlled by operational level commanders. Soviet armies usually consisted of four to as many as twelve divisions, while the corresponding Western formation—the corps—usually was limited to two to four divisions. A typical Soviet army or front commander had at his disposal far more assets than did a typical Western corps or army commander. In addition to controlling tank and motorized rifle divisions, Soviet army and front commanders controlled large numbers of fixed-wing aircraft; attack helicopter regiments; combat engineer regiments or brigades; air defense brigades; artillery regiments, brigades, or divisions; and even independent tank regiments. This allowed them maximum flexibility in deciding where to concentrate assets for the decisive blow (see Table 3.1).

Soviet Military Doctrine in the Arab World: The Cases of Egypt, Syria, and Iraq

As Soviet military involvement in the Middle East deepened during the 1960s and 1970s, the Soviets tried to impart their approach to war to their Arab clients, with mixed results. Some Arab armies tried to imitate the Soviets, while others selected only those elements they found useful. The experiences of Syria and Egypt in the 1973 October War, and that of Iraq during the 1991 Gulf War, provide three different examples of how Soviet military methods affected Arab combat operations and influenced Arab military performance. Overall, it is not possible to establish a direct relationship between the adoption of Soviet methods and Arab success or failure on the battlefield.

Syria, 1973

Of the three armies examined, Syria's most closely resembled the Red Army, and most closely followed Soviet practices. Of the three, the Syrians in 1973 also fared worst in combat. The problem, however, lay less in their adherence to Soviet methods than in their application of those methods in a rigid, mechanical manner never intended by the original authors. This factor, plus the incompetence of Syrian troops and their tactical commanders, led to the failures of October 1973.

Organizationally, the Syrian armed forces generally resembled the Red Army. On the eve of the 1973 war, the Syrian army consisted of three partially mechanized infantry divisions, two armored divisions, three independent armored brigades (including the Defense Companies, a praetorian guard force), five special forces battalions, and one airborne battalion.[20] The Syrian army was highly mechanized. Its ground order of battle was more or less evenly divided between mechanized and armored formations. Syrian tactical formations—for its army, air force, and air defense forces—were largely patterned after their Soviet counterparts. Syria subordinated large numbers of military assets directly to the general staff in Damascus, including several independent armored brigades, special forces battalions, and additional artillery and rocket launcher battalions.[21]

Syria's objective in 1973 was to retake the Golan by force. Syria (unlike Egypt) did not have a political-military strategy that saw war as a means to initiate a political process. It was not unrealistic for Syria to believe that it could recover the Golan in war, whereas Egypt's only alternatives were either a series of costly wars or using a war to create momentum toward a negotiated return of the Sinai.[22]

At the operational level, the Syrian plan for retaking the Golan relied heavily on Soviet concepts. The Syrian army sought to break through Israeli lines in two sectors, envelop the Israeli forces defending the Golan, and seize and hold positions along the Jordan River and the Sea of Galilee, thereby securing the entire Golan within the first twenty-four hours of the war. The assault was led by three mechanized divisions, followed by a second operational echelon of two armored divisions to exploit any penetrations. Syria conducted several largely unsuccessful airmobile operations against various objectives on the Golan, to disrupt Israeli command, control, communications and intelligence, and mobilization efforts. Syria also launched a number of FROG rockets at a key air base in northern Israel, in an effort to disrupt Israeli air operations.[23]

[20]U.S. Defense Attache Office (USDAO) Tel Aviv, Intelligence Information Report (IIR) 6 849 0094 74, "Egyptian and Syrian Armies and Their Activities during the October 1973 Yom Kippur War," 10 Apr. 1974, p. 11 (declassified under the Freedom of Information Act).

[21]The organization of Syrian forces in the 1970s and early 1980s was dictated, to a great extent, by Syrian president Hafiz al-Asad's fear of a military coup. For a number of years after the 1973 war, Asad refused to add new divisions to the army's order of battle or to create an intermediate echelon of command between division commanders and the Syrian general staff, fearing that corps or army commanders could threaten his grip on power. With the consolidation of his rule during the 1970s and the Israeli invasion of Lebanon in 1982, Asad created a number of new divisions and even permitted the formation of two to three corps commands in the early to mid-1980s.

[22]After 1973, however, Asad showed greater sophistication in using the military instrument as a political tool in Lebanon and vis-à-vis Israel. See, for instance, Lt. Col. Daniel Asher, "Ha-Plisha Ha-Surit LeLevanon: Mahlachim Tzvaiim KeMachshir Medini" ("The Syrian Invasion of Lebanon: Military Moves as a Political Tool"), Ma'arachot (June 1977): 7–16.

[23]For concise Arab and Israeli assessments of the fighting on the Golan, see: Lt. Col. Al-Haytham al-Ayyubi, "Al-Qital 'ala al-Jabha al-Suriyya" ["The Fighting on the Syrian Front"], in Lt. Col. Al-

The Syrian attack broke down at the tactical level, a product of the ineptitude of Syrian troops and tactical commanders. Many bridging units tasked to span Israeli antitank ditches were located at the rear of the advancing Syrian formations and were not in position to facilitate the movement of the assault force when it arrived at the ditches.[24] Syrian division commanders led their attacks against the dug-in Israelis with pure tank battalions that had not been reinforced with infantry—a violation of the Soviet emphasis on combined arms—leaving them easy prey for Israeli tank crews and anti-tank teams.[25] Most of the airmobile assaults were conducted during daytime—when they were most vulnerable—and were consequently thwarted by Israeli air or ground forces. Syria lacked accurate surface-to-surface missiles that could have enabled them to disrupt Israeli mobilization efforts. Of greatest importance, Syrian infantrymen, tank crews, and fighter pilots were not very skillful. They were outmaneuvered and outfought by the Israelis.

The Syrians implemented a caricature of the Soviet model that lacked the flexibility and adaptability inherent in the Soviet approach. In the Red Army, tactical commanders had the latitude to choose among several courses of action in order to accomplish a mission. Syrian tactical commanders seemed loath to depart from their original plans, and plodded forward in the face of stiff resistance, often failing to maneuver against the enemy. Syrian infantry frequently remained mounted during the assault, even as their vehicles were picked off one by one by Israeli tanks. Except for a few instances, the Syrians did not aggressively employ dismounted antitank teams to neutralize Israeli armor during the assault, even though the Israeli tanks on the Golan had limited infantry support.[26] Lack of flexibility and initiative enabled small Israeli units to consistently maul far larger Syrian forces.

Lack of initiative had dire consequences for Syria's operational plan. For reasons that are still unclear, Syrian tactical commanders who had broken through

---

Haytham al-Ayyubi, gen. ed., *Al-Mawsu'a al-'Askariyya* [*The Military Encyclopedia*] (Beirut: Arab Institute for Studies and Publishing, n.d.), vol. 1, pp. 709–16); Bassam al-'Assali, "Harb al-Jawlan" ["The War in the Golan"], *Al-Difa 'al-Islami* 4, no. 13 (Apr.–June 1985): 20–26; and Lt. Col. (Res.) Tzvi, "Ma'arachah Hatkafit Surit BeRamat Ha-Golan" ["The Syrian Assault on the Golan in the Yom Kippur War"], *Ma'arachot* (Jan.–Feb. 1989): 21–29.

[24]Jerry Asher with Eric Hammel, *Duel for the Golan: The 100-Hour Battle that Saved Israel* (New York: William Morrow, 1987), pp. 90–91; Col. Trevor N. Dupuy, *Elusive Victory: The Arab-Israeli Wars 1947–1974* (New York: Harper and Row, 1978), pp. 445–47.

[25]Chaim Herzog, *The War of Atonement* (London: Weidenfeld and Nicholson, 1975), p. 75; Charles Wakebridge, "The Syrian Side of the Hill," *Military Review* (Feb. 1976): 28–29.

[26]Dupuy, *Elusive Victory*, pp. 455, 590; Wakebridge, "The Syrian Side of the Hill," pp. 28–29. Syrian antitank teams did, however, play a role in harassing Israeli tank laagers at night in the first days of fighting, and in defending the Sa'sa' salient against advancing Israeli forces during the latter phase of the war.

the Israeli lines failed to press their advantage and exploit their gains. On the second day of the war, despite enjoying an overwhelming local numerical advantage, Syrian forces stopped but a few kilometers short of the crucial bridges over the Jordan River. Through lack of initiative, Syrian tactical commanders failed to grasp a victory that was within their reach.[27]

Egypt, 1973

Although the Egyptians leaned heavily on their Soviet mentors to get them back on their feet after the catastrophe of June 1967, and to help them turn their armed forces into a professional organization capable of taking on the Israelis, the Egyptian military in 1973 was hardly a carbon copy of the Red Army. Cairo's senior military commanders recognized quickly that aspects of Soviet military doctrine could prove useful, but that it could not solve all of Egypt's operational and tactical problems. They explicitly acknowledged that the shortcomings of Egyptian soldiers and tactical commanders required them to modify key elements of Soviet doctrine. Egypt's initial victories and subsequent reverses during the October War demonstrated both their successes in borrowing and adapting Soviet practices, as well as the inability of Soviet methods to transcend critical Egyptian weaknesses.

The Egyptian army did not resemble the Red Army nearly as closely as the Syrian army did. The Egyptian army consisted of five infantry divisions, three mechanized infantry divisions, two armored divisions, five independent armored brigades, three independent infantry brigades, one naval infantry brigade, one airborne brigade, two air assault brigades, and six commando groups.[28] The Egyptian army of 1973 was mechanized to only a limited degree. The main assault formations were infantry divisions (long after the Red Army had abandoned the "straight leg" infantry division)—each augmented by an armored brigade. Armored formations constituted only about one-quarter of the total ground order of battle. On the other hand, Egyptian air and air defense forces closely resembled their Soviet counterparts.

The most important difference between the Egyptian and Soviet models in

---

[27]Several explanations for these failures have been offered: (1) advanced Syrian tank units outran fuel and ammo resupply convoys that had been disrupted by Israeli air strikes; (2) having reached their geographic objective for the day, the Syrians halted to consolidate their gains and await additional orders that never came; (3) fearing that their flank was exposed to an Israeli counterattack, the vanguard of the Syrian force halted to allow the rest of the force to catch up (which it never did). Herzog, *War of Atonement*, p. 104; Dupuy, *Elusive Victory*, p. 456. Regardless of which explanation is correct, the key consideration is that Syrian tactical commanders lacked the initiative to seize the paramount objectives of their war plan (the bridges on the Jordan) when they were within their grasp. This was an unforgivable violation of Soviet military doctrine, which stressed seizing such war-winning opportunities when they presented themselves.

[28]USDAO Tel Aviv IIR 6 849 0094 74, p. 2.

1973 lay at the operational level. Egypt's plan for crossing into the Sinai owed lit-tle—and in fact ran contrary to—Soviet operational principles. When the Egyp-tians and their Soviet advisors started planning an operation to cross the canal, they envisioned a grand, Soviet-style assault and breakthrough along the canal line followed by a vast exploitation to retake all of Sinai. The plan was code-named Granite. Following the appointment of Lieutenant General Ahmad Isma'il as minister of war in October 1972, however, the Egyptian general staff drastically revised its plans, recognizing that the limitations Egyptian forces displayed in previous conflicts made such an ambitious operation a pipe dream. They scrapped the original Granite plans and secretly developed a new operational concept without any Soviet input (which in fact they hid from both the Soviets and their Syrian allies), called High Minarets. It became the basis for Operation Badr—the actual crossing operation implemented in 1973.[29]

High Minarets was based on Egyptian planning assumptions and an Egyptian operational concept that departed in significant ways from Soviet operational doctrine.[30] The Egyptian plan called for seizing and consolidating a bridgehead on the eastern side of the Suez Canal, followed by an "operational pause."[31] Egypt would transition to the defense and Egyptian infantry—heavily reinforced with antitank weapons and armor—would destroy Israeli armored counterattacks.[32] Although the possibility of resuming the offensive in the future was left open, no detailed plans were written for subsequent offensive operations, no Egyptian field commanders were briefed about such an eventuality, nor was such an operation rehearsed. The method of strategic offensive coupled with an operational-tactical defensive, and reliance on attrition to wear down an enemy, rather than on ma-neuver to defeat him decisively, was contrary to the Soviet approach, which stressed offensive operations and retention of the initiative. However, Egyptian president Anwar Sadat's strategy in 1973 was a limited-war approach; he felt that Egyptian forces needed only to gain a toehold in the Sinai in order to initiate

[29]Lt. Gen. Saad El Shazli, *The Crossing of Suez* (San Francisco: American Mideast Research, 1980), pp. 36–37, 111; Ahmed Fakhr, "Sadat and the Transformation of Egyptian National Security," in Jon B. Alterman, ed., *Sadat and His Legacy: Egypt and the World, 1977–1997* (Washington, DC: Washington In-stitute, 1998), p. 68.

[30]Col. (Res.) Daniel Asher, "Me 'Horaah 41' Le-'Tahrir 41': MeTorat Lechima Mitzrit—LeMilcha-mah" ["From 'Order 41' to 'Liberation 41': From an Egyptian Combat Doctrine—to War"], *Ma'arachot* (Sept.–Oct. 1993): 46–53.

[31]The Egyptians thus reinterpreted the Soviet concept of the "operational pause." In Soviet military thought, a commander might order an operational pause in order to reorganize and reconstitute his forces in anticipation of the resumption of the offensive. In the Egyptian case, the operational pause was intended to facilitate the transition to the defense and the defeat of Israeli counterattacks.

[32]Each Egyptian infantry division—in addition to its four organic tank battalions, one BMP battal-ion, and one ATGM battalion—was reinforced with a tank brigade, an antitank gun (SU-100) battalion, and an ATGM battalion, and hundreds of other antitank weapons taken from second echelon units and army level reserves. Shazli, *Crossing of Suez*, pp. 225, 258.

postwar negotiations with the Israelis from a position of strength. His army did not need to seize more than a few kilometers of the Sinai.[33]

High Minarets featured a number of other concepts at variance with core tenets of Soviet doctrine. Egypt's infantry, not its armored divisions, were cast to play the decisive role in the October War. Five infantry divisions would cross the canal, consolidate the bridgehead, and beat back the Israeli counterattacks with antitank weapons. The mechanized and armored divisions were held in reserve to lend weight and firepower to any threatened sector and to defend against an Israeli crossing of the canal. This decision made good military sense given the poor performance of Egypt's mechanized units in tank combat with the Israelis, relative to the comparatively good performance of their infantry in defending fixed positions. Egyptian antitank units were not organized into antitank reserves to protect the flanks or rear of friendly forces against enemy counterattacks as prescribed by Soviet tactical thought, but were integrated into the assaulting infantry divisions and were to be key to the success of Operation Badr.[34] The canal crossing itself diverged from Soviet principles for water-crossing operations. These principles emphasize crossing along a broad front, at multiple points (which the Egyptians did) at night (the Syrians succeeded in insisting that the war start in the early afternoon), followed as soon as possible by a breakout from the bridgehead. Instead, the Egyptians consolidated their bridgehead and transitioned to the defense.[35] Except for a relatively modest number of strikes launched on the first day of the war, the Egyptian air force was held back from the front, used mainly to supplement the coverage of the ground-based SAMs and antiaircraft artillery (AAA), which bore the brunt of the air defense effort. Egyptian attack aircraft conducted limited strikes against Israeli command and control facilities and staging areas, but the air force never tried to disrupt Israeli operations systematically throughout the depth of Sinai. Finally, Egypt's conventionally armed Scud missiles were used for strategic deterrence and signaling, since they lacked the accuracy to serve as battlefield support weapons when mounting conventional warheads.

The genius of the Egyptian war effort lay in this unique operational plan. The severe limitations of Egyptian tactical formations were its Achilles' heel. The solution devised by Ahmad Isma'il and his staff was an approach that worked around this problem.

---

[33]See Heikal, *Road to Ramadan*, p. 184.

[34]Col. (Res.) Daniel Asher, "Ha-Neged-Tank KeMa'anah: Tichnon Haf'alat Ha-Emtza'im Neged-Tank Al-Yadei Ha-Mitzrim BeMilchemet Yom Ha-Kippurim" ["The Antitank Weapon as Response: The Employment of Antitank Weapons by the Egyptians during the Yom Kippur War"], *Ma'arachot* (Feb. 1996): 6–10.

[35]Avraham (Bren) Adan, *On the Banks of the Suez* (San Francisco: Presidio, 1980), pp. 64–65.

In the twenty-five years prior to the October War, Egyptian forces had been crippled by the consistently poor performance of their junior officers. To compensate for past difficulties with combined arms, initiative, and improvisation at tactical levels, Cairo's high command came up with a novel approach. The general staff scripted the entire operation down to the last detail. Every action of every squad and every platoon was detailed at every stage of the operation by general staff planners. Isma'il and Egypt's chief of staff, Sa'd al-Din Shazli, decreed that every Egyptian soldier should have only one mission, and that he should learn to perform that mission by heart. Full-scale mockups of the Israeli fortifications, the terrain on the east bank of Suez, and the canal itself were constructed and used by the Egyptian units to learn their missions. Operations were rehearsed incessantly, until every member of every unit knew exactly what he was supposed to do at every step of the attack. The entire offensive was rehearsed as a whole thirty-five times before the assault. Egyptian soldiers and officers were encouraged to memorize a series of programmed steps, and during the actual canal-crossing operation, junior officers were expressly forbidden from taking actions not specifically included in the general staff plan.[36] Because of this detailed scripting and rote memorization of tasks, the general staff planners were able to write combined arms coordination, tactical maneuver, and synchronized movement into the operations order, obviating the need for tactical commanders to innovate or take initiative—at least as long as things went according to plan.

The weakness in this approach was that it was practicable only as long as the Egyptians were able to stick to their rigidly planned and exhaustively rehearsed operation. At first, surprise and tremendous numerical advantages allowed Egypt to dictate the course of battle. (At the start of the war, the Egyptian assault forces held a 12:1 advantage in manpower, a 5:1 advantage in tanks, and a 20:1 advantage in artillery over the Israelis defending the Bar-Lev line.)[37] However, friction, the fog of war, mobilization of Israel's reserves, and adjustments in Israeli tactics brought the Egyptian advance to a halt short of its objectives on the fourth day of the war, despite Cairo's spectacular early advance. The Egyptians were able to penetrate only eight to twelve kilometers into the Sinai; they had hoped initially to seize a bridgehead some fifteen to twenty kilometers deep.[38]

Without a detailed script, Egyptian tactical formations floundered. The clearest illustration came on 14 October (the eighth day of the war), when Sadat

[36]Mohammed Heikal, "An Interview with Lt. General Ahmed Isma'il," *Journal of Palestine Studies* 3, no. 2 (winter 1974): 217–19; Herzog, *The War of Atonement*, pp. 34–37; Charles Wakebridge, "The Egyptian Staff Solution," *Military Review* (Mar. 1975): 6–7; Edgar O'Ballance, *No Victor, No Vanquished: The Yom Kippur War* (San Rafael, CA: Presidio, 1978), pp. 27–30, 338.
[37]Herzog, *The War of Atonement*, pp. 34–37; O'Ballance, *No Victor, No Vanquished*, pp. 27–30.
[38]Gamasy, *October War*, p. 139.

bowed to pressure from Syria to resume the offensive over the objections of General Ahmad Isma'il. Since the general staff had not planned for a renewed offensive, nor had their tactical formations been able to learn it by heart through endless repetitive rehearsals, the assault Egypt mounted on 14 October bore little resemblance to the canal-crossing operation. The Egyptian generals had insufficient time to prepare to resume the offensive against the Israelis. They had not brought up in sufficient numbers many of the assets needed to support a new attack—including combat engineer, artillery, and mobile air defense systems. They lacked sufficient space on the east bank of the canal to array the forces they would need to bring over from the west bank to support a renewed attack. Because of difficulties in improvising such an operation, what was intended as a pair of concentrated divisional attacks to break through and envelop Israeli forces opposing them, devolved into nine brigade-sized attacks at separate points along the breadth of the front line.[39] Within hours the Israelis had beaten back the Egyptian offensive, destroying 265 tanks and 200 other armored vehicles. Only 40 Israeli tanks were put out of action, and of those only 6 were total losses.[40]

What was so striking about the Egyptian failure on 14 October was how little it resembled the crossing of the Suez just days before. Without the detailed plans of the general staff, Egyptian tactical commanders fared poorly. The Egyptian artillery barrage covering the attack was huge, employing at least five hundred guns, but completely ineffective. Egyptian artillery was unable to deliver accurate fire without the prepared fire plans they had employed during the canal crossing.[41] Egyptian armor attacked Soviet-style, in waves, but made no effort to maneuver. The Egyptian T-55s and T-62s literally drove straight at the Israelis. In the words of one Israeli brigade commander, "they just waddled forward like ducks."[42] Egyptian forces were chewed up by superior Israeli tank gunnery, and—in the south, where Egyptian forces had ventured beyond the SAM umbrella—by the Israeli air force.[43]

[39] Adan, *Banks of the Suez*, p. 237.

[40] For accounts of the offensive of Oct. 14, see: Adan, *Banks of the Suez*, pp. 232–42; Dupuy, *Elusive Victory*, pp. 486–91; Gamasy, *October War*, pp. 276–80; Herzog, *War of Atonement*, pp. 205–6.

[41] Dupuy, *Elusive Victory*, pp. 488–89.

[42] Maj. Gen. Amnon Reshef, interview, Sept. 1996. Reshef was a brigade commander in Mendler's (later Magen's) *ugdah* (roughly, division), facing the Egyptian Third Army.

[43] Maj. Doron, "Krav Ha-14 BeOctober BeGizrat Vadi Mabuk" ["The Battle of 14 October in the Wadi Mabuk Sector"], *Ma'arachot* (Nov. 1978): 23, 26–27. The Egyptians were also hindered by the mechanistic implementation of Soviet doctrine. Before 1973, Soviet doctrine called for tank platoon leaders to designate a single target, which the entire platoon would fire at until destroyed. The platoon leaders would then identify a new target to engage. The Soviets calculated that, given the gunnery skills of their crews, it normally would take three salvoes from the platoon (nine rounds) to kill an enemy tank at the preferred range of engagement. Rather than seeing this as a general guide for action, the Egyptians turned it into a hard-and-fast rule and taught all of their tank platoons to fire three rounds at the designated target and then proceed to the next. Egyptian gunnery skills were so poor that often none

Iraq, 1991

Of the three armies under consideration, the Iraqi army in 1991 was least influenced by Soviet ideas and practices. Despite lingering resentment over the British colonial legacy, the Iraqi military clung to British military organizational forms and tactics, though it modified them with certain elements of Soviet, French, and U.S. organization and tactics.[44] Most of the Soviet advisors in the country maintained certain complex systems (such as the air defenses) and provided weapons instruction.[45] Like the Egyptians, Iraqi soldiers and junior officers lacked the skills needed to implement foreign doctrines as intended. They performed best when they devised indigenous tactics and techniques that conformed to their own (limited) range of abilities, based on past experiences against the Kurds, Israelis, and Iranians.[46] The Iraqi military achieved its greatest successes when it implemented its indigenous approach, and suffered its worst defeats when circumstances or a skilled adversary prevented it from doing so.

During the Iran-Iraq War (1980–88), Iraqi military organization reflected British, Soviet, and French influences. Although Iraqi tank platoons always consisted of three tanks (per Soviet practice), Iraqi companies sometimes consisted of three platoons (the Soviet model) and sometimes four (the British model). Battalions might consist of three, four, or even five companies. Divisions had no set table of organization and equipment: a divisional headquarters frequently controlled more than half a dozen brigades (sometimes as many as a dozen); an armored divisions headquarters might command mostly leg infantry brigades, while leg infantry divisions headquarters might command more armored bri-

---

of the three salvoes hit home. Nevertheless, because they had been taught to fire three rounds and then move on, tank platoon leaders would generally shift fires to the next target even though they had not actually destroyed the first one. In this way, the Egyptians lost a great many tank duels to the Israelis. Interviews with retired Israeli military officers, 1992–94.

[44]Ofer, ed., *Tzva 'Iraq BaMilchemet Yom Ha-Kippurim* [*The Iraqi Army in the October War*] (Tel Aviv: Ministry of Defense Publishing House, 1986), p. 33; National Training Center (NTC), *The Iraqi Army: Organization and Tactics*, Handbook 100–91, Jan. 3, 1991, pp. 1, 35, 37, 110, 112, 117, 128, 133, 152, 159, 173; Belzyudnyy, "I Taught Saddam's Aces to Fly," p. 62. The Iraqis basically retained the old World War II version of British doctrine (this was the last time that Iraqi troops had been trained by the British), and many of their manuals captured during the 1991 Gulf War were translations of old British Sandhurst manuals. However, while the doctrine of the Iraqi army remained overwhelmingly British, its execution was uniquely Iraqi.

[45]For instance, according to a Soviet air force advisor based in Iraq prior to the 1991 war, "[M]any Iraqi pilots had been trained in Western countries, France or England. . . . Correspondingly, the organization of the units as well as the tactics of air combat followed Western models very distinct from ours. But ours was a very narrow, specific task and we focused all our attention on instructing the Iraqis in piloting techniques, without imposing our own notions about air force tactics." Ibid., p. 62.

[46]William Staudenmaier, "Iran-Iraq, (1980–)," in Robert Harkavy and Stephanie Neuman, eds., *The Lessons of Recent Wars in the Third World* (Lexington, MA: Lexington Books, 1986), vol. 1, p. 218; and John S. Wagner, "Iraq: A Combat Assessment," in Richard Gabriel, ed., *Fighting Armies in the Middle East* (Westport, CT: Greenwood, 1983), p. 67.

gades than many armored divisions. Brigades were reassigned constantly from one division to another.[47] Iraqi air defenses were patterned on the Soviet model, though a French firm (Thomson-CSF) had designed Iraq's KARI national integrated air defense system, which consisted of Soviet and French AAA, SAMs, and radars.

On the eve of Operation Desert Storm in 1991, the Iraqi army looked even less like the Red Army than had its Egyptian or Syrian counterparts in 1973. The Iraqi army consisted of sixty-six divisions, including eight armored divisions, five mechanized infantry divisions, fifty-two infantry divisions, one special forces division, and about thirty independent brigades of various kinds.[48] The Iraqi army was, in quantitative terms, largely an infantry army. Its backbone, however, resided in three heavy divisions of the Republican Guard and seven of the regular army. These had borne the brunt of offensive operations during the Iran-Iraq War. The large number of infantry divisions were used mainly to hold the border against Iran.

Baghdad's concept for fighting the U.S.-led coalition during the 1991 Gulf War diverged from a number of Soviet doctrinal precepts. The Iraqis intended to remain largely on the defensive; Baghdad attempted neither to gain air superiority over the battlefield nor to seriously challenge coalition air operations over its own air space. Instead, Iraq tried to absorb coalition air strikes during the air campaign and attrit coalition air forces by the use of SAMs and AAA, while withholding its air force for the latter (presumably decisive) stages of the war. On the ground, the Iraqi concept was to use dug-in infantry to absorb the coalition attack, and to counterattack with Republican Guard and regular army heavy divisions once the main coalition effort had been identified. Rather than seek to destroy coalition ground forces, Iraq's strategy was to fight the coalition to a standstill, inflict unacceptable losses, and compel them to accept a negotiated end to the fighting. Iraq's strategy and its operational concept, flawed in retrospect, aimed at broader political objectives.

At the tactical level, the Iraqis least resembled the Soviets. They used mostly British and indigenous doctrine in their ground forces. Where they did attempt to employ Soviet doctrine, they experienced the same problems as the Egyptians and Syrians; the rigidity and incompetence of their tactical formations precluded them from implementing Soviet methods as intended. The Iraqi air force had embraced Soviet-style ground-controlled intercept procedures. Iraqi ground-controlled interception (GCI) officers vectored Iraqi fighters flying defensive counterair missions against enemy aircraft, positioning them for combat. Once

---

[47]NTC, *The Iraqi Army*, pp. 5, 11.

[48]Michael Eisenstadt, *Like a Phoenix from the Ashes? The Future of Iraqi Military Power* (Washington, DC: Washington Institute, 1993), p. 84.

engaged, Iraqi pilots were taught to rely on British and French air-to-air tactics, which placed a premium on independent action and allowed pilots considerable latitude to engage the enemy. However, Iraqi pilots proved unable to perform Western dogfight tactics, and thus leaned heavily on GCI guidance (which was frequently inept). When coalition jamming disrupted GCI direction, Iraqi pilots failed to react, flying straight ahead or fleeing battle. U.S. fighter pilots were astonished at Iraqi pilots' slow and clumsy maneuvering, simplistic tactics, and fundamental mistakes. Iraqi fighters were shot down almost effortlessly. Several were killed because their GCI controllers guided them into the ground.[49]

Over the course of its long war with Iran, Iraq had greatly modified its tactical doctrine. Like their Egyptian counterparts, the Iraqi general staff concluded that efforts to train Iraqi units to employ tactical maneuver, conduct combined arms operations, and aggressively seize battlefield opportunities had consistently failed. In response, they began to rely almost exclusively on heavily scripted set-piece offensives. Baghdad began planning both counterattack and offensive operations in minute detail.[50] Detailed plans were given to the Republican Guard and the handful of competent regular army divisions. For months before an operation, these units would rehearse their specific missions repeatedly in training areas where the terrain in the planned area of operations had been replicated. According to the commander of the Republican Guard's Hammurabi Armored Division, the guard regularly trained for ten to fourteen hours without pause, day after day, in preparation for the series of offensives to take back the al-Faw peninsula in April 1988.[51] Eventually they reached the point where the entire operation could be performed by rote. The same approach was used in air operations: Iraqi Mirages flew rigidly prescribed and rehearsed flight profiles when attacking Iranian oil targets or tankers in the Persian Gulf.[52]

[49]Michael R. Gordon and Lt. Gen. Bernard Trainor, *The Generals' War* (Boston: Little, Brown, 1995), pp. 104–5; U.S. Air Force, *Gulf War Air Power Survey (GWAPS)*, vol. II, *Part I: Operations* (Washington, DC: Government Printing Office, 1993), pp. 75–77. For the travails of one Soviet pilot sent as an adviser to Iraq to teach air-to-air operations, see Belzyudnyy, "I Taught Saddam's Aces to Fly," pp. 62–63. For descriptions of Gulf War aerial combat, see U.S. Air Force, *GWAPS*, vol. II, *Part II: Effects and Effectiveness* (Washington, DC: Government Printing Office, 1993), pp. 119–30.

[50]"Saddam's al-Faw Anniversary Meeting; Part II," *Iraqi News Agency*, 20 Apr. 1993, translated in the U.S. government's Foreign Broadcast Information Service, FBIS-NES-93-075, 21 Apr. 1993, pp. 23–24.

[51]"Saddam's al-Faw Anniversary Meeting, Part IV," *Al-Thawra* (23 Apr. 1993): 2–3, translated in FBIS-NES-93-081, 29 Apr. 1993, p. 30.

[52]Capt. Michael E. Bigelow, "The Faw Peninsula: A Battle Analysis," *Military Intelligence* (Apr.–June 1991): 16; Major James Blackwell, *Thunder in the Desert* (New York: Bantam, 1991), pp. 56–57; Aaron Danis, "Iraqi Army Operations and Doctrine," *Military Intelligence* (Apr.–June 1991): 12; *GWAPS*, vol. II, *Part I, Operations*, p. 64; Stephen C. Pelletiere and Douglas V. Johnson, *Lessons Learned: The Iran-Iraq War* (Carlisle Barracks, PA: Strategic Studies Institute, 1991), pp. 47–49; and interviews with General Bernard Trainor, June 1994. The Iraqi solution was almost identical to that devised by the Egyptians prior to the October War. There were considerable numbers of Egyptian military officers attached to

This approach had worked well for the Iraqis against the Iranians in 1988 and the Kuwaitis in 1990. It failed against the U.S.-led coalition in 1991. Against weaker foes, the Iraqis could seize and hold the initiative and control the course of battle. The Iraqis made sure they had overwhelming advantages in firepower and numbers and limited their operations in both time and space to minimize the possibility that battlefield realities would force them to diverge from their carefully scripted plans. The Americans, however, would not play by those rules. Just as the Israelis had overwhelmed the Egyptians and Syrians through maneuver and a rapid pace of operations, the Americans used these same advantages—plus a massive technological edge—to prevent the Iraqis from sticking to their carefully designed and rehearsed plans. The most obvious example of this was the coalition's "left-hook," the main attack by the U.S. VII Corps, which flanked the western end of the Iraqi line and attacked the Republican Guard from the west, as opposed to the south as expected. This completely threw off Iraqi plans, and forced Iraqi field commanders to fall back on their limited repertoire of improvisational skills.

The Battle of Wadi al-Batin on 26 February illustrated these problems. The Tawakalna 'ala Allah Mechanized Division of the Republican Guard was deployed in reserve in southern Iraq west of Wadi al-Batin, which marks the western border of Kuwait. The Tawakalna had planned and practiced to counterattack southward, against the expected U.S. assault up the wadi. Late on 25 February, the division learned that the U.S. VII Corps was bearing down on it from the west. The Tawakalna was ordered to occupy hasty defensive positions facing westward to screen the retreat of the Iraqi army out of Kuwait. During the afternoon of 26 February the division was attacked by three U.S. heavy divisions and an armored cavalry regiment. The Republican Guards fought hard but extremely poorly. With ample time, Iraqi units normally established a defense-in-depth. The Tawakalna deployed in line abreast, with no depth to their positions whatsoever. When the Americans attacked, Iraqi tactical commanders made almost no effort to counterattack or maneuver against the Americans to get flank and rear shots, even though that is a central element of British doctrine. The division was wiped out in just over twelve hours.[53]

---

the Iraqi armed forces as advisers in the latter part of the Iran-Iraq War, and the Egyptians claim that they taught the Iraqis to script their operations. The Iraqis insist that they hit upon the same method without any input from the Egyptians. Indeed, they claim that Egyptian personnel were assigned only to training commands and so did not have contact with the Iraqi generals who actually formulated this approach.

[53]Lt. Col. Peter S. Kindsvatter, "VII Corps in the Gulf War," *Military Review* (Feb. 1992): 26–34; Gordon and Trainor, *The Generals' War*, pp. 390–95; Brig. Gen. Robert H. Scales, Jr., *Certain Victory: The United States Army in the Gulf War* (Washington, DC: Office of the Chief of Staff, U.S. Army, 1993), pp. 261–91.

The Red Herring of Red Army Influence

In the cases considered here, there was no apparent correlation between the extent of reliance on Soviet equipment and advisors, and the degree to which that country's military adopted Soviet organizational forms, concepts, and practices. Egypt hosted a very extensive Soviet military presence until a year before the 1973 war but was selective in its approach to Soviet doctrine, showing great innovative flair, imagination, and creativity in planning for the 1973 war. Syria relied heavily on Soviet organizational forms and tactical and operational concepts, even though its relationship with the Soviets was much less extensive than that of Egypt. The Soviets maintained a significant advisory presence in Iraq through Desert Storm with only a modest impact on Iraqi thought and practice.

Organization

Soviet influence on the armed forces of Egypt, Syria, and Iraq was greatest at the organizational level. All three countries used, to some extent, modified Soviet tables of organization and equipment as the basis for their units, which were largely equipped with Soviet or Soviet-type weapons. From squad to division in the ground forces, and particularly in the air defense forces, Soviet organizational forms were often copied, though frequently modified when appropriate. The main exceptions were praetorian units such as Syria's Defense Companies and Iraq's Special Republican Guard, each organized in tailor-made fashion by local leaders to fulfill regime-defense roles.[54]

Soviet doctrine was intended to be implemented by a modern, mechanized military that only an industrial power could afford to raise and maintain. Translating the Soviet model to the Middle East necessarily entailed adaptations. Most of the Soviet Union's Middle Eastern allies lacked the money to fully mechanize their forces (through the acquisition of modern infantry fighting vehicles, self-propelled artillery, vehicle mounted antitank weapons, and mobile SAM and AAA systems) and the material and human resources (i.e., large numbers of skilled technicians and operators) to support and man such an army. The Soviets were also unwilling to provide their Arab allies with the full panoply of weapons necessary to create a truly modern force.[55]

---

[54]Michael Eisenstadt, "Syria's Defense Companies: Profile of a Praetorian Guard Unit," unpublished manuscript, 1989.

[55]For much of the Cold War the Soviets tried to discourage their Arab clients from going to war with Israel, fearing that such a conflict could spark a confrontation with the United States (as nearly happened during the 1956 and 1973 wars). Following Anwar Sadat's assumption of the presidency in 1970, the Soviet Union repeatedly refused to transfer to Egypt the arms that Cairo thought it needed to go to war with Israel. The Soviets finally relented in early 1973 and transferred modern SAMs and MiG fighters, heavy bombers, and Scud missiles. Tensions between Moscow and Cairo over this issue were a con-

Tactics

Arab reliance on Soviet tactics varied widely across the cases. The Syrians embraced the Soviet model almost completely, with few modifications, despite unique requirements and circumstances. Egypt adopted many Soviet tactics, but modified them to fit Egyptian circumstances. Egypt's greatest successes came from taking the germ of a Soviet idea—limiting the command responsibilities of junior officers—and devising a new method that better suited Egyptian needs: detailed scripting and rote memorization of set-piece moves to compensate for shortcomings in tactical leadership. Its greatest reverses occurred when circumstances prevented them from employing their preferred method. Iraq leaned on Soviet tactics least, although there was considerable Soviet influence on Iraqi air and (especially) air defense operations. Iraq's experience with scripting paralleled Egypt's: Iraqi forces did well when they were able to rely on their uniquely "Arab" approach of carefully scripted, set-piece operations, but failed when they were unable to do so.

Operations

Arab armies generally did not adhere to Soviet operational principles. While the Syrian war plan in 1973 reflected a certain Soviet influence, the Egyptian war plan in 1973 and the Iraqi war plan in the 1991 Gulf War diverged from Soviet operational thought, which saw the defensive as an expedient to be adopted only as a prelude to offensive operations. The war strategies of both Egypt in 1973 and Iraq in 1991 hinged on the success of an operational defensive. Both Egypt and Iraq developed these concepts on their own, based on professional analysis of their own strengths and weaknesses vis-à-vis their enemies and an assessment of their war aims.

None of the Arab states adopted the Soviet emphasis on attaining air superiority, retaining the initiative, or attacking the enemy throughout his operational depth. The Arabs discarded those and other Soviet operational-level principles as irrelevant to their circumstances. The main operational-level problem that the Soviets faced in Europe—the possibility of nuclear escalation on the battlefield—was not a problem for their Arab clients. Israel's adversaries realized that it would use nuclear weapons only if its existence were threatened. As a result, the need to destroy the enemy's (tactical) nuclear weapons, such an important part of Soviet doctrine, was absent from Arab war planning in 1973. Since Egypt and Syria did not have to worry that concentrating their forces at breakthrough points would

---

stant leitmotif of Soviet-Egyptian relations during this period. See Heikal, *Road to Ramadan,* pp. 117–18, 156–57, 160, 163–64, 173, 183.

leave them vulnerable to nuclear strikes, the need to concentrate and disperse forces rapidly was not as crucial as it was for the Soviets. However, like the Soviets, Egypt and Syria were fighting an enemy that relied heavily on its mobilization potential for victory. Consequently, they had to accomplish their wartime objectives during the initial phase of the war, before Israel could mobilize its citizen army.

Arab Culture and Soviet Military Practices

Looking at modern Arab military history, there is a consistent pattern of failings at tactical command levels that emerges as the crux of Arab difficulties in implementing Soviet doctrine. The Egyptians, Syrians, and Iraqis lacked the capacity to implement key elements of the Soviet model, particularly its emphasis on combined arms coordination, adaptation of standardized "battle drills" to specific combat conditions, fast-paced operations, combat maneuver, and deep thrusts into enemy territory. Arab armies repeatedly demonstrated the lack of tactical leadership and command and control capabilities needed to wage this kind of war.[56]

These problems of tactical leadership derive principally from facets of Arab society. Every culture fosters a certain set of values through formal and informal education that tends to condition its members toward certain distinctive patterns of behavior; the dominant Arab culture fosters patterns of behavior that have had a decisive impact on the battlefield performance of Arab armed forces. In particular, Arab culture tends to promote conformity with group norms over innovation and independent thinking, and a rather severe deference to authority, which discourages initiative among subordinates. Arab culture also tends to promote the avoidance of shame at all costs, which discourages acceptance of responsibility and encourages the manipulation of information to conceal acts that could lead to embarrassment and humiliation.[57] The impact of these factors on training and operations can be judged by the following anecdote offered by a Soviet air force advisor who served in Iraq before the 1991 Gulf War:

[56]Kenneth M. Pollack, *The Influence of Arab Culture on Arab Military Effectiveness*, Ph.D. dissertation, Massachusetts Institute of Technology, Feb. 1996.

[57]Ibid., pp. 37–65. Most sociologists, psychologists, and anthropologists recognize the validity of such generalizations within and across societies, while simultaneously acknowledging that regional, class, national, and other differences are also important. While all the traits described above may not apply to any specific Arab individual, they do describe the societal mean around which individual behavior tends to cluster. Halim Barakat, *The Arab World: Society, Culture, and State* (Berkeley: University of California Press, 1993); Hisham Sharabi, *Neopatriarchy: A Theory of Distorted Change in Arab Society* (Oxford: Oxford University Press, 1988); Sania Hamady, *The Temperament and Character of the Arabs* (New York: Twayne, 1960).

As is known, in the East it is the custom to respect elders, and not only persons who are older in age but also in position or status. Take an ordinary situation: we were taking off in a two-man trainer to work on elements of combat training. In front was the aircraft of the squadron commander, and I was flying with his deputy. "Now," I said to him, "make a turn. We are going to intercept him." The deputy squadron commander, a bright fellow, soon was coming in on the tail of his commander. But I could feel that my "student" was beginning to reduce speed. "Go on," I said, "attack the target!" And he re-plied: "I cannot. That is my commander." [I responded:] "What in the devil does a commander mean in combat training? Carry out the order." Generally, the squadron commander was safely "downed." In truth, the commander for a week would not speak to me as he was insulted. But when he was put up for promotion, to wing commander (we had trained the squadron as required), he arrived after two weeks, embraced me and said, "Thank you brother, for the science I learned." This is how the story ended.[58]

These patterns of culturally conditioned behavior have crippled the tactical performance of Arab militaries. Junior officers demonstrate little initiative, cre-ativity, flexibility, or capacity for independent action in combat. This leaves Arab ground and air forces incapable of improvising ad hoc operations or conducting maneuver warfare. Twentieth-century Arab armies have consistently been plagued by passivity, dogged adherence to the letter of an order, inability to react to unforeseen circumstances, and a sluggish pace of operations. Because Arab junior officers and enlisted personnel often dissemble, exaggerate, and obfuscate in order to conceal mistakes and unpleasant news, Arab militaries frequently op-erate on the basis of unreliable information, with dire consequences in battle.[59]

Consequently, Arab armed forces could not exploit the potential offered by Soviet doctrine. Arab officers had been reared, taught, and trained to behave in a manner inappropriate to the demands of the Soviet system. Arab cultural pro-clivities discouraged subordinates in a hierarchy from taking initiative, innovat-ing, or acting independently. They encouraged passivity, precise adherence to "school solutions," and inaction in the absence of specific orders. For similar rea-sons, the Iraqis (and the Saudis and Jordanians as well) experienced problems implementing British (and U.S.) doctrine, which also called for high degrees of tactical initiative, innovation, flexibility, and maneuver. In many ways the Iraqis experienced greater difficulty executing British doctrine than did the Syrians and Egyptians implementing Soviet doctrine, because British methods required greater tactical flexibility and aggressiveness than did Soviet practices.

By the same token, the Arabs quickly took to certain aspects of Soviet doc-

[58]Bezlyudnyy, "I Taught Saddam's Aces to Fly," p. 63.

[59]While there are alternative explanations for the relative ineffectiveness of Arab militaries—socioeconomic underdevelopment or civil-military tensions (i.e., praetorianism)—neither of those factors is as important as Arab culture. For an in-depth discussion of the relative importance of these various influences on Arab military effectiveness, see Pollack, *The Influence of Arab Culture*, pp. 83–135, 535–80, 661–752.

trine, because some features of the Soviet military system conformed to accustomed patterns of Arab behavior. The Soviet military system's emphasis on centralization, detailed planning of military operations, and the careful limitation of the decision-making of junior officers all fit comfortably with Arab societal norms.[60] These similarities to patterns of behavior characteristic of Arab armies have led many observers to conclude that reliance on Soviet methods was the root of Arab problems.[61] That assumption is incorrect. Arab armed forces have manifested these tendencies regardless of whether they relied on Soviet methods, or had any contact with the Soviets at all. Arab culture rather than Soviet doctrine was the key. Today, after twenty years of U.S. training and retraining, the Egyptian military continues to manifest the same problems with tactical leadership and the same predilection for scripted operations they did during the era of Soviet assistance. Although some U.S. military personnel continue to blame these problems on Soviet influence, most now recognize that Egyptian difficulties had little to do with the Russians. In the words of one U.S. military officer: "[A]ll the Soviets ever did was bring 'science' to what the Egyptians like to do anyway."[62] Many U.S. military personnel with lengthy service in Egypt note that problem tendencies in the military are reflected in all walks of Egyptian life, demonstrating their roots in Egyptian society, rather than Soviet training.[63]

[60]On the limitations on tactical decision-making imposed by the Soviet military system, see John Erickson, *Soviet Ground Forces: An Operational Assessment* (Boulder, CO: Westview, 1986); Capt. Jonathan M. House, *Toward Combined Arms Warfare: A Survey of Twentieth Century Tactics, Doctrine and Organization* (Fort Leavenworth, KS: U.S. Army Combat Studies Institute, 1984); and John J. Mearsheimer, *Conventional Deterrence*, (Ithaca: Cornell University Press, 1983), pp. 183–88 and 193–98.

[61]Thus, in an article concerning lessons learned from the U.S. Central Command BRIGHT STAR 83 exercise, two U.S. Army officers observed that "(t)he Egyptians follow the Soviet doctrine of centralized decision-making and are quite bureaucratic in their hierarchy. Rarely is a major decision made below brigade level, and staff decisions routinely require general officer approval before they can be acted upon. Highly structured operations schedules 'drive the train'; even battalion commanders cannot modify them without the approval of higher headquarters. And once briefed to a higher Egyptian authority, a decision or an agreement is difficult to change." Lt. Col. Wolf D. Kutter and Maj. Glenn M. Harned, "Interoperability with Egyptian Forces," *Infantry* (Jan.–Feb. 1985): 15.

[62]Interview with U.S. military officer, Feb. 1997.

[63]In a 1994 article concerning the BRIGHT STAR 94 exercise, a U.S. army officer noted, among the lessons learned: "The Egyptian's approach to training exercises differs radically from U.S. training doctrine. . . . The training events for the Egyptians are scripted demonstrations with set times for maneuver, close air support, etc. The U.S. expected a 'free play' maneuver exercise." Thus, scripting remains a key element of the Egyptian style of warfare. In addition, the author noted: "The Egyptians did not have the authority to establish a Joint Task Force [JTF] with Egyptian forces below the Chief of Services. The Egyptian Task Force Commander had no authority or coordination capability to employ Egyptian Air, Special Forces, or Naval Forces. [Consequently, t]he U.S. JTF commander had to conduct individual coordination with each Egyptian component commanders [sic]. This organizational difference resulted in much more time and effort to coordinate combined operations than originally anticipated." The author attributed these Egyptian idiosyncrasies to "cultural/military differences." Capt. Jay Stefaney, "Operation BRIGHT STAR 94," January 1994, Center for Army Lessons Learned, "News from the Front!" at http://call.army.mil/products.nftf/jan94/pt5jan94.htm.

These points suggest that the Soviet military system conformed in some ways to accustomed Arab practices because of certain broad similarities in behavioral norms fostered by Russian (or Soviet) culture and Arab culture. At least at first blush, discussions of the Soviet military system, and of Russian/Soviet culture, seem to support this notion. Although Soviet culture and Arab culture are different in many ways, they share a number of similarities. A sociological study conducted in the 1970s compared three thousand recent Soviet immigrants to an American control group and found that Russians expected initiative, direction, and organization to come from higher authority to a far greater extent than the did Americans.[64] It appears that the dominant Arab and Russian/Soviet cultures shared similar notions that change and action should come from the top of a hierarchy and be transmitted downward, that subordinates should not exercise much independent judgment, that creative approaches were generally to be avoided, and that power should be concentrated in the hands of those at the top of a hierarchy. Although important differences exist, it may well be that Egypt, Syria, and other Arab states felt some degree of comfort with certain elements of the Soviet military system because it was shaped by the forces of a Russian/Soviet culture that shared many characteristics of their own.

For this reason, Arab reliance on Soviet methods was sometimes helpful to Arab militaries. The Soviets found simple but effective solutions to many of the problems resulting from their own culturally driven behavior, solutions that became central elements of the Soviet military system.[65] To the extent that Arab culture manifests similar behavioral traits, Arab adoption of Soviet methods proved helpful to Arab militaries. The best example of this was the scripting of operations by the Egyptians in 1973 and by the Iraqis in the latter half of the Iran-Iraq War. Although both Baghdad and Cairo took this practice to an extreme never envisioned by the Soviets—essentially eliminating all tactical improvisation—it allowed Egypt and Iraq to conduct limited, set-piece offensives that achieved narrow but important military and political goals.

Arab culture was the ultimate key to the diffusion of Soviet military doctrine in the Arab world. Arab armies that relied on Soviet equipment and doctrine did best when they borrowed (and adapted) aspects of Soviet doctrine that best suited their needs and capabilities. They did worst when they attempted to apply Soviet tactics and operational concepts that required skills and capabilities that their armies lacked.

[64]Cited in Arthur Alexander, *Decision-Making in Soviet Weapons Procurement*, Adelphi Paper 147/8 (London: IISS, 1978/79), p. 28.
[65]Donnelly, *Red Banner*, pp. 29–43, 195, 211.

Soviet Doctrine and Arab Armies

The impact of Soviet military doctrine on Arab armed forces has been exaggerated in a number of ways. The resemblance between the Red Army and the Arab militaries was often purely physical. Arab states made conscious decisions not to adhere rigidly to Soviet practices. Soviet doctrine, particularly at the operational level, often was not applicable to the circumstances of Arab armed forces. On the other hand, the Soviets brought a rigorous, comprehensive approach to thinking about, planning for, and waging war, after the disaster of 1967 left Egypt and Syria casting about for ways to undo the consequences of that war. The Soviets taught the Egyptians and Syrians how to train, plan, and put together large-scale military operations both tactically and logistically, and, most important, how to think systematically about waging war. The scripting of combat operations that provided an Arab solution to Arab problems with tactical leadership was an innovation derived ultimately from Soviet methods of command and control, with their emphasis on detailed planning of operations.[66]

Arab reliance on Soviet doctrine varied considerably from state to state. None of the Arab states had the ability to apply it properly. Because they lacked those capabilities, the Arabs fared best not when they tried to execute Soviet practices, but when they adapted Soviet methods to their own cultural and organizational predilections and operational needs.

Lessons on the Diffusion of Military Doctrine

The experiences of the armed forces of Egypt, Syria, and Iraq in applying Soviet military doctrine during the Cold War offer some broader lessons about the spread of military technology and doctrine from one society to another. Throughout the latter half of the twentieth century, both superpowers worked assiduously to arm and assist allied militaries. Frequently, Western and Soviet analysts believed that the impact of such training was the decisive factor in regional military balances. History has demonstrated otherwise. The case of Arab militaries and Soviet doctrine is one such example.

These cases demonstrate first and foremost the importance of culture in the diffusion of military doctrine. A society's culture helps determine what skills and behavioral predilections the nation's manpower will bring to military service. A military doctrine created by one nation will inevitably reflect the dominant cultural traits of its society, and thus it may not "fit" another military. Other nations will, at the very least, have to adapt the doctrine to suit their own set of skills and values to make it function effectively. In some cases, it may prove nearly impos-

[66]Pollack, *The Influence of Arab Culture*, pp. 658–60.

sible for one nation to employ the military doctrine of another. The Arab states are a good example of this, as their own cultural patterns of behavior did not provide them with the skills demanded by Soviet (or Western) doctrines of the Cold War era.

Militaries sometimes overlook the problem of cross-cultural diffusion. The Soviet and U.S. armed forces both seem to have assumed that their military doctrines were "universal" and so could be employed successfully by any other military. By contrast, the British appear to have been more sensitive to cross-cultural differences—hence their willingness to try to adapt European military practices to foreign cultures. (See, for example, John A. Lynn's chapter on the Indian sepoys in this volume.) It may be that the universalist credos of both the United States (the humanist/Enlightenment tendency to focus on similarities, rather than differences between peoples) and the former Soviet Union (the fundamental assumption underpinning socialist internationalism—that class, not nationality, determines values and molds ideology and culture) resulted in a tendency to downplay the role of culture, and fostered the dogmatic belief that military doctrines were universally applicable.

Arab experience with Soviet doctrine during the Cold War makes clear that the diffusion of military doctrine can be heavily conditioned by cultural factors. But neither culture nor warfare is static. Arab culture will evolve in accord with its own logic while also assimilating foreign influences. Likewise, new forms of warfare will emerge to render obsolete those that dominated the battlefields of the Cold War. It is impossible to predict how the two will interact. Arab culture was not a good fit for the forms of conventional maneuver warfare dominant during the Cold War. This need not be so in the future.

CHAPTER FOUR

Cooperative Diffusion through Cultural Similarity

The Postwar Anglo-Saxon Experience

THOMAS-DURELL YOUNG

> Men are not tied to one another by papers and seals. They are led to
> associate by resemblances, by conformities, by sympathies. Nothing is
> so strong a tie of amity between nations as correspondence in laws,
> customs, manners and habits of life. They have more than the force of
> treaties in themselves. They are obligations from the heart.
> —Edmund Burke[1]

During the February 1998 crisis over the question of the inspection of potential
sites of illegal weapons of mass destruction in Iraq, a rather curious and largely
unnoticed development occurred. As the Clinton administration endeavored to
build international and domestic consensus to force Saddam Hussein to allow
the UN arms inspectors to visit restricted sites, an ad hoc coalition of forces was
quietly deployed to the Gulf. In addition to Oman, ships and aircraft from Britain
and Canada were joined by army special forces and aircraft from Australia and
New Zealand.[2] Thus, at a time when stalwart NATO allies such as the Federal Republic of Germany and Portugal would acquiesce only to the use of U.S. bases on
their territory, and France, an ally with a nuanced, albeit intimate, relationship
with Washington, actively opposed the use of force against Iraq,[3] these four
"Anglo-Saxon" countries were willing to use military forces alongside those of
the United States. Significantly, while one might understand why Washington
and London, permanent members of the U.N. Security Council, would act to

---

The views expressed in this essay are those of the author and do not necessarily reflect the official policy
or position of the Department of the Navy, Department of Defense, or the U.S. government.

[1]Edmund Burke, *First Letter on a Regicide Peace*, VI, 155.

[2]See *The Canberra Times*, 21 Feb. 1998.

[3]See *The Washington Post* and *The New York Times*, 11 Feb. 1998.

enforce Security Council resolutions, why would Canada and Australia, let alone New Zealand, move to support this policy with armed force?

This recent historical anecdote provides a suitable example of how the current writer speculated some years ago that U.S. alliance relationships might develop in the post–Cold War world.[4] Had a force of arms come to pass during this crisis in the Gulf, this manifestation of the pursuit of common political objectives would have also demonstrated how this unique and informal coalition of states could conduct combined operations with a high degree of interoperability. But while the casual observer would understand how Britain and Canada, as NATO members, would be able to do this, it is less obvious how they, in combination with Australia, let alone New Zealand, which has ostensibly lost direct military contact with U.S. armed forces since 1985, could achieve close interoperability on the battlefield. After all, there is not a single collective defense treaty that binds together the five countries, and moreover, one would think that the singular lack of geographic propinquity would obviate the development of close military ties. This would be particularly the case with armies which, due to their nuanced nature (in comparison with ships and aircraft), almost defy any attempt at integration into multinational formations. Indeed, even where there is a peacetime stationing and planning relationship, as has been the case among NATO Central Region allies, effecting interoperability among armies has proven to be very difficult.[5] Yet, the fact is that the armed forces of these five countries have long enjoyed very close service-to-service ties among themselves. Indeed, one of the primary objectives of this informal coalition has been to ensure that the five countries' defense forces could operate effectively alongside their allied counterparts should the need arise.

Two important and directly related issues arise when examining this unique relationship among these five countries, which are of direct application to the topic of diffusion of military innovation. First, the quiet and all but exclusive Anglo-Saxon deployment of armed forces to the Gulf in February 1998 demonstrates a high degree of political cohesion and a similar outlook on international affairs among these five countries. The reason for this similarity in views is arguably attributed to the five countries' Anglo-Saxon heritage, which is reflected in their sharing of common law and their similar governing institutions.

Second, this close, if not necessarily formalized/institutionalized political, relationship has allowed the creation of a bureaucratic "culture" that has even fos-

---

[4]See my essay, "Whither Future U.S. Alliance Strategy: The ABCA Clue," *Armed Forces and Society* 17, no. 2 (winter 1991): 277–97.

[5]For an explanation of the plethora of problems faced by NATO in achieving interoperability among bi- and multinational formations, see my monograph, *Multinational Land Formations and NATO: Reforming Practices and Structures* (Carlisle Barracks, PA: Strategic Studies Institute, 1997).

tered and encouraged the transmission of the most difficult to absorb elements of military innovation: the conceptual basis for future warfare and its attendant required capabilities. Unlike the mere transfer of technology, either for the purpose of gaining new capabilities or to effect standardization, the Anglo-Saxon case has addressed itself to conveying the intellectual foundations of how best to address future warfare. While it is true that special arrangements exist for the sale of some of the most sensitive high technology weapon systems among these five countries, it is rather the all but free exchange of new ideas, concepts, and supporting data at almost all levels (i.e., strategic to tactical) that makes this relationship a powerful force in the diffusion of military innovation. In fact, despite the end of the Cold War, and the end of a common threat, this relationship among these five countries has actually grown closer, particularly among the five armies.

The purpose of this essay is to assess this "unique relationship" with a view toward examining and explaining how cultural and linguistic similarities can provide an effective transmission vehicle of innovation. The foundation for this cooperation has been provided by the existence of collective defense treaties, with general political guidance to the armed forces to "cooperate." An examination of this relationship reveals that its unique means of allowing for the diffusion of military innovation has been largely based upon cooperation carried out among similar services, achieved through contacts at all levels, with the important accompanying lack of overriding "national" positions. Moreover, while there are provisions for the exchange and transmission of the latest technologies, these programs tend to focus more on the issue of "intellectual" diffusion. Particular attention in this essay will be directed to the armies' standardization program to demonstrate how the countries' armies have been able to achieve a continuous dialogue and exchange of innovative ideas with the aim of maintaining interoperability.

This essay is organized in the following manner. First, a brief overview of the political and security bases that support the "ABCA" relationship among defense forces will be presented. Second, the service programs established to effect interoperability, save the armies, will be examined. Third, the armies' program, which has been probably the most important and significant of these activities, will be described and assessed in depth. Fourth and finally, the entire relationship will be assessed and recommendations for improving this relationship will be proffered.

Political Bases for Military Cooperation

We begin with a brief discussion related to basic alliance theory. Classical alliance theory generally holds that alliances among states are the product of common threat perceptions. Another school of thought, immortalized in the opening quotation by Edmund Burke, argues that alliances can be formed based on shared traits between states. Hans Morganthau described this form of alliance formation as "ideological solidarity," which while relatively rare among alliances, does have historical precedent.[6]

The origins of the peacetime quinquepartite and "informal" alliance that has developed among the five Anglo-Saxon countries' armed services can be traced to World War II. Given the gravity of the global conflict, strong interwar feelings of antipathy were set aside and a unique solution to strategy formulation was found in creating the Combined Chiefs of Staff between Britain and America.[7] Perhaps most important to achieving victory in this conflict and continued cooperation after the end of hostilities was, according to a senior British intelligence officer, the establishment of very close wartime intelligence ties between Washington and London. After 1945 these ties were maintained through ad hoc exchanges and regular liaison agreements.[8] Perhaps the most important nonintelligence aspect of these arrangements was the signing of "security agreements" between the United States and the United Kingdom, later expanded to include Australia, Canada, and New Zealand, which established standard security classifications and rules regarding its protection.[9] The establishment of such "ground rules" for the exchange of security information resulted in the ability of the armed services to cooperate with their counterparts in a decentralized and (perhaps most important) informal manner.[10]

However, as societies based on law, collective defense treaties have been a necessary basis for such cooperation to become institutionalized and regularized—notwithstanding the fact that they have not always preceded such cooperation.

---

[6]Hans Morganthau, *Politics among Nations: The Struggle for Power and Peace* (New York: Alfred A. Knopf, 1973), pp. 183–84.

[7]Indeed, it is often forgotten that the origins of the modern-day "Joint Staff" were based in the creation and operation of the Combined Chiefs of Staff. See Vernon E. Davis, *The History of the Joint Chiefs of Staff in World War II. Organizational Development.* Volume I: *Origin of the Joint and Combined Chiefs of Staff* (Washington, DC: Historical Division, Joint Chiefs of Staff, 1972).

[8]See Michael Herman, *Intelligence Power in Peace and War* (New York: Cambridge University Press, 1996), pp. 202–3.

[9]For a description of these arrangements, see my work, *Australia, New Zealand and the United States Security Relations, 1951–1986* (Boulder, CO: Westview, 1992), pp. 162–63.

[10]It is interesting to note that the Republic of South Africa is not included in this coalition, particularly given that it, too, was a wartime ally of the United States. Obviously, the adoption of Apartheid led the United States in the early 1950s to distance itself from the National Party regime, particularly after the election sweep by Daniel Malan in 1953. However, apparently during the years prior to 1953, the Truman administration worked to effect close military cooperation with Pretoria. Washington's principal aim was securing supplies of uranium. See Thomas Borstelmann, *Apartheid's Reluctant Uncle: The United States and Southern Africa in the Early Cold War* (New York: Oxford University Press, 1993), esp. pp. 125–31, 184–93.

Anglo-American security cooperation that developed during the postwar era has been well researched and documented by scholars and analysts on both sides of the Atlantic.[11] Leaving aside cooperation conducted under the auspices of NATO, bilateral defense cooperation between defense establishments and services has grown to include many security-sensitive areas, including nuclear weapons and submarine nuclear-propulsion research and development.[12]

Following the end of World War II, the three "Old Dominions" of the British Empire/Commonwealth (Canada, Australia, and New Zealand) entered into formal peacetime collective defense arrangements with the United States. For Canada, this took the form of the establishment of the Military Cooperation Committee in 1946, whose legal and political basis was established by the 1940 Ogdensburg agreement declared by President Franklin D. Roosevelt and Prime Minister William Lyon Mackenzie King. Ottawa and Washington expanded their mutual security commitments when each became signatories to the North Atlantic Treaty in 1949.[13] Australia and New Zealand, by virtue of the ANZUS Security Treaty with the United States, signed in 1951, gained official allied status in the eyes of Washington.[14] (The latter's security relationship was formally "suspended" in 1986, albeit a bit of a thaw in defense cooperation has taken place since the mid-1990s.)[15]

A prime motivation behind the continuation and certainly the deepening of the conditions allowing for the cooperative diffusion of innovation has been the existence of cultural and linguistic similarities, and the sharing of common law,[16] as opposed to total agreement in threat perceptions, let alone uniform conformity regarding each and every diplomatic objective. These factors have provided the bases for continued intimate collaboration during periods of diplomatic contretemps and international crisis. Two examples are offered to support this argument. First is the fact that the New Zealand Defence Force has retained membership in quinquepartite military cooperation fora following the break in formal bilateral U.S.-New Zealand defense and security ties in 1985 over the issue

---

[11]For example, John Baylis, ed., *Anglo-American Relations since 1939: The Enduring Alliance* (New York: Manchester University Press, 1997).

[12]John Baylis, *Anglo-American Defence Relations, 1939–1984: The Special Relationship*, 2d ed. (New York: St. Martin's Press, 1984).

[13]For an excellent assessment of the U.S.-Canadian defense relationship, see, Canada, Parliament, House of Commons, "NORAD 1986," Report of the Standing Committee on External Affairs and National Defence, Ottawa, Feb. 1986, pp. 3–15.

[14]Joseph G. Starke, *The ANZUS Treaty Alliance* (Carlton, Vic.: Melbourne University Press, 1965).

[15]New Zealand, Ministry of Defence, *The Shape of New Zealand's Defence: A White Paper* (Wellington, 1997), p. 20. For details on the extent of the termination in U.S. defense ties with New Zealand, see Young, *Australia, New Zealand and the United States Security Relations*, pp. 212–19.

[16]It is interesting that John Baylis describes the defense relationship emerging out of World War II as the "Common-law Alliance." See Baylis, *Anglo-American Defence Relations*, pp. 1–8.

of port access for U.S. warships.[17] Wellington's continued participation in these bodies has mitigated against its complete isolation from the United States in defense matters stemming from its antinuclear policies, which is clearly in the Western alliance's best interest. Second, the British campaign to regain possession of the Falkland Islands in 1982 provides a revealing example of the close relationship between the defense bureaucracies in Washington and London. One published source claims that while Secretary of State Alexander Haig was attempting to mediate the crisis, the Department of Defense began to assist the Ministry of Defence prior to the announcement of official U.S. diplomatic support for Britain due to the existence of a multitude of bilateral cooperative agreements between the two bureaucracies and armed services.[18]

The peculiar relationship that has developed among these five countries has encouraged the diffusion of military innovation with clear objectives, if indeed employing subtle means to achieve them. Five broad characteristics of this relationship can be discerned and point to the important role played by similarities in cultures and language:

1. The existence of collective defense treaties gives defense bureaucracies the necessary broad political and legal freedom to conduct peacetime defense cooperation. In the words of an Australian Army brigadier describing how the ANZUS treaty affects tactical level cooperation: "The treaty does not limit us to certain actions but instead provides the room for like-minded people to function together. ANZUS works on the tactical level because it feels right."[19]

2. The overriding objective in encouraging the diffusion of military innovation among these five countries has been to achieve the illusive objective of interoperability. A high degree of interoperability is crucial if the armed forces of these countries can expect to operate alongside one another in wartime, particularly in high-intensity operations.

3. Encouraging the sales of state-of-the-art technological innovation and the standardization of kit in its most strict sense have not proven successful in the diffusion of innovation among these five defense forces. Rather, in accepting the political realities that govern defense procurement, the five countries have endeavored to pursue the diffusion of innovation through the standardization of

[17] *The Dominion Sunday Times* (Wellington), 13 Nov. 1988.

[18] Max Hastings and Simon Jenkins, *The Battle for the Falklands* (New York: W. W. Norton, 1983), p. 142.

[19] See Peter Leahy, "ANZUS: A View from the Trenches," *Joint Force Quarterly*, no. 17 (autumn–winter 1997–98): 88. This insightful essay is drawn from then-Brigadier Leahy's full statement and answers to questions posed to him in a parliamentary committee hearing. The full text of his testimony can be found in Australia, Parliament, The Joint Standing Committee on Foreign Affairs, Defence and Trade, Defence Sub-Committee, "ANZUS after 45 Years: Seminar Proceedings, 11–12 Aug. 1997," Canberra, House of Representatives, Sept. 1997, pp. 102–9.

tactics, techniques, and procedures. In other words, the five armed forces have endeavored to create and maintain "intellectual interoperability."

4. With one major exception, the achievement of all of the above has been delegated to the individual services, vice defense forces per se.

5. The ability to maintain close collaborative ties and exchange information has produced an environment whereby cooperation extends to the sensitive area of the promulgation of doctrine and collaboration in future force development requirements.

The programs and agreements, which act to diffuse innovation, all exist at the departmental and armed service levels and therefore, according to traditional treaty law, do not formally bind the governments of these five countries in any way.[20] Nevertheless, the implications of these programs, referred to in this essay as the "ABCA" fora, have had a strong influence for many years on the training, development, organization, and interoperability capabilities of these five defense forces. These arrangements are: the Air Standardization Coordinating Committee; ABCA navies arrangements and programs; the Combined Communications Electronics Board; the Technical Cooperation Committee; and the ABCA Armies Standardization Program.

The implications of these programs for their participants have been significant apropos developing the capability to conduct combined military operations. At the services' level, these programs have provided the basis for continuing peacetime cooperation among the five countries' armed services in the areas of combined operations, mutual logistic support, and cooperation in, and coordination of, defense scientific research and development. It is interesting to note, en passant, that many of the ABCA standardization efforts actually preceded the creation of NATO and continue to operate outside of that organization, although there has been a long record of information exchange between them and NATO.[21] The long-standing and informal working relationship that has typified the ABCA programs has acted to diffuse military innovation in a manner not possible in NATO, because of its diverse national cultures, languages, and formalistic nature of conducting business. For example, practically all the charters of the ABCA

[20]For example, one such operational arrangement that exists among the Australian, British, New Zealand, and U.S. navies drew public criticism in a high-level report in Australia because of the arrangement's use to justify the Royal Australian Navy's force structure. In other words, allied, vice national considerations had long been used by the RAN to justify the type and numbers of warships. The report's author pointed out that such "arrangements" do not fall under traditional treaty law and therefore are not legally binding upon governments. See *Review of Australia's Defence Capabilities*, Report to the Minister for Defence by Mr. Paul Dibb (Canberra: Australian Government Publishing Service, 1986), p. 47.

[21]NATO Information Service, "Relationship between NATO and Regional Bodies in Standardization, Bruxelles," n.d.

programs provide for the routine exchange of information between members of this exclusive club, up to and including the SECRET level: a condition that does not exist in NATO.

The Development of ABCA Programs

The genesis of the ABCA standardization programs can be traced to the very early postwar years, when Washington, London, and Ottawa were becoming increasingly anxious about the worsening relationship between the Western democracies and their former wartime ally, the Soviet Union. Senior military leaders from the United States, Britain, and Canada, drawing on their experience from both world wars, fully appreciated the difficulty of attaining the capability to conduct successful coalition warfare. Given the increasingly bellicose behavior of the Soviet Union in world affairs, it was logical for these former allies to attempt to retain a high degree of interoperability among their respective defense forces.[22] Their foresight was vindicated in 1950 when these three allies found themselves once again (with Australia and New Zealand) fighting alongside each other in Korea.

Specifically, in 1946, the Chief of the British Imperial General Staff, Field Marshal Bernard Montgomery, during a visit to North America, recommended that the United States, Britain, and Canada "cooperate closely in all defense matters; discussions should deal not only with standardization, but should cover the whole field of cooperation and combined action in the event of war."[23] By late 1946, press reports from London stated that these three countries were considering the feasibility of standardizing the weapons, tactics, and training of their armed forces.[24] Indeed, in 1947 a standardization agreement between the armies of the three countries was signed, followed by a similar accord effected among their respective air forces in 1948. Of importance to future standardization efforts was the agreement reached in November 1948 under which a standard thread pattern was adopted for all nuts and bolts, the "Unified American-British-Canadian Screw Thread."[25] Impetus was given to this inter-allied standardization movement through a directive issued by the U.S. Joint Chiefs of Staff in July 1949 to the U.S. Armed Services to initiate standardization programs with their British and Canadian counterparts.[26]

---

[22]Edward C. Ezell, "Cracks in the Post-War Anglo-American Alliance: The Great Rifle Controversy, 1947–1957," *Military Affairs* 38, no. 4 (Dec. 1974): 138–39.

[23]John B. McLin, *Canada's Changing Defence Policy, 1957–1963: The Problems of a Middle Power in Alliance* (Baltimore: Johns Hopkins University Press, 1967), pp. 9–10.

[24]Ezell, "Cracks in the Post-War Anglo-American Alliance," p. 138.

[25]Forster Lee Smith, "Canadian-United States Scientific Collaboration for Defense," *Public Policy* 12 (1963): 307.

[26]Ibid., p. 318.

From this modest beginning, the ABCA standardization programs have expanded to include almost all areas of defense activity, notwithstanding the evident failure of many of them to succeed in accomplishing the formal standardization of weapon systems among themselves. Rather, where these programs have had an important influence has been in the area of achieving and maintaining interoperability, through the diffusion of particularly intellectual innovation as it relates to current and future warfare requirements (operational and technical). Activity in these programs has not diminished since the end of the Cold War, and, indeed, in some key cases, cooperation has intensified. A concerted effort recently has been placed on these programs not only to maintain their efforts to achieve interoperability but also to explore new areas of cooperation. Significantly, these programs, based in the Washington, D.C., area, coordinate their activities in multiforum meetings to a higher degree than before and have recently widened their focus to address "joint" interoperability issues.

A brief description and analysis of each of these standardization programs reveals that these fora cover almost all services and functional areas where interoperability is crucial to combined operations. Due to the unique nature of "armies," the ABCA Armies program will be addressed separately from and following the other standardization arrangements.

Air, Naval, Communications, and R&D Programs[27]

Air Standardization Coordinating Committee (ASCC). The standardization process between the air forces of the five Anglo-Saxon countries (to include the membership of the U.S. Navy and the Royal Navy) is the ASCC.[28] Membership in the committee comprises officers of flag rank who meet annually to resolve any outstanding policy issues and to approve the annual report of the ASCC Management Committee, which is located near Washington.[29]

The program's mission is "to ensure member nations are able to fight side by side as airmen in *joint* and combined operations."[30] In acknowledging these realities, ASCC attempts to minimize obstacles to operational cooperation among its members, by enabling cross-servicing of aircraft, conducting justifiable logistic

[27]For an excellent official source on all of the ABCA fora, see American, Australian, British, Canadian and New Zealand International Military Standardization Fora, *Washington Staff Handbook*, 4th ed., Aug. 2001.

[28]The ASCC was formed by the air forces of the United States, Britain, and Canada in Jan. 1948. The RAAF and RNZAF joined the ASCC as full members in 1964 and 1965, respectively.

[29]*The Air Standardization Coordinating Committee* (Washington, DC: ASCC Management Committee, 1984), p. 1.

[30]Emphasis added. The addition of "joint" was effected at the 50th annual meeting of the ASCC Management Committee, Mar. 1998. See *ASCC News*, Newsletter #5, ASCC Homepage: http://www.hq. af.mil/xo/xxo/xoox-iso/ascc/news.htm.

support, and generally promoting a rationalization of resources.[31] The members of the committee are also signatories to the "Master Agreement for the Exchange of Equipment for Test Purposes," which provides for the loan of equipment for testing and evaluation purposes by the ASCC members at no cost, and often at short notice.[32] ASCC standardization objectives are normally reached by the negotiation of Air Standards among the five air forces. There are currently more than 350 Air Standards and 60 ASCC Advisory Publications.[33] Close coordination is maintained with the NATO Military Agency for Standardization and, indeed, NATO AIR Standardization Agreements (STANAGS) are automatically dispatched to Australia and New Zealand.[34]

The process by which Air Standards are reached (which must be approved by unanimous agreement) includes: exchange of information in approved areas; adoption of standard or similar methods and procedures; tactics, techniques, equipment, and terminology; the establishing the design of equipment for cross-servicing of aircraft.[35]

Information eligible for exchange in the ASCC extends up to the SECRET level.[36] In the early 1980s, the ASCC members faced a bit of an identity crisis and recognized that the continued lack of materiel standardization between themselves placed the program's future relevance in doubt. One of the recommendations of the 33d meeting of the ASCC Management Committee was to take a fresh look at the program with the aim of possibly lowering costs by reducing the number of projects and working groups under its sponsorship. The RAND Corporation, commissioned to study the problem, issued a report in 1982 arguing that the ASCC could revive much of the impetus of its early years by working toward "sufficient" standardization (i.e., interoperability), instead of standardization in the strictest sense of this term.[37] Subsequent to the release of this study, the number of engineer working groups was reduced, and achieving interoperability is now the primary objective of the ASCC.

ABCA Navies. The ABCA Navies have a number of disparate agreements and arrangements to encourage interoperability. To a large extent, peacetime cooperation between these navies predates that of the ABCA Armies, inasmuch as the

[31] *The Air Standardization Coordinating Committee*, pp. 1–2.

[32] Ibid., p. 7; and New Zealand, *Appendix to the Journal of the House of Representatives (AJHR)* H. 4 *Defence*, p. 29.

[33] The ASCC also issues *Information Publications, ASCC Instructions,* and the *ASCC Handbook 1998.* See, "ASCC Document Listing," ASCC Homepage.

[34] See *The Air Standardization Coordinating Committee, ASCC Handbook 1998,* p. D–1.

[35] Ibid., p. 2.

[36] New Zealand, Ministry of Defence, *Background Briefs for Minister of Defence,* p. 2. "ANZUS," Wellington, July 1984, p. 6.

[37] John K. Walker, *Responses-Synopsis of Questionnaires and Some Observations: Air Standardization Coordinating Committee* (Santa Monica: RAND Corporation, Apr. 1982), pp. 5, 29.

Royal Navy and the U.S. Navy began exchanging classified information regarding Japan in July 1937.[38] Very close wartime cooperation was followed by cordial peacetime navy-to-navy relations that culminated in an extensive array of service-level operational agreements and arrangements. The parties to the ABCA Navies arrangements belong to the Information Exchange Project, which enables the exchange of technical data in areas of common interest—for example, undersea and electronic warfare.[39] Where it is found that a NATO STANAG would be of benefit to the Royal Australian Navy (RAN), there are procedures whereby a similar ABCA NAVSTAG is established for the benefit of the RAN through the sponsorship of the Royal Navy.

Operational procedures are addressed between the five navies in a quinquepartite agreement that governs naval and maritime exercises held in the Pacific and Indian oceans among the five navies and their respective maritime air units. This agreement, the Combined Exercise Agreement (commonly known as COMBEXAG), was initially a bilateral agreement between the Commander, Far East Fleet, Royal Navy, and the Commander, U.S. Navy's 7th Fleet. The RAN informally began using the document in September 1964 and became a formal participant with the document's release in 1966. The Canadian Forces joined the agreement in 1978, and the Royal New Zealand Navy (RNZN) at a later date became a participant in COMBEXAG. To a large extent, the COMBEXAG merely formalized existing arrangements between the five navies in the area of operational procedures. The document itself is essentially a planning manual that ensures that the five navies and associated maritime air contingents can conduct combined maritime operations in the greater Pacific region.[40]

In the area of naval communications, the CAN-UK-US NAVCOMMS Board was created in 1960 to resolve signals incompatibilities among the British, Canadian, and U.S. navies. Australia became a member of the Permanent Board in 1966 and New Zealand, which had associate status, became a full member in 1980. Following New Zealand's accession, the arrangement adopted its present nomenclature, the AUSCANNZUKUS Naval C4 Organization.[41] The mission of the or-

[38]Richard A. Harrison, "Testing the Water: A Secret Probe towards Anglo-American Military Cooperation in 1936," *International History Review* (May 1985): 217.

[39]ABCA Armies, *Standardization Program* (Canberra: Australian Army, n.d.), p. E–1 (Annex E).

[40]Personal correspondence, Vice Admiral Michael Hudson, RAN, Chief of Naval Staff, 18 Nov. 1986; and Major General D. Huddleston, CF, Associate Assistant Deputy Minister (Policy), Department of National Defence, 20 May 1987.

[41]The AUSCANNZUKUS Naval C3 Organization is directed by a committee that meets every November in Washington, a Technical Working Group (which meets twice a year), and the Permanent Support and Coordination Group comprising the Washington naval attachés from Australia, Britain, Canada, New Zealand, and U.S. Navy delegates. The Permanent Steering Committee meets twice a month with the mission to standardize communications equipment procedures. See *Washington Staff Handbook*; and Robert Howell, "Aus-Can-What?" *Signal* (Sept. 1982): 36–37.

ganization is "to provide a seamless information infrastructure to enable Allied commanders at any level access to information as required to accomplish their assigned tasks."[42]

An official release of the NAVCOMMS organization argues that the success of the program is clearly evident since first "the five Allied navies can, and do, communicate, and thereby operate, at sea." Second, over time the organization has developed a methodology that identifies potential impediments to communications interoperability at an early stage of development. Even if problems cannot be resolved, there are procedures whereby alternative solutions can be explored by the board and its members.[43]

Combined Communications Electronics Board (CCEB). In the area of national defense communications and electronics, the Combined Communications Electronics Board is tasked with coordinating common communications and electronics matters (particularly standardization issues) that are of mutual interest to two or more members. The board also coordinates communications and electronics issues with other ABCA standardization programs.[44] The CCEB is the oldest standardization forum, established in July 1942. It was dissolved in 1949 (or better, "absorbed" into NATO activities) but was re-established on a peacetime basis in 1951.[45] Its membership then consisted of the United States, Britain, and Canada (with Australian and New Zealand participation when appropriate) and was then called the Joint Communications-Electronics Committee. Australia became a full member on 18 December 1969, and New Zealand on 20 September 1972, when its present name was adopted.

The aim of the CCEB is to maximize the effectiveness of combined operations. In other words, it is to advance combined military communications-electronics interoperability when problematic issues are referred to it by one of its members. In this respect, the role of the CCEB is to consider any military "communications and information systems in support of command and control" (CE). To accomplish this ambitious objective, the CCEB develops combined operation CE policies, doctrine, operating methods, and procedures. The CCEB is responsible for the establishment of the content, format, distribution, and release policy of the

[42]The organization attempts to achieve its mission by (1) promoting interoperability among member nations by adopting standards and agreements to maximize operational compatibilities, (2) exchanging information on issues of interoperability, (3) providing a forum to highlight issues to national authorities, and (4) using national resources to cooperatively coordinate studies to resolve long-term, complex, interoperability matters. See "Overview," AUSCANNZUKUS Homepage; http://auscannzukus-navalc3.hq.navy.mil/overview.htm.

[43]*AUS-CAN-NZ-UK-US Naval Communications Organization* (Washington, DC: Permanent Secretariat, NAVCOMMS Organizations, n.d.).

[44]Howell, "Aus-Can-What?" p. 35.

[45]*CCEB Publication 1: Organization, Mission and Functions* (Washington, DC, n.d.).

thirty-three Allied Communications Publications (ACPs) and their general supplements. These important publications are used in more than ninety nations.[46]

The Technical Cooperation Program (TTCP). The body most intimately involved in the diffusion of scientific and technological innovation is The Technical Cooperation Program. TTCP was one of the many by-products resulting from the launch of the Soviet Sputnik satellite on 4 October 1957. Fearing a sense of scientific inferiority in the Western alliance, British prime minister Harold McMillan, during a visit to Washington, issued a joint public declaration with President Dwight Eisenhower on 25 October stating that both countries should pool their defense science information and coordinate future defense R&D projects in order to avoid needless and costly duplication.[47] The government of Canada immediately endorsed this Declaration of Common Purpose, thereby forming the Tripartite Technical Cooperation Program. The program's nomenclature was changed to its current usage when Australia joined in July 1965. New Zealand gained admission in October 1969.[48]

TTCP, as established by its Declaration of Common Purpose, recognizes that no single member has the resources to conduct research in all areas of defense science by itself. The program provides to its members the means of acquainting themselves with the defense science activities of their counterparts. In providing this conduit for diffusing technological innovation, each country is able to plan its activities in cognizance of the efforts of others. Given diminishing defense budgets in Western countries, the ever-growing complexity of defense science and its technological application, the value of TTCP has grown.[49]

Under TTCP, there are two subcommittees, one of which is solely concerned with atomic-related defense R&D, of which Britain and the United States are members. The other is called the Non-Atomic Military Research and Development Subcommittee, to which all five countries belong.[50] TTCP, it should be stressed, is a "program" and is not a corporate body. Therefore, it does not have any resources or projects under its direct sponsorship.[51] Rather, the program is headed by the respective heads of the defense science establishments of the five countries, administered by seconded representatives ("Washington Deputies"),

[46]See Thom McCabe, "The Combined Communications Electronics Board . . . What Is It?" 2 Sept. 1997, www.dnd.ca/commelec/vol31/ccebe.htm.

[47]Smith, "Canadian-United States Scientific Collaboration for Defense," pp. 309–10.

[48]For an informative description of the creation and development of this program, see The Technical Cooperation Program Document, Subcommittee on Non-Atomic Military Research and Development, *Policies, Organization and Procedures in Non-Atomic Military Research and Development (POPNAMRAD)* (Washington, DC: Dec. 1997), pp. 1–2. (The TTCP Homepage also provides a wealth of information on the program and extent of its activities. See http://www.ttcp.osd.mil/.)

[49]Ibid., pp. 2–3.

[50]See ibid., pp. 5–7, for a description of TTCP's organization.

[51]ABCA Armies, *Standardization Program*, p. D–1 (Annex D).

and served by a small secretariat in Washington. TTCP acts to facilitate "the definition and initiation of joint complementary research studies of defense problems
of mutual concern."[52] Thus it acts as a conduit by which military scientific and
technological innovation can be known to the other program members. Research
studies can, in principle, cover the entire range of military-related R&D topics. As
an illustration of the value of TTCP to the Australian defense science community
in particular, the Australian government has estimated that without TTCP, its Defence Science and Technology Organisation would have to be doubled in size to
maintain its current level of scientific support to the Australian Defence Force.[53]

ABCA Armies

Armies, by their very nature, defy attempts at their integration into bi- or
multinational formations. Unlike their naval and air counterparts, armies have
their own sui generis characteristics that mitigate against their integration with
other armies. In contrast, ships and aircraft can be thought of as individual and
discrete platforms carrying weapons and possessing capabilities that can be integrated in their entirety in combined operations.[54] Armies, conversely, are combined arms teams made up of various subset formations, each of which may have
different mission-essential tasks assigned to them that must be accomplished if
the parent unit is to achieve its mission.[55] For example, it is conceivable that for
each tactical maneuver, a commander may need to assign subordinated units
new missions and tasks, reassign forces, or task-organize ("fragment") subordinate forces. Thus, to achieve interoperability, it is essential that standardization of
procedures and materiel must be effected at much lower organizational levels
than is the case with navies and air forces in order for armies to be capable of
providing mutual support. Given that armies can be defined largely as an amal-
gam of individuals, efforts at standardization and interoperability face considerable challenges under the best of circumstances. An exception to this truism has
been the success of the ABCA Armies program in overcoming many of these difficulties through the diffusion and exchange of technological and organizational
innovations.[56]

---

[52]Australia, Parliament, *Defence Report 1970*, Parliamentary Paper No. 171 (Canberra: Australian Government Publishing Service, 1970), p. 9. (Note: The Australian Washington deputy represents New
Zealand.)

[53]Hon Bronwyn Bishop, MP, Minister for Defence Industry, Science and Personnel, "Speech for the
1997 Technical Cooperation Program Awards," Canberra, 24 July 1997.

[54]See Roger H. Palin, "Multinational Military Forces: Problems and Prospects," Adelphi Paper No.
294 (London: The International Institute for Strategic Studies, Apr. 1995), pp. 52–54.

[55]See Department of the Army, *FM 100–5, Operations* (Washington, DC, June 1993), pp. 2–2/2–3.

[56]For a very candid view of the high degree of interoperability between the U.S. and Australian armies, see Leahy, "ANZUS: A View from the Trenches," pp. 86–90. (Note: Brigadier Leahy was Com-

The ABCA Armies Standardization Program was initially established with the signing of the "Plan to Effect Standardization" in 1947 among the American, British, and Canadian (ABC) armies. The aim of this particular agreement was to ensure that there should be no doctrinal or materiel obstacles to complete cooperation between these three armies in time of conflict. This accord was replaced by the "Basic Standardization Agreement" in 1954.[57] Following the deployment of Commonwealth forces from Australia, New Zealand, and Britain to Malaysia during the period of Confrontation with Indonesia, the Australian Army was invited to join the ABC armies' forum in 1963. Australia accepted this invitation to join the arrangement on 18 January 1963,[58] and the Basic Standardization Agreement became the ABCA Armies Program in 1964. New Zealand subsequently gained associate membership through Australia's sponsorship in 1965.[59] Both the Australian and New Zealand armies' decision to join this allied standardization forum proved timely, because both subsequently deployed forces to Vietnam alongside the U.S. Army in 1964.

As they now stand, the stated objectives of the ABCA Armies program are to achieve not only the fullest cooperation and collaboration but also the highest possible degree of interoperability through both material and nonmaterial standardization, and also to obtain the greatest possible economy by the use of combined resources and effort.[60]

"Material standardization" under this arrangement is not defined as constituting the strict standardization of weapon systems themselves. The ABC and ABCA Armies' programs are replete with examples of the inability (or unwillingness) of these armies to come to agreement concerning the joint acquisition of equipment, even in the most basic areas:

1. U.S. and British officials attempted without success in the early 1950s to adopt a common rifle for their armies, a singular lack of standardization in such a basic weapon that continues even today.[61]

2. During the 1960s, the U.S., British, Canadian, and Australian armies entered into a cooperative agreement to support the research, development, and production of a secure tactical trunk communications system called "Project Mallard"

---

mander Combined Army Force during the TANDEM THRUST '97 field training exercise that comprised some twenty-seven thousand personnel—the largest field training exercise held in Australia since the end of World War II. A Battalion of the 25th Infantry Division [Light] was "chopped" OPCON to Brigadier Leahy during the exercise.)

[57] ABCA Armies, *Standardization Program: Information Booklet* (Washington, DC: Washington Standardization Officers, 13 Feb. 1987), p. 1–1.

[58] Australia, *Commonwealth Parliamentary Debates*, House of Representatives, vol. 118, 22 May 1980, p. 3199.

[59] ABCA Armies, *Standardization Program*, p. 1–1.

[60] "Foreword," *Army Research, Development and Acquisition Magazine* (Jan.–Feb. 1982): 1.

[61] Ezell, "Cracks in the Post-War Anglo-American Alliance," pp. 138–41.

with interoperability among the four armies as one of the system's primary objectives. While a considerable amount of advanced R&D was carried out on this project, it ultimately failed to reach the production stage because of cost overruns.[62]

3. As late as 1996 during a corps level command post exercise (CPX) CASCADE PEAK 96, some incomprehensible incompatibilities of equipment were discovered: 105mm tank ammunition, JP8/diesel fuel, and (for Heaven's sake) even fuel nozzle sizes![63]

Hence, because of the political sensitivity of equipment standardization, which requires purchases of foreign systems or domestic manufacture under license, from the late 1960s onward the ABCA Armies Program has redirected its principal efforts to the goal of doctrinal and procedural standardization.[64] The realization of this reality resulted in the program's de facto acceptance of the NATO concept of the different levels of standardization: compatibility, interoperability, interchangeability, and commonality.[65]

One publication[66] on the ABCA Armies Program lists its many advantages to its members along with methods by which standardization is achieved under the program:

1. Standardization Lists: The Standardization Lists contain a listing of the ABCA Armies R&D projects of interest to two or more armies.

2. Cooperative R&D: The ABCA Program provides an army with the means of matching requirements with other armies.

3. Loans of Equipment: The program provides an army with the opportunity to borrow equipment from other armies for its own test and evaluation if the loan is in the interest of standardization. When appropriate, these items can be tested to destruction. Loans are generally at no cost to the borrowing army.

4. Defense Sales: [A] sure method of achieving standardization.

---

[62]Australia, Parliament, *Defence Report 1969*, Parliamentary Paper No. 136 (Canberra: Australian Government Publishing Service, 1969), p. 32.

[63]American, British, Canadian, Australian Armies, *ABCA Exercise CASCADE PEAK 96, Post Exercise Report* (Washington, DC: ABCA Primary Standardization Office, 20 Jan. 1997), p. 13, "Lessons Learned." Note: "I (US) Corps was the lead nation headquarters, with 25 (US) Infantry Division (Light), 3 Brigade/2 (US) Infantry Division, 1 (CA) Division and 1 (AS) Brigade under command. The principal exercise objective was to evaluate standardization achieved in C^3I^2, EW and logistics." See p. vii.

[64]This was, interestingly enough, the initial objective of the 1947 "Plan to Effect Standardization." The products of the early postwar standardization program among the U.S., British, and Canadian armies were titled "Standardization of Operations and Procedures," or "SOLOGS." SOLOGS provided the foundation in many areas during the early standardization efforts of NATO. See James Huston, *One for All: The United States and International Logistics through the Formative Period of NATO* (Newark, DE: University of Delaware Press, 1984), p. 218.

[65]See *AAP–6, NATO Glossary of Terms and Definitions* (English and French), Jan. 1995; and American, British, Canadian, Australian Armies, *Standardization Program Information Booklet* (Washington, DC: Primary Standardization Office, 1994), p. 5.

[66]ABCA Armies, *Standardization Program*, pp. 15–17.

5. Quadripartite Standardization Agreements (QSTAGs): Armies may participate in formal agreements on common equipments and procedures, called QSTAGS. QSTAGs record the degree of standardization achieved and to be maintained for any item of equipment and agreement to standardize on operational, logistic, administration, and technical procedures. When applicable, QSTAGs are offered to national air forces and navies, which may also accept them as a binding agreement. Approximately 20 percent of QSTAGs are all but copies of NATO STANAGs, thereby tying Australia and New Zealand particularly closely to NATO technical standards.[67] A key finding of CASCADE PEAK 96 regarding QSTAGS is revealing: the Corps Commanding General in his after action report stated emphatically that they "work."[68]

6. Quadripartite Advisory Publications (QAP): There are several specialized functional or technical areas within the ABCA Program in which the standardization of procedures and processes for materiel and nonmateriel items are not possible. When the identification of these national procedures can be an aid to mutual understanding, they can be published as QAPs. There are more than two thousand QSTAGs and QAPs.[69]

7. Exchange of Ideas: The program provides a forum for the continuing exchange of ideas and thoughts among the scientists, developers, and army staffs.

The method by which information is exchanged under the ABCA Armies Program includes correspondence, Quadripartite Working Groups (which cover functional military branches—e.g., infantry, artillery, air defense artillery, etc.), Special Working Parties, Information Exchange Groups, and the exchange of Standardization Representatives located in each of the member countries.[70] The international management board, which oversees the activities of this program, is staffed by the Washington Standardization Officers who are field grade officers stationed near Washington, DC. Moreover, approximately every twenty-four months there is a general meeting of Army officials at the vice chief of staff/deputy chief of staff level of all five partners for the purpose of providing direction and establishing guidelines for future standardization efforts called, in keeping with the program's anatidae flavor, TEAL.[71]

---

[67]The existence and extent of QSTAGS are often transparent in the U.S. Army because of the manner in which field manuals and technical manuals are developed. QSTAGS are integrated directly into these documents and the fact that they are "ABCA" QSTAGS is not often acknowledged.

[68]American, British, Canadian, Australian Armies, "ABCA Exercise CASCADE PEAK 96, Post Exercise Report," p. 13. The corps commanding general was LTG C. G. Marsh, USA.

[69]See *Washington Staff Handbook*, p. 1.

[70]Ibid., p. 20.

[71]"Foreword," *Army Research, Development and Acquisition Magazine*, p. 1. "TEAL" actually takes its name from the British-sponsored Tripartite Conference on Tactics held in 1957 whose theme was tactics, equipment, and logistics; ergo, "TEAL."

This particular program has also attempted to coordinate the orientation of force development of the five armies through its "Armies Combat Development Guide." This classified publication is continuously updated and reissued every five years. The document assesses the outlook for global security for the West over the next ten years and identifies the likely combat requirements and examines available innovations to meet these requirements of its members. From this publication, "Quadripartite Objectives" are developed, which in turn provide direction to the numerous ABCA Army Quadripartite Working Groups in their efforts to formulate "Quadripartite Working Group Concept Papers."[72] Derivative from these studies are national "Army Objective/Requirements Documents" that are regularly circulated to other armies for comment.[73] From these coordinating efforts, the five armies are able to formulate their own objectives and plan for midterm military capabilities at the conceptual stage of development in conjunction with their allies, while making them cognizant of innovations existing in other armies.[74]

There is little question that, because of the immense size of the U.S. Army in comparison to its Anglo-Saxon counterparts, it predominates in many functional areas. Indeed, in recognition of this simple reality, the ABCA program openly acknowledges U.S. predominance in higher-level formations (i.e., corps) and concentrates its efforts on interoperability through the exchange of innovations at the level of lower maneuver formations (i.e., division, brigade, battalion).[75] Also, in terms of technological and operational innovations, the influence of the U.S. Army should not be underestimated. For example, the Australian, British, and Canadian armies[76] have liaison officers at Ft. Hood and Ft. Monmouth to monitor developments in the U.S. Army's FORCE XXI project. FORCE XXI seeks to leverage information-age technology through intensive and extensive experimentation with technologies and organizational concepts.[77] This project has resulted in the planned

---

[72]J. F. Koek, "A Guide for International Military Standardization—ABCA Armies' Operational Concept, 1986–95," *Defence Force Journal* (July–Aug. 1977): 51–54.

[73]ABCA Armies, *Quadripartite Standard Operating Procedures* (Falls Church, VA: Primary Standardization Office, 9 Mar. 1981), pp. 6–2, 7–1.

[74]Stuart K. Purks, "The ABCA Standardization Program," *Army Research, Development and Acquisition Magazine* (Jan.–Feb. 1982): 15.

[75]Leahy, "ANZUS: A View from the Trenches," p. 88.

[76]Although New Zealand has enjoyed an improved defense relationship with the United States since the mid-1990s, significant restrictions remain, to include the proscription against the presence of liaison officers at U.S. defense facilities. Nonetheless, despite this limitation, the New Zealand Army has been able to draw upon many of the lessons and developments from the U.S. Army's FORCE XXI through other liaison fora. See *Army Development 2015(AD2015) (Formerly known as Tekapo Manoeuvres)*. About the New Zealand Army, see http://www.govt.nz/army/about.htm.

[77]For a brief explanation of the FORCE XXI program, see "United States Army Posture Statement FY99," presented to the Committees and Subcommittees of the U.S. Senate and the House of Representatives, 2d Session, 105th Congress, Feb. 1998, pp. 28–29.

redesign of the U.S. Army's six armored/mechanized divisions[78] and has strongly influenced in particular the Australian Army's earlier plan to reorganize its entire army along RMA-inspired lines in its "Army 21 programme."[79]

Finally, despite the fact that the ABCA Armies Program is not a formal treaty organization (and as such does not, for instance, conduct contingency planning), it was decided in the mid-1980s to review the process of the program. As a result of this review, the armies determined the need to conduct a series of combined exercises to evaluate the state of interoperability, to determine shortcomings, and to validate and assess existing operating concepts and doctrine. The first ABCA command post exercise, CPX CALTROP TYRO, was held at Fort Ord, California, in November 1987. Following that, for the first time since the end of the Korean War, units from these four armies, comprising fifty-five hundred troops (constituting four maneuver battalions with support units), participated in a brigade-level field training exercise, FTX CALTROP FORCE, at Fort Hunter-Liggett, California, between 15 March and 1 April 1989.[80] Since then, field and command post exercises have been held on a biennial basis, with each army in turn hosting the event.

Conclusion

That the ABCA fora and the special relationships they represent are unique as regards the diffusion of military innovation is an understatement. This is particularly the case when one examines how formalistic the United States has been in its approach to its peacetime alliance relationships since the end of World War II. What is even more unusual is that the vitality of these arrangements has little to do with mutually held threat perspectives. In the post–Cold War era the ABCA arrangements have continued and some have even increased in intensity. The clear objective in all of the arrangements described and assessed in this essay has been the difficult task of achieving that elusive capability of effecting combined operations through maintaining peacetime programs whose object is to achieve a high degree of interoperability. Attaining the ability to conduct combined operations requires maintaining a constant dialogue among member services with the objective of remaining abreast of changes in operating procedures and the latest technological and organizational innovations developed by the members.

---

[78]See *The Washington Post*, 9 June 1998.

[79]See *Restructuring the Australian Army: A Force Structure for the Army of the Future*, DPUBS: 24432/96 (Canberra: Directorate of Publishing and Visual Communications, 1996).

[80]Scott R. Gourley, "CALTROP Force an Exercise in Co-operation," *Pacific Defence Reporter* (July 1989): 63; *The Age* (Melbourne), 13 Mar. 1989; and "ABCA's CALTROP FORCE Ends," *Jane's Defence Weekly*, 22 Apr. 1989, p. 694.

Three characteristics of these fora warrant mention as they relate to the wider issue of the diffusion of technological and organizational innovation. First, notwithstanding the existence of the TTCP and the defense science and technology aspects of the ABCA fora, that which appears to be the principal strength of these programs is not their cooperative technological activities but rather their focus on developing compatible procedural and operating arrangements. The intellectual processes behind these innovations have encouraged the Anglo-Saxon armed services to remain in constant contact with their counterparts on practically all key military activities. Second, the activities of these fora do not appear to have any direct connection with their nations' threat perceptions, let alone deliberate planning processes. Even in the extreme case of the U.S.-New Zealand diplomatic contretemps that led to a break in bilateral defense ties, New Zealand's membership and status in these programs has remained remarkably unaffected. That the diffusion of innovation has increased in some of the programs since the end of the Cold War clearly indicates a deeply shared commonality in the five countries' global objectives (i.e., an expansion in the adherence to Western norms, and failing that opposition to challenges to Western prestige and favorable regional political balances). Third, size and resources do not appear to be a consideration in continued access to, and the exchange of, innovation. Rather, as long as participating states bring some unique operational experience or new technological innovations to exchange, their bona fides are accepted by the other members.

That said, the special diffusion relationship among these five Anglo-Saxon countries is not without some noticeable weaknesses. The overwhelming size of the U.S. armed forces and the dissimilarities in technological capabilities has created problems in the area of maintaining meaningful cooperation. More precisely, that the U.S. armed forces have embraced the concept of the "revolution in military affairs" and have undertaken ambitious modernization programs has caused concern among many U.S. allies that they will be unable to maintain interoperability with the United States in future. Concern has been expressed both in terms of anxiety that allies will be unable to keep up with U.S. advances, as well as being incapable of contributing to the development of these innovations because of limited resources.[81] Given the generally shared trend in the Western Alliance of diminished defense budgets, which mitigates against innovation and modernization, there is the potential for the creation of a substantial capabilities gap between the United States and its allies. Notwithstanding the legitimacy of this point, the U.S. armed forces are not unaware of this important problem and

[81]See the comments made by General Klaus Naumann, German Army, Chairman of the North Atlantic Military Committee, in *The Washington Post*, 6 July 1997; and comments by the Netherlands vice-chief of Defence Staff, Lieutenant General A. van Baal, in *Jane's Defence Weekly*, 11 July 1997, p. 17.

are endeavoring to maintain their ability to operate alongside forces that are less technically advanced—both allies and their own reserve components.[82]

A further weakness in the ABCA framework of arrangements and programs has been its inability to transcend its "combined operations" orientation and address effectively what has become one of the leading military innovations since the end of the Cold War—joint (service) operations.[83] Arguably, the ABCA programs, despite recent provisions for holding multifora ABCA meetings,[84] have yet to overcome some of their postwar service parochialism—that is, concentration on combined operations. In this respect, it is surprising to note that NATO is more advanced in developing draft joint operations doctrine in a combined setting than the ABCA programs.[85] It would not make sense to suggest creating an entirely new body of "Anglo-Saxon" joint doctrine. Unilateral joint operations are challenging enough without adding yet another body of doctrine to what is a growing doctrinal morass. Nonetheless, consideration should be given to the creation of an informal body of ABCA joint doctrine development officials/centers to ensure that they are aware of other ABCA efforts and findings. Ideally, this could mitigate against the emergence of substantive differences in national approaches to conducting joint operations.

Notwithstanding these problems or weaknesses, the fact remains that the ABCA fora are a unique example of how the diffusion of innovation can transcend national borders. What is remarkable is how these allies allow an all but free flow of information of military technical and organizational innovations, as well as its *exclusive* nature. Other key U.S. allies neither are members nor participate directly in ABCA activities. In other words, the development of, for example, Allied Communications Publications, is an exclusively Anglo-Saxon activity, but whose usage is worldwide, to include key U.S. allies. To be blunt, to date, such important Western allies as France, the Federal Republic of Germany, and Japan are simply not given the breadth of access to information and low-level, but important, decision-making opportunities by the United States as are its Anglo-Saxon brethren. Thus, in view of this unique relationship among these five countries, there can be no doubt that cultural similarity can be an effective, if not indeed subtle, means of facilitating the diffusion of innovations.

[82]See the interview with General William Hartzog, Commander, U.S. Army Training and Doctrine Command, *Jane's Defence Weekly*, 1 Oct. 1997, p. 32ff.

[83]It is interesting to note that the vacuum in combined joint doctrine is being formally addressed in the Multinational Interoperability Council, a body created in 1996 that consists of the ABCA nations plus France and Germany. It remains to be seen whether this body will supplant ABCA efforts in coalitional, let alone combined joint, doctrine. See *Washington Staff Handbook*, pp. 21–22.

[84]See *ASCC News*, Newsletter #4, ASCC Homepage.

[85]See, for example, *Allied Joint Operations Doctrine*, AJP–1(A), NATO UNCLASSIFIED, Change 1, Sept. 1997.

PART II

Managing and Controlling Diffusion

CHAPTER FIVE

Reflections on Mirror Images

Politics and Technology in the Arsenals of the Warsaw Pact

CHRISTOPHER JONES

The diffusion of military technology from the Soviet Union to its Warsaw Pact (WP) allies highlights the process of "coercive" diffusion and efforts to shape and tightly control the military practices of other states. In coercive diffusion, forms and practices are imposed on dependent organizations. Direct imposition may facilitate standardization, but it also produces distortions because diffusion is driven by the goals of the "supplier" state and is less likely to be sensitive to the social, political, and cultural contexts of "recipient" states, or to the goal of military effectiveness. Soviet decisions about the size and structure of Warsaw Pact ground forces show that dissemination of technology and hardware was used as a tool for political purposes.

The Soviets were virtually the sole supplier of most East European military technology, either by export or licensing. The distribution of Soviet technology within the Warsaw Pact was influenced by the desire for political, rather than military, effectiveness. The most important dimension of diffusion was in ground force technology—the offensive weapons that stood at the heart of mounting an offensive against the West. The Soviets provided most of their allies with token quantities of state-of-the-art tanks. The exception was the East Germans, who received the largest quantity of state-of-the-art weaponry, even though they were the smallest of the non-Soviet Warsaw Pact (NSWP) armies. The East Germans were also the only national units trained and equipped close to the level of Soviet forces. What accounts for this distribution of military technology?

Domestic political considerations in East Germany and other Warsaw Pact states dictated the size of Warsaw Pact forces and the distribution of Soviet mili-

tary equipment to the non-Soviet members of the Warsaw Pact. The highest Soviet priority was to outfit six East German divisions of the National People's
Army so that they could be paired as symbolic allies of the Soviet divisions in the
German Democratic Republic (GDR). The primary Soviet goal was to deny the
armed forces of the Federal Republic of Germany (FRG), the *Bundeswehr*, a claim
to be the sole heir of German national military tradition. A secondary goal was to
use the East German military as political camouflage for the extremely large internal security forces of the GDR. These goals meant that as the military leader of
the Warsaw Pact, the USSR had a vested interest in both promoting as well as
controlling the spread of new technologies and practices within the military
sphere.

In comparison to the other East European military forces, the National People's Army (NVA) of East Germany came closest to matching the equipment of
Soviet forces in Eastern Europe. The NVA also maintained a much higher state of
military readiness and much closer integration with Soviet forces. The overall
quality of the NVA allowed it to compete with the West German army as a symbol of German national military identity. Over the course of the Cold War, the
NVA served as the GDR's principal proof that the GDR was a legitimate German
state: like West Germany, it had its own army.[1]

During the Cold War the six ground forces divisions of the NVA constituted a
relatively small force compared with other non-Soviet armies of the Warsaw Pact,
particularly when available human and economic resources are compared (Table
5.2). The Soviets appear to have kept the NVA just large enough to pass muster as
a symbolic national force and small enough so that in the event of war the consequences of a possible NVA collapse would be tolerable. Even though the NVA
was comparatively better equipped than the other East European armies, its divisions were never brought up to the armament standards of the nineteen Soviet
divisions in East Germany. Only two of the six NVA divisions were tank divisions. There were eleven Soviet tank divisions in the nineteen-division Group of
Soviet Forces in Germany (GSFG).

The main military-political consequence of the Soviet decision to create and
arm the NVA was the necessity of creating the Warsaw Pact, so that the NVA
could have military allies symbolically comparable to those of the *Bundeswehr*.
These East European forces did not have to compare themselves with those of a
"West Poland" or "West Hungary." As a result, they were armed to a lower tech-

---

[1]This claim expired following the night of 9–10 Nov. 1989, as GDR citizens tore down the Berlin
Wall. The NVA disintegrated shortly afterward. The Warsaw Pact had collapsed by the spring of 1990,
though it remained a formal organization until the summer of 1991. See Dale R. Herspring, *Requiem for
an Army: The Demise of the East German Military* (Lanham, MD: Rowman and Littleman, 1998).

nical standard than the NVA, maintained much lower levels of readiness, and were less integrated with Soviet forces.

The principal device for conferring military credibility on the NVA was to drill and train NVA personnel to the standards set by the GSFG. Neither internal nor external observers could contest the intensity of the NVA training and exercise programs. But the first purpose of such programs may have been political, not military. These programs were Potemkin drills in Potemkin skills. To fulfill its political mission, the NVA did not have to make a major contribution to the overall number of Warsaw Pact divisions. Its six standing divisions served three main purposes: (1) to compete with the *Bundeswehr* as a symbol of state sovereignty; (2) to serve as hosts for the GSFG as ally rather than occupier; and (3) to camouflage the relatively much larger numbers of uniformed personnel (border troops, Ministry of the Interior troops, and other special units) assigned to the internal front. The internal forces were focused on the real security threat in East Germany, as demonstrated by the events of 1989—and 1953 as well.

After providing East Germany with a politically useful military force of minimum size, the next Soviet priority was to outfit the Polish, Czechoslovak, and, to a much lesser extent, Hungarian forces with weapons to enable their elite units to assume symbolic status both as allies of in-country Soviet garrisons and as "Northern Tier" partners of the NVA and the GSFG (later renamed the Western Group of Forces). In Poland in particular, the military also became a symbol of national identity and bestowed a degree of legitimacy on the local Communist Party/state apparatus.

The Soviets outfitted the Bulgarian forces with the minimum armaments necessary to make a show of alliance participation. The Albanians, who remained formal members of the WP until 1968, ceased purchasing Soviet weaponry after 1960. About the same time, the Romanians sought to minimize the Soviet weaponry in their arsenal. This policy complemented a territorial defense strategy intended to maximize their military and political independence of the USSR. Remaining in the Warsaw Pact allowed Nicolae Ceausescu to pose before his own people as a defiant anti-Soviet nationalist skillfully deflecting the permanent threat of a Pact intervention. WP membership also won him special consideration from both NATO and the USSR that he would not otherwise have obtained.

The symbolic political functions of the WP demanded that the Pact convince NATO observers that the Warsaw Pact should be taken seriously as a military alliance—otherwise the East Europeans would never take it seriously. The Soviets persuaded the West by means of massive investments in the conventional and nuclear forces of the USSR. The Soviets could then use the responses of the Atlantic alliance to justify an offensive military posture for the Warsaw Pact, backed up by the threat of nuclear weapons. This offensive posture, formally revealed in 1961 by WP commander Andrei Grechko, facilitated three Soviet goals: (1) persuading West Germany to accept the inner-German *status quo* as permanent; (2) configuring Soviet forces for rapid and massive military interventions to rescue failed communist regimes in East Europe; and (3) preempting East European capacities for organizing national defense of national territory by national means on the models of Yugoslavia (never a Pact member), Albania, and Romania. These three states were independent of the USSR, frequently challenged Soviet foreign and security policies, and denied use of their national armed forces to the Warsaw Pact, either for interventions in East Europe or for offensive campaigns against NATO.

It is the contention of this chapter that the sizing of Warsaw Pact ground forces and internal security forces, and also the distribution of tanks, were functions of the political assignments, rather than military missions. This is not to say that the Soviets may have hoped that in the long run, the East European forces could eventually reach a level of political reliability necessary to make a genuine military contribution as steadfast allies. But the Berlin Wall fell before the Soviet *maskirovka*[2] could be replaced by a genuine military alliance. For the duration of the Cold War, the contributions made by the East European forces were largely political. They posed as military additions to the Soviet forces in the Western Theater. They served as hosts for Soviet garrisons and for the joint exercises that maintained a Soviet capability for military intervention. The low military value of these forces to the Soviets is reflected in the technology transfer policies of the Soviets to the Pact members. (See Tables 5.1–5.7.)

The National People's Army in the *Bundeswehr* Mirror

The explanation for the creation of the GDR National People's Army lies in the political consequences of NATO's decision to develop a German national military force, the *Bundeswehr*. The decision on German rearmament came after a five-year Western debate over German rearmament involving Bonn, Paris, London, Washington, and other NATO capitals. This debate saw the French parliament reject in 1954 a plan for an integrated multinational European army under the auspices of a European Defense Community. One historian argues that the 1955 decision to bring a German military force into NATO was the reluctant

---

[2]See the entry for *maskirovka* on p. 175 of N. V. Ogarkov, ed., *Sovetskaia Voennaia Entsiklopediia*, vol. 5 (Moscow: Voenizdat, 1978). The two-page entry lists a bibliography of six Soviet scholarly works on the subject. The encyclopedia entry reads in part: "Strategic *maskirovka* is achieved by the decision of the supreme high command and includes measures for the preservation of secrecy in the preparation of . . . measures for the disorientation of the adversary relative to the actual intentions and actions of one's armed forces."

TABLE 5.1
WP Ground Forces—East European and USSR, 1988–89
Category A Divisions

Country	Population	Personnel in Cat A Divisions	Cat A/B/C Divisions	Category A Divisions	All Ground Forces Personnel Cats A/B/C
GDR	16,593,000	73,720	6	6	120,000
GDR Divs, Cats A/B/C			2 Tk, 4 MR	2 Tk, 4 MR	
Czechoslovakia	15,676,000	49,555	10	4	145,000
Czechoslovak Divs Cats A/B/C			5 Tk, 5 MR	1 Tk, 3 MR	
Poland	38,122,000	107,435	13 Divs 2 Brigades	8 Divs 2 Brigades	230,000
Polish Divs Cats A/B/C			5 Tk, 8 MR 1 Airborne Brigade 1 Marine Brigade	5 Tk, 3 MR 1 Airborne Brigade 1 Marine Brigade	
Hungary (1989)	10,628,000	0	(1985) 6	0	84,000
Hungarian Divs Cat A/B/C			(1989) 5 Tk Brigades 10 Mechanized Rifle Brigades	0	
Bulgaria	9,078,000	0	8	0	157,000
Bulgarian Divs Cats A/B/C			8 MR	0	
NSWP Divs personnel, Cats A/B/C (except Romania)			43		729,000
NSWP Cat A Divs personnel (except Romania)		230,710		18	
Sov Cat A Divs in EE		565,000		30 15 Tk, 15 MR	565,000
Sov Divs in western military districts of USSR			65	15 Cat A/B[a]	
GSFG Divs		380,000		19 11 Tk, 8 MR	380,000
NGF Divs		40,000		2 1 Tk, 1 MR	40,000
CGF Divs		80,000		5 2 Tk, 3 MR	80,000
SGF Divs		65,000		4 2 Tk, 2 MR	65,000
NVA as percentage of all NSWP Divs(except Romania)				6 of 43	14%
NVA as percentage of all NSWP Cat A Divs (except Romania)				6 of 18	33%

TABLE 5.1—*cont.*

Country	Cat A/B/C Divisions	Category A Divisions	All Ground Forces Personnel Cats A/B/C
Sov Force Groups in EE as percentage of Sov Cat A Divs European Theater (Western and Southwestern TVDs)[b]		30 of 45	66%
GSFG as percentage of Sov Cat A Divs in EE		19 of 30 (before 1981, 20 of 30)[c]	64.5% 66%
Sov Cat A as percentage of all WP Cat A Divs		30 of 48	62.5%
NSWP Cat A Divs as percentage of all Cat A WP Divs in EE		18 of 48	37.5%
NSWP Cat A/B/C Divs as percentage of all Sov and NSWP Divs (except Romania)	43 of 138 (43 + 30 + 65)		31.2%
GDR-NVA as percentage of Cat A NSWP Divs (except Romania)	6 of 18		33%
GDR-NVA as percentage of GSFG	6 of 19		31.5%
Sov Force Groups in EE as percentage of all Sov Cat A/B/C Divs in European Theater (Western and Southwestern TVDs)[b]	30 of 95		31.5%

SOURCE: *The Military Balance, 1988–89.*

ABBREVIATIONS: Tk=Tank Division; MR=Mechanized Rifle Division; EE=Eastern Europe; Sov=Soviet; WP=Warsaw Pact; NSWP=Non–Soviet Warsaw Pact; Div=Division; Cat=Category; GSFG=Group of Soviet Forces in Germany; NGF=Northern Group of Soviet Forces in Poland; CGF=Central Group of Soviet Forces in Czechoslovakia; SGF=Southern Group of Soviet Forces in Hungary.

[a]*The Military Balance, 1986–87,* p. 37 "Perhaps 15 of the 65 divisions in the West and Southwest European USSR are Category A (70–100 percent ready) or Cat .B (50–75 percent ready). Implication: the remaining are Cat C (about 20 percent ready).

[b]*The Military Balance, 1986–87,* p. 37. TVD=Theater of Military Action.

[c]In 1981, Brezhnev announced the "withdrawal" of one Soviet division from East Germany as an opening bid in the NATO-WP negotiations over the deployment of intermediate-range nuclear forces. Some Western observers claimed that there was little net reduction in personnel and weaponry in the GSFG as a result of the "withdrawal" of one Soviet division, because of alleged transfers of personnel and equipment to the remaining nineteen divisions.

TABLE 5.2

External Front of the Warsaw Pact
Ground Combat Divisions

	GDP Estimate	Per Capita GDP Low	Per Capita GDP High	Ground Force Divisions	Available GDP per Standing National Division Low	Available GDP per Standing National Division High	Population per National Division
Bulgaria pop. 9,078,000	$30.4–$67.7bn	$3,348	$7,457	8	$3.8bn	$8.64bn	1,134,750
Czechoslovakia pop. 15,676,000	$69.5–$140.5bn	$4,433	$8,962	10	$6.95bn	$14bn	1,567,600
GDR pop. 16,593,000	$94-$182bn	$5,665	$10,968	6	$15.6bn	$30bn	2,765,500

TABLE 5.2—*cont.*

	GDP Estimate	Per Capita GDP		Ground Force Divisions	Available GDP per Standing National Division		Population per National Division
		Low	High		Low	High	
Hungary pop. 10,628,000	$28–$95bn	$2,634	$8,938	6	$4.6bn	$15.8bn	1,1771,300
Poland pop. 38,122,000	$98–$260bn	$2,570	$6,820	13 Divisions 2 Brigades	$7.5bn	$20bn	2,087,100
Romania pop. 22,400,000	$68.1–$135bn	$2,857	$5,663	10	$6.8bn	$13.5bn	2,240,000
WP Average 1	$64–$146.7bn	$3,584.50	$8,134.66	8.83	$7.54bn	$16.99bn	2,068,600
WP Average 2 w/o GDR	$58.8–$139.6bn	$3,168.40	$7,568.00	9.4	$5.93bn	$14.38bn	1,929,200
Average 2 as % of GDR	68–80%	63%	74%	147%	48%	56%	69.9%
GDR as % of Average 2	146–124%	158%	134%	67%	206%	176%	144%

SOURCE: *The Military Balance, 1988–89.*

TABLE 5.3

Tank Inventories of NSWP Ground Forces, 1988–89

	Cat A Tank Div	Cat A MR Div	Cats.B/C	T-80 Tanks (in Soviet Force Groups)	T-72 Tanks	T-54/55 Tanks	T-34 Tanks
Tanks required per standard Soviet Division (Div)	328[a]	222[a]					
GDR	2	4					
Tanks needed for 2 tanks and 4 MR Divs Cat A	656	888	0				
1544 tanks needed for 6 Divs					1544 needed to match quality of GSFG Tks		
Actual inventory all tanks: 2850				0	350	1700	800
Czechoslovakia	1	3	6 (3 Tk, 3 MR)				
Tanks needed for 1 tank and 3 MR Divs, Cat A, plus 3 tank Divs, 3 MR Divs, Cats B/C	328	666	1650				

TABLE 5.3—*cont.*

	Cat A Tank Div	Cat A MR Div	Cats.B/C	T-80 Tanks (in Soviet Force Groups)	T-72 Tanks	T-54/55 Tanks	T-34 Tanks
2644 tanks needed for 10 Divs					2644 needed to match quality of CGF Tks		
Actual inventory all tanks: 3,400				0	500	2900	0
Poland	5 plus 1 Airborne Brigade 1 Marine Brigade	3	5 (all MR Divs)				
Needed for 3 tank and 5 MR Divs, Cat A, plus 5 MR Divs, Cat B/C	1640	666	1110				
2306 tanks needed for 13 Polish divisions					2306 needed to match quality of NGF Tks		
Actual inventory all tanks: 3950				0	350	3400	200
Hungary	0	0	before 1987: 6 Divs (1 Tk, 5 MR) after 1987: 15 brigades				
1438 tanks needed for 5 MR Divs, 1 tank Div, all Cat B/C	0	0	1438		1438 needed to match quality of SGF Tks		
Actual inventory all tanks: 1335				100 PT-76	135 or 60[b]	1200 or 1300 or 1400[b]	0
Bulgaria, 1985–86[c]	0	0	before 1987: 8 MR Divs 5 Tk brigades after 1987: 13 brigades				
1776 tanks needed for 8 MR Divs (no figure for tank brigades)							
Actual inventory all tanks: 2550					200	1200 or 1500[d]	850

TABLE 5.3—*cont*

	T-62 Tanks	PT-76 Tanks	T-80 Tanks	T-72 Tanks	T-54/55 Tanks	T-34 Tanks
Total for NSWP tank inventories	125	100	0	1595	10,700–12,000	2350

For Comparison to NSWP Soviet Tanks: Soviet Tanks in Syria, Libya, and Iraq

	T-72 Tanks	T-54/55 Tanks	T-62 Tanks
Syria	950	2,100	1,000
Libya	180	1,800	0
Iraq	500	2,500	1,000
TOTAL	1630	6,400	2,000

SOURCE: *The Military Balance, 1988–89.*
[a]*U.S. Army Field Manual FN-2-3 The Soviet Army: Forces and Equipment.*
[b]Discrepancy in estimates, *The Military Balance, 1986–87, 1988–89.*
[c]*The Military Balance, 1985–86.*
[d]Discrepancy in estimates, *The Military Balance, 1989–90, 1988–89.*

TABLE 5.4

Tank Requirements—Groups of Soviet Forces in East Europe, 1988–89

	Tanks Needed	T-80 Tanks	T-72 Tanks	T-64/62 Tanks	T-54/55 Tanks
Soviet Cat A Tank Divisions in East Europe; 328 tanks per Tank Division					
16 Tank Divisions	5248				
Soviet Cat A Mechanized Rifle Divisions in East Europe 222 tanks per Mechanized Rifle Division					
14 MR Divisions	3108				
Total Soviet tanks needed in East Europe	8356				
Soviet Tank Inventory	Available: 53,300	2500	9000	22,500	19,300
Tanks needed for GSFG	5384	?	?	?	?
Tanks needed for NGF	550	?	?	?	?
Tanks needed for CGF	1322	?	?	?	?
Tanks needed for SGF	1100	?	?	?	?

SOURCE: *The Military Balance, 1988–89.*

TABLE 5.5

Internal Front of the Warsaw Pact National
Paramilitary Forces

	Regular armed forces personnel	Paramilitary personnel	Paramilitary w/ militia	Armed forces as % total population	Paramilitary as % of armed forces	Paramilitary w/ militia as % of armed forces
Bulgaria 9,078,000	157,000	22,500	172,500	1.70%	14%	109%
Czechoslovakia 15,676,000	197,000	11,000	131,000	1.20%	5.50%	66%
GDR 16,593,000	172,000	92,500	592,500	1.03%	53%	344%
Hungary 10,628,000	99,000	16,000	76,000	0.90%	16%	76%
Poland 38,122,000	406,000	87,000	437,000	1.06%	21%	107%
Romania 23,836,000	179,000	40,000	290,000	0.70%	22%	162%
Average 1 with GDR				1.09%	21%	144%
Average 2 w/o GDR				1.11%	15.7%	104%

SOURCE: *The Military Balance, 1988–89.*

NOTE: Militia organizations consisted of Communist Party members supplied with weaponry from Ministries of the Interior and Ministries of Defense. Militias had paramilitary tasks and sometimes participated in Warsaw Pact military exercises.

TABLE 5.6

East German and West Germany Military Forces in 1988–89

	GDR	GDR as % of FRG	FRG
GDP (US $ billion)	94–182 (estimate)	8–16%	1119.73
Population	16,800,000	27.6%	60,931,000
18–22 years old as % of population	7.0%		7.4%
Population of 18–22 years old	1,188,000	26.2%	4,519,000
Personnel in Ground Forces combat divisions	73,720	28.03%	260,800
Personnel in Ground Forces combat divisions as % of population	0.0044%		0.0043%
Armed forces—total	172,000	35.2%	488,700
Armed forces as % of population	1.02%	127.5%	0.80%
Conscripts	94,500	42.45%	222,600
Armed forces reserves	323,000	37.9%	852,000
Reserves as % of population	1.92%	138%	1.39%
Total regular and reserves	495,000	36.9%	1,340,700
Total Armed forces plus reserves plus paramilitary (1)	587,500	43.1%	1,361,700
Total Armed forces plus reserves plus paramilitary (2)	1,087,500	79.86%	1,361,700
Navy personnel	15,000	41.6%	36,000
Navy reserves	21,000	80.7%	26,000

TABLE 5.6—*cont.*

	GDR	GDR as % of FRG	FRG
Naval vessels	56	28.8%	85
Air Force personnel	37,000	34%	108,700
Air Force reserves	30,000	28.8%	104,000
Combat aircraft	330	54.7%	603
Ground Forces personnel	120,000	36.1%	332,000
Conscripts	71,500	40.6%	175,900
Conscripts as % of regular forces	41.5%	115%	35.9%
Reserves	500,000	69.7%	717,000
Ground Forces divisions	6		12
	4 MR @ 12,695 persons each		4 MR @ 21,700 persons each
	2 Tk @ 11,470 persons each		6 Tk @ 22,000 persons each
			1 AB @ 21,000
			1 Mtn @ 21,000
Personnel in Ground Forces combat divisions	73,720	28.03%	260,800
Paramilitary (without Workers' Militia)	92,500	440%	21,000
Border Guard	47,000[a]	235%	20,000
Coast Guard	0		1,000
Berlin Guard	7,000		0
Interior Ministry	22,000[b]		0
Workers' Militia full-time	3,000		0
Workers' Militia mobilized	500,000[c]		0
Paramilitary with Workers' Militia	592,500	2,821%	21,000
Paramilitary as % of personnel in Ground Forces combat divisions	125%	1,552%	8.05%
Paramilitary as % of total regular military (without Workers' Militia)	53.8%	1,251%	4.3%
Paramilitary as % of total regular military (with Workers' Militia)	344%	8,000%	4.3%
Regular Armed Forces as % of paramilitary with Workers' Militia	29.02%	1.2%	2,327%

SOURCE: *The Military Balance, 1988–89.* GDR Society for Sports and Technology (450,000 members) not included in these tables.

[a]Border Guard trained by Soviet Advisors.

[b]People's Police Alert Force: 12,000; Transport Police: 8,500; Civil Defense Force: 15,000.

[c]Worker's Militia frequently participated in Warsaw Pact exercises.

TABLE 5.7

Warsaw Pact Ratios of Distrust: Ratios of Career Cadre/Paramilitary/"Militia Organizations" to Conscripts in Regular Armed Forces

	USSR	GDR	Czech.	Poland	Hungary	Bulgaria	Romania
Regular forces total	5,050,000	167,000	204,000	340,000	105,000	162,000	189,500
Conscripts in regular forces	3,300,000	92,000	117,000	190,000	58,000	94,000	109,000
Career cadre in regular forces[a]	1,250,000	75,000	37,700	150,000	47,000	68,000	80,500

TABLE 5.7—*cont.*

	USSR	GDR	Czech.	Poland	Hungary	Bulgaria	Romania
Paramilitary forces[a]	450,000	74,000	11,000	85,000	15,000	23,000	37,000
Militia[b] (party members)	NA	500,000[c]	120,000	350,000	60,000	150,000	1,550,000[d]
Paramilitary/ career cadre total	1,700,000	149,000	98,500	235,000	62,000	91,000	117,500
Paramilitary/career cadre in regular armed forces plus militia total	NA	649,000	213,500	585,000	122,000	241,000	NA[e]
Soviet forces in country[f]	4,274,000	380,000	80,000	40,000	65,000	0	0
Ratio of Distrust 1[g]	0.4	1.6	0.84	1.23	1.0	.97	1.07
Ratio of Distrust 2[h]	NA	7.05	1.87	3.07	2.1	2.56	NA[i]
Ratio of Distrust 3[j]	NA	11.18	2.5	3.28	3.22	2.56	NA[k]

[a]Calculated from *The Military Balance, 1983–84*. This table adapted from Teresa Rakowska, in Harmstone and Jones, *The Warsaw Pact: The Question of Cohesion*, Vol. 2: *The Greater Socialist Army*, (Ottawa: Department of National Defense, 1985), p. 325.

[b]Various names: Citizen's Militia, Workers Militia. These were Communist Party–based organizations with equipment supplied by Ministries of Defense and Interior.

[c]William J. Lewis, *Warsaw Pact Forces: Arms Doctrine, Strategy* (New York: McGraw Hill, 1982), p. 73. GDR Citizens' Militia organized in 15,000 combat units.

[d]Romanian Territorial Defense Forces include 900,000 Patriotic Guard, 650,00 Youth Home Defense Forces, plus supplemental civilian organizations.

[e]Not applicable: Romanian militia under territorial defense command outside WP control system. No participation in joint exercises of Warsaw Pact. No plans to support WP conventional forces. Romanian system organized for guerrilla resistance to large-scale conventional invasion. No Soviet links to internal security system, which was under Ceausescu's personal control. See "Osobaia pozitziia Rumynskogo rukovodstva" ["The Special Position of the Romanian Leadership"], in A. I. Grikov (former Chief of Staff, WP), *Su'dba Varshavskogo Dogovora* [*The Fate of the Warsaw Pact*] (Moscow: Russkaia kniga, 1998), pp. 74–80.

[f]*The Military Balance, 1983–84*.

[g]Ratio of career cadre plus paramilitary (border troops, internal security troops, other special units) to conscripts in the regular armed forces.

[h]Ratio of Distrust 2: Ratio of career cadre plus paramilitary (border troops, internal security troops plus special units) plus Militia (party-based organizations) to conscripts in the regular armed forces.

[i]Not applicable: Romanian militia under territorial defense command outside WP control system. See note e above.

[j]Ratio of Distrust 3: Ratio of career cadre plus paramilitary (border troops, internal security, special units) plus Militia (party-based organizations) plus Soviet troops in country to conscripts in regular armed forces.

[k]Not applicable. See note e above.

acceptance by all parties—particularly by Paris—of a policy of "dual containment," directed not only against the external Soviet military threat but also at integrating a democratic Germany into the structures of Western military and economic cooperation.[3]

Bringing a first-class German army into NATO would also undercut the do-

[3]Thomas Alan Schwartz, *America's Germany: John J. McCloy and the Federal Republic of Germany* (Cambridge: Harvard University Press, 1991). See chap. 5, "The Dilemmas of Rearmament" and chap. 8, "The Skeleton Key: Military Integration."

mestic political base in the FRG for right-wing nationalists, for neutralists, and for officials contemplating a revival of the Rappallo era of Soviet-German military cooperation.[4] And equally important, the Christian Democrats were able to prove that their policy of cooperation with the Allies ultimately resulted in the *de facto* restoration of German sovereignty, symbolized by the *Bundeswehr*'s leading role in NATO's conventional forces. *De jure* restoration of full sovereignty required an end to the four-power rights in Berlin.

The NATO solution to the German question was to create twelve West German divisions plus a formidable tactical air force and modest naval force, but not a German general staff. The German divisions were placed in a multinational force, under NATO command, with an American as supreme commander. With the creation of the *Bundeswehr*, U.S., British, and French forces stationed in the FRG became allies rather than occupiers. And Kurt Schumacher, head of the German Social Democrats, could no longer mock Konrad Adenauer, leader of the Christian Democrats, as "the chancellor of the Allies."

The NATO decision confronted Moscow in the spring of 1955 with the political problem of deciding whether or not to create an East German army out of the Barracked Police, an internal security force whose primary purpose was to buffer the Group of Soviet Forces in Germany from direct contact with civil disturbances.

In 1953 the Barracked Police had failed to put down an uprising, and the GSFG, under the command of Andrei Grechko, had to go into action. That event, coupled with the savage conduct of Soviet soldiers toward German women in the early years of the occupation,[5] may have permanently undermined the possibility of genuine military cooperation among Soviet and East German conscript personnel. The absence of an East German army in early 1955 may have reflected in part such Soviet misgivings, although the Soviets had clearly been outfitting the Barracked Police with armaments appropriate for a national army. Even after the creation of the NVA, the commander of the GSFG retained until the unification of Germany in 1990 the legal right to declare martial law, suspend the GDR constitution, and rule the GDR directly.

The establishment of the *Bundeswehr* required the Soviets in late 1955 to go through the motions of creating and arming a relatively small East German army, consisting of six ground forces divisions supported by token air and naval forces. Not to have created the NVA would have been to bestow on the FRG an exclusive right to the German national military tradition, a powerful symbol of sovereignty and legitimacy. This was particularly true in the Prussian *lander* that constituted

---

[4]Ibid.

[5]Norman Naimark, *The Russians in Germany: A History of the Soviet Zone of Occupation* (Cambridge: Harvard University Press, 1994), passim.

the territory of the GDR. Such a situation would have reinforced the claim that the Federal Republic made in its Basic Law of 1949 to be the sole legitimate German government. The "Hallstein doctrine" of the 1950s and 1960s maintained this claim by refusing to establish diplomatic relations between the FRG and any state that recognized the GDR (with the exception of the USSR). The Soviet decision to create an East German military was an effort to legitimize not only the GDR but also the Socialist Unity Party of Germany (SED). The NVA also changed the nominal status of the GSFG from an occupying army to a fraternal ally.

The ground forces of the NVA were proportional in size to the corresponding size of the ground forces of the *Bundeswehr*. The number of personnel in the six ground forces combat divisions of the NVA was about 0.0044 percent of the GDR population. The comparable figure for the FRG is 0.0043 percent (Table 5.6). But by every other measure, the number of personnel in other service branches and other units of the NVA was disproportionately larger than the corresponding entities of the *Bundeswehr* (Table 5.6). In other words, the size of the NVA ground forces was minimized to the lowest level necessary to provide a symbolic rival to the *Bundeswehr*. The internal security forces of the NVA were enormously larger than the size of the corresponding forces in the FRG (Table 5.6).

If the GDR Workers Militia is included in the category of paramilitary forces, then the total GDR paramilitary forces constituted 344 percent of the regular armed forces. In the FRG, the comparable figure was 4.3 percent. Put another way, the GDR had 592,000 personnel on the internal security front; the FRG had 21,000. The GDR put substantially greater proportion of its population into the regular armed forces than the FRG (1.02 percent of the population in the GDR, versus 0.80 percent in the FRG—that is, 25 percent more than the FRG), not counting the paramilitary forces assigned to the internal front in both countries. Yet in terms of the six ground forces personnel (73,720), the GDR put into the field constituted 28.03 percent of the number of personnel that the FRG put into combat divisions (260,800). The population of the GDR was about 27.5 percent that of the FRG.

Hence the conclusion: using the FRG as a standard, the GDR put a larger percentage of its population in military/paramilitary uniforms than the FRG, but shortchanged the Warsaw Pact in terms of ground forces personnel assigned to the external front (Table 5.2). At the same time, the GDR devoted far greater resources to the internal front (Table 5.6). The events of 1989 demonstrated that such allocation of manpower resources correctly addressed the primary security threat to the GDR.

The internal security forces of the GDR were also proportionately much larger than those of the other Warsaw Pact forces (see Table 5.5). Table 5.7 examines "ratios of distrust" in the Warsaw Pact armies, measured as the ratios of para-military forces to regular armed forces, the ratio of cadre personnel to conscripts, and the ratio of in-country Soviet troops to regular military personnel. Tables 5.6 and 5.7 indicate that the GDR was in a class by itself in terms of the high ratios of distrust toward conscript personnel.

The figures in Tables 5.1 and 5.6 also suggest that the ratios between the ground forces personnel in the NVA of the German Democratic Republic and the *Bundeswehr* of the Federal Republic of Germany were the bases for a series of concentric ratios within the Warsaw Pact.

The six divisions of the NVA amounted to roughly one-third of the nineteen divisions of the GSFG (Table 5.1). The six NVA divisions constituted roughly one-third of the total (eighteen) of all non-Soviet Warsaw Pact Category A divisions. The nineteen divisions of the GSFG constituted roughly two-thirds of all Soviet divisions based in Eastern Europe (thirty—that is, nineteen in the GDR, two in Poland, five in Czechoslovakia, and four in Hungary). The thirty Soviet divisions in Eastern Europe constituted roughly two-thirds of all Soviet and War-saw Pact Category A divisions in Eastern Europe (forty-eight). The Soviet divisions in Eastern Europe also constituted roughly two-thirds of the roughly forty-five Soviet Category A divisions in the entire European theater (western military districts of the USSR plus Eastern Europe) (Table 5.1). By building Warsaw Pact force ratios around the NVA, the Soviets were also building force ratios around the *Bundeswehr*.

According to one of the participants/analysts in the long and inconclusive WP-NATO talks on Mutual and Balanced Force Reductions (MBFR) during the 1970s and early 1980s, every single WP proposal was geared toward setting specific limits on the ground forces of the *Bundeswehr*.[6] According to this expert, the So-viets were relatively unconcerned about the size of the British, U.S., and other forces in the MBFR zone. But Moscow focused intently on the size of the West German forces. This consistent and intense focus may have been tribute not only to the quality of the *Bundeswehr*. It may also have been a response to the fact that the force ratios of the Warsaw Pact were indirectly geared to the size of the *Bundeswehr* ground forces through the internal WP force ratios built around the NVA.

The NVA in the Soviet Mirror

During the 1960s, the Socialist Unity Party, under the banner of preparing for combat with NATO, emphasized the acquisition and mastery of Soviet military

[6]John G. Keliher, *The Mutual and Balanced Force Reduction Talks* (New York: Pergamon, 1982), pp. 62–70.

technology as a symbol of national pride and socialist modernism.[7] The NVA, like its putative Prussian ancestor, was held up as the "school of the nation"—a model technological institution for the rest of East German society. It was also, from the Leninist perspective, the model socialist institution: ideological, hierarchical, and disciplined.[8]

The NVA also served as a symbol of Soviet–East German solidarity and collaboration. The GDR constitution contained an article stating that the two fraternal armies would stand "shoulder to shoulder forever."[9] Virtually all studies of the NVA note the embedding of NVA divisions into larger Soviet-controlled formations.[10] According to an officer at the GDR Institute of Military History, in 1969 the NVA and GSFG drew up a plan for intensifying the existing contacts between parallel corresponding GSFG and NVA units.[11] During the 1970s the number and frequency of such contacts grew extensively, linking divisions, regiments, and smaller units.[12] In December 1973 the SED Central Committee issued a new charter for such contacts, the "regiment next door" program, which called upon NVA units "to work shoulder-to-shoulder with the regiment next door in combat training and army life. Each serviceman must participate in the strengthening of combat cooperation, learn from the glorious Soviet Army and imitate it."[13] By 1975 the number of agreements for such contacts had reached 570.[14]

The Soviets placed military hardware in the GDR in excess of the NATO estimates for East Germany. There was in fact enough to outfit an additional five divisions.[15] They also prepared an inventory of medals to be awarded to NVA personnel in combat against NATO, and new street signs for West German towns.[16]

[7]Dale R. Herspring, *East German Civil-Military Relations: The Impact of Technology, 1949–72* (New York: Praeger, 1973). The main argument of this volume was that the NVA was the spearhead of a nationwide effort to emphasize modern technology across all sectors of East German life—education, industry, research, and living standards.

[8]Douglas A. Macgregor, *The Soviet-East German Military Alliance* (New York: Cambridge University Press, 1989), p. 15.

[9]Quoted in E. F. Ivanovskii et al., eds., *Na boevom postu: kniga o voinakh Gruppy sovetskikh voisk v Germanii* [*At the Battle Station: A Book about the Soldiers of the Group of Soviet Forces in Germany*] (Moscow: Voenizdat, 1975), pp. 273–74.

[10]Macgregor, *The Soviet-East German Military Alliance*, p. 110.

[11]Col. Prof. Reinhard Bruehl, "Brotherhood in Arms of the Fraternal Armies—Reliable Guarantor of Socialism and Peace," *Horizont*, no. 9 (1967), translated in Joint Publications Research Service, no. 67020 (March 1976), p. 6.

[12]Dr. Col. NVA G. Jokel, Dr. Col. NVA T. Nelles, and K. U. Kuebke, "The Development of Combat Cooperation between the GDR NVA and the Soviet Army during the 1970s," *Voenno-istoricheskii zhurnal*, no. 7 (1978), translated in Joint Publications Research Service, no. 72066 (18 Oct. 1978), p. 45.

[13]Ibid.

[14]Bruehl, "Brotherhood in Arms," p. 8.

[15]Michael Boll, "By Blood, Not Ballots: German Unification, Communist Style," *Parameters* 24, no. 1 (spring 1994): 67–70.

[16]Ibid., pp. 70–71.

Such preparations were consistent with the offensive doctrine and posture of the WP. But such evidence may be balanced by other indicators, which suggest that the NVA was a "Potemkin" force: the apparent lack of command training for the very highest level NVA officers, the low state of morale in the NVA because of poor living conditions for the rank and file, the excessive (by Western standards) time devoted to guarding weapons depots, the corruption and personal extravagance of the high command, the secretive and distrustful relations between superiors and subordinates at every level of the NVA, from private to general.[17] General Jeorg Schoenboehm, the officer responsible for merging the NVA into the *Bundeswehr*, found that the vast majority of the NVA officer corps did not have the professional military ethic necessary for the combat standards of the *Bundeswehr*.[18] In a 1992 monograph, General Joachim Goldblatt, former chief of the NVA agency for armaments, observed: "From today's knowledge it is bitter to have to say that the NVA was a product and an instrument of Soviet policy, that it together with its state [the GDR] was superfluous and allowed to collapse as Soviet power politics fell apart."[19]

A Thought Experiment: Replace NVA Ground Forces Personnel with Soviet Soldiers

Perhaps the 73,720 combat personnel of the six NVA divisions were exactly what Moscow required to meet military requirements for well-trained, combat-ready, politically reliable military personnel on the NATO front. All six were Category A divisions. Category A divisions were those at the highest level of readiness. In terms of military reliability against NATO, it probably would have been more effective simply to bring in 73,720 Soviet soldiers from the fifty Category B and C divisions in the western military districts of the USSR. Within the Warsaw Pact, Category B divisions are 50–75 percent fully manned and equipped. Category C divisions were equipped at about 20 percent of requirements. The six NVA divisions were equipped as two tank divisions (some 11,470 personnel and 328 tanks per division) and four mechanized rifle (MR) divisions (12,695 personnel and some 220 tanks per division). In contrast to the 2:1 ratio of MR divisions to tank divisions in the NVA, the GSFG had eleven tank divisions, eight MR divisions, and one artillery division, plus an additional five independent tank regiments. Failure to field six East German divisions would not have tipped the military balance in Central Europe against the Soviets, especially given the fact that

---

[17]Joerg Schoenbohm, *Two Armies and One Fatherland: The End of the Nationale Volksarmee* (Providence, RI: Berghan, 1996), pp. 32–38.

[18]Ibid.

[19]Quoted in Herspring, *Requiem for an Army*, p. 133.

the NVA fielded only two tank divisions. But a failure to provide East Germany with an army to match the *Bundeswehr* would have constituted a political assault on the GDR. Keeping the NVA relatively small was sufficient for Soviet political purposes. In the event of war with NATO, the role of the NVA would be relatively small—thus minimizing the risks if the NVA rapidly disintegrated, as it did after the fall of the Berlin Wall in 1989.

The Warsaw Pact in the NVA Mirror

In deciding to give the East Germans a national military, the Soviets also had to decide whether to give the GDR a military alliance to match the one that the FRG had joined. The NATO alliance treated the Federal Republic as a sovereign state and equal partner. Not to give the GDR an alliance would require continuation of a bilateral treaty relationship with the USSR. That could never be presented as an alliance of equals. And not to create a multilateral response to the Atlantic alliance (albeit some six years after the NATO treaty) would have been to mock the "internationalism" of the Soviet bloc.

But to create a multilateral military counterpart to NATO also created an extended version of the problem of arming the National People's Army: creating alliance standards for arming the existing East European armies. Before 1955 these armies were partly staffed by high-ranking Soviet officers (the most spectacular example being the Polish defense minister, Konstantin Rokossovky, a former Soviet marshal). In addition to turning commands over to national officers, the standards of armament, size of ground forces, and readiness had to appear credible to both internal and external observers. And these forces had to be sized and outfitted for interaction with the Soviet garrisons in their host countries.

Apart from the creation of the NVA, the best example of this dynamic concerns the establishment of the Central Group of Forces (CGF), a five-division Soviet occupation force placed in Czechoslovakia after the 1968 WP invasion. This force, in addition to provoking NATO alarms about Soviet troops in a state bordering Germany, had to establish "fraternal" unit-to-unit relations with Czechoslovak divisions that had not previously had to compare themselves with Soviet troops stationed in their country or to train and exercise together. The result was to upgrade four Czechoslovak divisions (of ten altogether) to Category A standards—and thus to confront NATO with nine Category A divisions that trained together (five Soviet, four Czechoslovak). And yet, the initial purpose of the CGF was to deal with internal security problems in Czechoslovakia. But it could do so only under the rubric of training to fight NATO. And NATO had to view these nine Category A divisions as an offensive threat. A similar problem arose in Hungary after 1956. The Soviets doubled the size of the Southern Group of Forces to four divisions and undertook a major overhaul of the Hungarian army.

The consistent pattern in Warsaw Pact states other than the GDR was to have relatively large regular forces characterized by much lower levels of readiness and lower per capita inventories of state-of-the-art tanks. That is, the NVA ground forces were the only NSWP forces always at full strength—as in the case of the *Bundeswehr*. (Both German armies had reserve forces, as did all the NSWP ground forces.) The question that arises is why all the other East European armies of the WP did not have to meet the readiness standards of the NVA, the GSFG, and of the Soviet force groups in Hungary, Czechoslovakia, and Poland—or of the Warsaw Pact's offensive doctrine.

The answer offered here is that there was no political equivalent to West Germany—a West Poland, a West Hungary, a West Czechoslovakia, a West Bulgaria. The non-German militaries of the Warsaw Pact could operate at lower levels of readiness and quality than the NVA because they did not have to compare themselves to the professional standards set by the competition between the *Bundeswehr* on one side and the NVA on the other.

There is an historical test for this argument: Germany was not the only Central European state divided into Eastern and Western occupation zones. Austria, incorporated into Germany by Hitler in 1938, was also divided. In the two-week interval between West Germany's entry into NATO and the creation of the WTO, the Soviets abruptly resolved long-deadlocked negotiations over the Soviet occupation zone in Austria. The Soviets withdrew their forces and agreed to the unification of Austria in exchange for two key Austrian pledges: no Austrian membership in NATO, and no Austrian political merger—*anschluss*—with the Federal Republic. Later a *de facto* third condition emerged—no Austrian membership in the European Economic Community (later the European Community).

Thus there was no possibility of a West Austrian army—and no possibility of NATO membership for the West Austrian army. Had there been an inner-Austrian competition over political legitimacy, there probably would have been an East Austrian army armed to the technical standards set by the West Austrian army. This would have created two sets of "German problems" in Europe and perhaps linked four "Germanies" into a single German problem stretching from the Baltic to the Tyrolean Alps. That in turn would have extended the central front southward and further escalated the cost of a Soviet arms race with NATO.

Instead, the Soviets opted for a united but neutral Austria. Stalin had dangled a similar possibility before the West Germans in 1952 at the high point of the Western debate over West German membership in NATO. A neutral Austria also meant the preemption of a "Southern Tier" arms race between the WP and NATO. Subsequently, the low levels of readiness and equipment in the Hungarian military helped avoid such an arms race. Of course, Soviet reluctance to arm the Hungarians with modern equipment may have also reflected subsequent Soviet doubts about the political reliability of Hungarian soldiers in the years after the 1956 Soviet invasion of Hungary.

The East German National People's Army (NVA) in the Warsaw Pact Mirror

The comparatively small six-division NVA ground forces constituted 33 percent of the entire Category A non-Soviet Warsaw Pact divisions. A Category A division is at 75 to 100 percent readiness. The contribution of the NVA to the Warsaw Pact was not in size—or even in the quality of its armament compared with Soviet forces.[20] The NVA contribution to the external front consisted of a high degree of readiness and a high degree of integration with the GSFG.[21] Put another way, the NVA did not constitute a distinct "national" combat force, but rather an "auxiliary" manpower pool, tightly controlled by the command structure of the GSFG.

The GDR lagged behind the other NSWP forces in terms of their contributions to the combat divisions of the Warsaw Pact by various indicators (Table 5.5). The average contribution (excluding the GDR) was 9.4 divisions. The GDR (pop. 16,593,000) contributed only 6 divisions. Only Hungary (pop. 10,628,000) made as small a commitment—6 divisions. The GDR was slightly below the Warsaw Pact average for total regular armed forces as a percentage of the total population (Table 5.5). The average per capita GNP of the WP states (excluding the GDR) was lower than that of the GDR (Table 5.5). For GDP per national division, calculations based on *Military Balance* information shows that the GDR had substantially more financial resources available per national division. This figure is a function of the relatively high GDP of the GDR and the very low number of East German divisions by WP standards (Table 5.5).[22]

Compared with the GDR, every other NSWP state made a greater numerical commitment to maintaining WP combat divisions and probably a relatively greater economic commitment to maintaining WP ground forces divisions. The point is that the GDR assigned most of its uniformed personnel to the internal security front rather than to the combat divisions serving on the external front (Tables 5.5, 5.7).

---

[20]See James M. Garret, *The Tenuous Balance: Conventional Forces in Central Europe* (Boulder, CO: Westview, 1989), p. 61: "Most tanks in Soviet forces are late model, but the majority of Polish and German tanks are not. . . . The Polish and East German units are acquiring T-72s, but most units have older Soviet tanks that are slower, have lighter guns and thinner armor and provide less stable firing platforms than the new models."

[21]Macgregor, *The Soviet-East German Military Alliance*, p. 10.

[22]International Institute for Strategic Studies, *The Military Balance, 1988–1989* (London: IISS, 1990).

Warsaw Pact Tank Forces

The tank forces of the Warsaw Pact armies were the key indicators of the technological relationship between Soviet and East European military forces. None of the NSWP ground forces in 1988–89 (Table 5.4) had equivalence in tank forces to their Soviet counterparts[23] (Tables 5.3 and 5.4). Proportionately, the NVA had more modern tanks per division than the Poles and Czechs. But in actual numbers, the NVA had fewer modern tanks than the Czechoslovak military (350 T-72 tanks in the NVA, compared with 500 in the Czechoslovak People's Army [CSPA]). And the Polish army had the same number of T-72 tanks as the NVA—350. The Polish Army had thirteen divisions, eight of which were Category A. The figure of 350 tanks was slightly more than the requirements for full modernization of one standard WP Tank Division. In contrast, Syria alone had about 950 T-72 tanks. The combined total of T-72 tanks in the arsenals of Syria, Iraq, and Libya appears to have exceeded that of the NSWP armies (Table 5.3).

The explanation may be that the token number of T-72 tanks in the WP enabled elite units to conduct symbolic exercises in which the NSWP forces were paired with Soviet partner regiments and divisions as "brothers-in-arms," outfitted with the same weaponry. Thus T-72 tanks may have been morale boosters and status symbols for elite units, particularly in the case of the Czechoslovak People's Army, which faced severe morale problems after the 1968 WP invasion.[24]

East European Armed Forces in the Soviet Mirror

The NVA regularly exercised on a bilateral basis with the GSFG. But it also exercised at the highest level on a multilateral basis with the Category A Soviet and NSWP divisions in Poland and Czechoslovakia.[25] The elite units of these forces had to be more or less comparable to the NVA lest it stand out as an exception to audiences in Eastern and Western Europe. The Hungarian forces—and especially the Bulgarian forces—were maintained at symbolic levels of advanced technology and of military readiness.[26]

[23]For a fuller discussion of this theme, see Richard Martin, "Warsaw Pact Force Modernization: A Second Look," in Jeffrey Simon and Trond Gilberg, eds., *Security Implications of Nationalism in Eastern Europe* (Boulder, CO: Westview, 1987).

[24]See Christopher Jones, "The Czechoslovak People's Army," in Jeffrey Simon, ed., *NATO-Warsaw Pact Force Mobilization* (Washington: National Defense University Press, 1989).

[25]Christopher Jones, *Soviet Influence in Eastern Europe: Political Autonomy and the Warsaw Pact* (New York: Praeger, 1981), pp. 119–21.

[26]For a standard survey of Warsaw Pact forces, see Martin Windrow, ed., Warsaw *Pact Ground Forces* (London: Osprey, 1987). Windrow notes on p. 27 concerning the Bulgarians: "Bulgaria is the USSR's most trusted ally, for racial, cultural, linguistic and political reasons. . . . Given such an ideal partner, it would be thought that the Bulgarians would play a major role within the WP, but they do not. . . . The

The single most important purpose of the NSWP armies was to ensure a So-viet capability for military interventions in their countries through the conduct of joint exercises. The Soviets did not have to rely on joint exercises to invade the GDR. The GSFG was the largest standing military force of the Cold War. But Po-land in 1955 had only two Soviet divisions based in the country; Czechoslovakia, none; Bulgaria, none. And Hungary had only two. The Soviets put four divisions in Hungary only after the uprising of 1956. There were still Soviet troops in Ro-mania in 1955, but the Romanians successfully argued that the signing of the Austrian State Treaty of 1955 had removed the legal basis for stationing Soviet troops in Romania.

The Soviets could maintain forces configured for the offensive and regularly practice invasions of Poland, Czechoslovakia, Hungary, and Bulgaria under the guise of practicing offensive actions against NATO. (In 1964 the Romanian lead-ership agreed to a WTO exercise that featured Soviet naval landings and para-troop landings inland, in addition to entry into Romania by other Pact forces. After that, Ceausescu refused any further participation in joint exercises.)[27] In 1968 the Soviet invasion of Czechoslovakia drew upon the practical experience of the 1963 Vltava exercise, a massive undertaking whose equipment could have formed a column longer than the East-West length of the country.[28]

Using the WTO as cover for practicing Soviet invasions required a certain number of East European Category A divisions and weaponry roughly compara-ble to Soviet Force Group weaponry. These were the elite Category A units that interacted with Soviet forces and often trained with the Soviet Force Groups based in their country.

Such units corresponded to the Soviet/WTO statements that declared that the Warsaw Pact forces under the command of the WTO commander-in-chief con-sisted only of certain designated units of East European forces—not all national forces.[29] This "Greater Socialist Army" met all the systemic Soviet requirements for East European forces: they served as hosts to legitimize both Soviet garrisons and Soviet troops on regular maneuvers. They linked Soviet forces to key ground forces divisions, in turn linked to the rest of the national armed forces. These linkages enabled the Soviets to preempt national capabilities for national defense

---

BNA (Bulgarian People's Army) is considered a marginal force of only limited combat value due to its limited training, which often takes second place to the demands of political indoctrination, and use as a labor force." This text dismisses the Hungarian military as follows: "In the eyes of the Soviets, the MN (Hungarian Army) is not a trustworthy army. . . . This lack of Soviet trust, its low levels of training, and its unfavorable geographic location are the primary reason the MN is provided only with limited quan-tities of older equipment" (p. 32).

[27]Jones, Soviet Influence, pp. 117–18.
[28]Krasnaia Zvezda, 21 Sept. 1966, p. 1.
[29]Sovetskaia Voennaia Entsiklopediia, vol. 5, p. 682.

by national means. They linked Soviet force groups through the national army to the large numbers of internal security formations of the East European states.

To be credible, the East European forces assigned to the Greater Socialist Army had to be more or less compatible technologically with Soviet forces in order to participate in joint military exercises. By orienting these forces to the conduct of offensive missions outside national territory, the Soviets preempted national planning for defense of national territory by national means. Further, the requirements of a westward offensive logically required subordination of national forces to a larger coalition requirements for command and control. The preparation and conduct of such exercises permitted Soviet commanders to monitor and control every aspect of these elite forces. This recalls the statement attributed to a high-ranking Soviet officer who defected to the West just before Marshal Grechko introduced the system of large-scale WTO joint exercises: "Soviet troop maneuvers will be conducted jointly with the troops of the people's democracies. During maneuvers they will be included in the operational command of the Soviet Army. This is necessary because we still do not trust them. They might turn their guns and run to the West."[30]

The professionalism required in the joint exercises—mainly because of the complex armaments and command systems—in turned bestowed professional prestige on the officers of these units. They prepared for such service by education in Soviet military academies. Over the years, the graduates of Soviet academies became a "greater socialist officer corps" with a vested interest in preserving the mechanisms of the Warsaw Pact *maskirovka*. The skills required of the East European members of the greater socialist officer corps were of an exceptionally high order: military, technological, linguistic, political, and academic. They were to take orders in Russian from the WP Staff and see that they were carried out in their national militaries. Such officers made it possible for the Soviet command to partially denationalize the security policies and defense ministries of the loyal WP states.[31]

The Warsaw Pact in the Romanian Mirror

Three communist states in East Europe—Yugoslavia, Albania, and Romania—completely rejected the Soviet doctrine formulated for the WP and developed competing nationalist military doctrines, adjusted for the peculiar circumstances of each country. The Yugoslavs broke with the USSR before the creation of the

[30]Quoted in Frank Gibney, ed., *The Penkovsky Papers* (New York: Doubleday, 1965), p. 245.
[31]Christopher Jones, "Agencies of the Alliance," in Teresa Rakowska-Harmstone and Christopher Jones, *The Warsaw Pact: The Question of Cohesion, Phase II*, vol. 1 (Ottawa: Department of National Defense, 1984), pp. 187–95.

Warsaw Pact. The Albanians had signed the Warsaw Treaty, but ceased any effective participation in the WP by the early 1960s. In 1968, Albania formally withdrew from the Pact to protest the invasion of Czechoslovakia. After 1964 the Romanians refused to participate in joint WP exercises, thus barring Soviet troops from returning to Romania, as they did in a 1963 exercise.[32] Although the Romanians condemned the invasion of Czechoslovakia, they remained formal members of the WP.[33] During the early 1960s the Romanians frequently called for the mutual decommissioning of both the NATO and WP alliances,[34] though during the post-1972 era of arms control negotiations they found that remaining in the Warsaw Pact gave them privileged access to NATO interlocutors.

The nationalist doctrines of these three Balkan states provided for "territorial defense" systems intended to function with locally made low-tech weapons. To the extent that Romania relied on foreign sources for military technology, these suppliers were West European and Chinese manufacturers.[35] In these Balkan concepts of warfare, there was no role for nuclear weapons or other high-tech forces. In the late 1950s and early 1960s a group of officers in the Polish defense ministry attempted to adapt such concepts to Poland. Although they failed, they left behind "territorial defense" forces later converted to WTO missions.[36] In Czechoslovakia in 1968 the staff of the Gottwald Military-Political Academy in Prague and the leadership of the Main Political Administration attempted to restructure Czechoslovak doctrine along the lines of territorial defense. After the invasion the Soviets disbanded the Gottwald Academy and founded a new, replacement institution located in Bratislava.[37] The practical focus of WP programs initiated by WP Commander Marshal A. A. Grechko in the early 1960s was to preempt such doctrinal heresies from spreading from Romania to other WP members.

The Romanian, Albanian, and Yugoslav doctrines of territorial defense were designed to deter Soviet intervention, and if necessary, fight a prolonged war of national liberation in the desperate hope that over time the West would provide material assistance and that Soviet troops would suffer a breakdown of morale. The Soviets encountered such a war in Afghanistan.

The Balkan territorial defense strategies did not plan to defend either territory or people. Instead, they were designed to generate loyalty to the national military-political leadership by ensuring a minimum level of civilian causalities. Draped in patriotic rhetoric, the Yugoslav, Albanian, and Romanian systems

[32]Jones, *Soviet Influence*, pp. 117–18.
[33]Ibid., pp. 191–97.
[34]Christopher Jones, "Romania," in Harmstone and Jones, *The Warsaw Pact*, vol. 2, pp. 354–57.
[35]Ibid., pp. 375–81.
[36]Teresa Rakowska-Harmstone, "Poland," in Harmstone and Jones, *The Warsaw Pact*, vol. 2, pp. 65–75.
[37]Jones, *Soviet Influence*, pp. 191–97.

trained their officer corps in techniques deliberately designed to shed civilian blood by provoking indiscriminate reprisals by the enemy military force. People's war doctrines also called for extensive propaganda capabilities—print and broadcast—to extract maximum political capital at home and abroad from enemy atrocities.

Such bloodshed was intended to inflame nationalism, encourage desires for revenge, and create popular support for leaders who promised reprisals against the common enemy. Such events would inevitably provoke condemnations from the United States and NATO—and perhaps even lead to covert aid to the anti-Soviet resistance. The Balkan doctrines of territorial defense thus attempted to leverage not only the resources of their own societies but also those of the NATO coalition. The Romanian and Yugoslav doctrines expected that over time such bloodshed would also demoralize Soviet soldiers and the Soviet home front. When confronted with such problems in Afghanistan, Gorbachev withdrew Soviet troops.

The best indicators of Romania's defection from the Warsaw Pact are its territorial defense doctrine, its independent defense posture, its refusal to participate in Warsaw Pact exercises after 1964, its refusal to send Romanian military officers to Soviet academies after the early 1960s, and its military technology.

Since the 1960s, the newer technology of the Romanian ground forces, air, and naval forces has been of non-Soviet manufacture—either Romanian, West European, or Chinese. Tanks played very little role in Romanian defense plans— evidently on the assumption that Soviet forces would be capable of destroying such tanks should they engage Soviet ground and air forces. As of 1965, the Romanians had acquired a collection of 1,060 Soviet T-34s from the 1930s, plus 757 T-55s as well as 30 T-72s.[38] The mainstays of the Romanian tank forces were Romanian-made weapons: some 556 TR-80 tanks, and some 414–580 of Romanian T-10 tanks. The Romanians acquired the 30 T-72 Soviet tanks from Israel, which had captured them in the 1973 war. Knowledge of the T-72's capabilities would have been of use for planning/practice defense against a possible Soviet invasion, even if the tanks came from the USSR. The same would apply to token quantities of other Soviet weapons in the Romanian inventory.[39]

[38]Lewis, *Warsaw Pact Ground Forces,* notes on p. 47, for the generally low quality of Romanian armament and training.
[39]The figures just cited are from the 1990–91 edition of the *Military Balance,* rather than the 1989–90 edition used for other comparisons above. The 1990–91 edition makes a point of saying that the information on Romania is considerably better than in any previous edition. By comparison, the 1989–90 edition listed for Romania the following entry: "Main Battle Tanks: 3200, including T-34, T-54/55, T-72 and M-77. By further comparison, the 1990–91 edition of the *Military Balance* lists for Bulgaria 1612 T-54 tanks (a higher figure than the 1989–90 edition), plus 80 T-62s (previously uncounted) and 334 T-72 tanks. So even if the 30 T-72 tanks in the Romanian arsenal were obtained directly from the USSR, they

In conclusion: The territorial defense doctrines of Yugoslavia and Albania were in place by 1960, when Marshal Grechko became commander of the Warsaw Pact. The Romanians appear to have adopted a territorial defense doctrine by 1964—and certainly by 1968, when Ceausescu publicly condemned the WP invasion of Czechoslovakia. But Ceausescu never sought to withdraw from the Warsaw Pact. Ceausescu had an interest in preserving the threat of a WP invasion against his country in order that he could pose as a nationalist. Membership also gave him a seat that he would not otherwise have had at various European forums.

Reflections on Marshal Andrei Grechko

This chapter has argued that the Warsaw Pact was a gigantic masquerade—a *maskirovka*. It is possible that Andrei Grechko provided the Soviet leadership with the institutional memory and the practical skills to combine the Warsaw Pact *maskirovka* with effective policy-making from 1955 to 1976, the year in which he died.

Grechko was commander of the GSFG from 1953 to 1957. His first experience after assuming command in June of 1953 was to put down the East German uprising that same month. In 1955 he was a firsthand witness to the creation of the Warsaw Pact, the NVA, and the Austrian State Treaty. He drew the NVA into regular joint exercises with the GSFG, beginning in 1956, the year of the upheavals in Poland and Hungary. From 1957 to 1959, he was commander of the Soviet Ground Forces, at a time when the ground forces eliminated the few remaining "national" (i.e., non-Russian) units in the Soviet army and prepared for NATO-like joint maneuvers with the WP forces. This was also a period during which hundreds of thousands of East Germans fled to the West, mainly through Berlin.

He assumed command of the WP in 1960, just prior to the Berlin Crisis of 1961 and the building of the Berlin Wall. He set in motion not only the system of joint offensive exercises but also the inclusion of nuclear weapons simulations in these exercises—backed up by Soviet nuclear systems deployed in East Europe and the western USSR. When Grechko became commander of the Warsaw Pact, both Albania and Romania had for all practical purposes defected from the Pact. But neither Poland nor any other state succeeded in escaping from the Warsaw Pact as Romania and Albania had.

After Grechko's appointment as WP commander in 1960, A. A. Epishev became chief of the Soviet Main Political Administration. In southern Poland and Czechoslovakia during 1944–45, Grechko had worked closely with Epishev, the

amount to about 9 percent of the Bulgarian inventory, despite the difference in population size (Bulgaria, 9,062,000; Romania, 23,295,000).

political commissar in charge of managing relations between Soviet personnel and the personnel of the Czechoslovak and Polish divisions originally recruited from Soviet prisoners of war. Epishev returned to military service after five years as Soviet ambassador to Romania during the late 1950s and one year in 1960 as Soviet ambassador to Yugoslavia.[40] A large part of Epishev's job as MPA chief was to manage the political reliability of Russian and non-Russian conscripts in the Soviet armed forces and the corresponding ethnic problem in the WP. Epishev became the principal theorist—or at least, the principal publicist—of the military-political component of Soviet/WP doctrine.[41] The thrust of his arguments was that doctrines of territorial defense, like those of Romania and Yugoslavia, threatened the unity and cohesion of the Warsaw Pact.[42] He and Grechko also argued that the dynamics of nuclear war demanded tight centralized control over all Warsaw Pact forces. When Grechko became the Soviet defense minister in 1967, he had left in place the system of joint exercises that facilitated the invasion of Czechoslovakia in 1968.

During his tenure as defense minister Grechko played a role in what may have been serious efforts to escape from the unintended consequence of 1955 decisions on creating the NVA and the Warsaw Pact. This was the subsequent nuclear arms race with the United States. That is, Brezhnev's policies of détente may have been attempts to moderate the confrontation between the two Germanies, to slow the U.S.-USSR nuclear competition, and to control the conventional arms race in Central Europe. Grechko had to adjust Soviet military policy accordingly while preserving the offensive *maskirovka* of the NVA and Warsaw Pact.

While Grechko was defense minister Moscow signed the Nuclear Nonproliferation Treaty of 1969, negotiated the SALT I Treaty of 1972 and the ABM Treaty, the Soviet-German treaty of 1970, and related 1971–72 treaties involving the two Germanies, Poland, Czechoslovakia, and Hungary. In 1973, Grechko became a full member of the Soviet Politburo, the first defense minister since 1957 to sit on the party's highest body. Grechko also supervised Soviet participation in the Mutual and Balanced Force Reductions Talks, which began in 1973 and lasted, with interruptions, until 1990. The 1975 Helsinki accords, built on the Soviet-German agreements of 1970–72, were negotiated while Grechko was still in office. Helsinki's "Basket One"—the military confidence-building measures—surely must have fallen under his review, as both defense minister and Politburo member. And so must have the progress in 1974 toward a Salt II agreement. In other words, Grechko was in a position to preserve the *maskirovka* of the Warsaw Pact

[40]Biographical information for A. A. Epishev in *Sovietskaia Voennaia Entsiklopediia*, vol. 3, pp. 311–12.

[41]A. A. Epishev, *Ideologicheskaia bor'ba po voennym voprosam* [*Ideological Struggle in Military Questions*] (Moscow: Voenizdat, 1974), pp. 108–9.

[42]Ibid., pp. 104–10.

even as the Soviets adjusted their policies toward NATO over the period from 1955 to 1975. Marshal Grechko died in office in 1976.

Had the NVA, the GDR, and the Warsaw Pact proven to be viable institutions—as many East German, West German, and NATO observers took them to be—the deception of creating the NVA and the Warsaw Pact would have proven a brilliant gamble. For the Soviet leadership to have made such a gamble was to take a calculated risk. But what was the political alternative in 1955? Or in 1961? The alternative was abandoning East Germany, as Gorbachev did in 1989. The alternative was setting in motion the disintegration of communist rule throughout the bloc. For the period from 1955 to 1976 the Soviet leadership appears to have assigned management of the USSR's high-risk German policy to a general with a demonstrated record of successfully commanding multinational military personnel in exceptionally difficult circumstances. He may also have compiled a record as a successful practitioner of détente, also in very difficult circumstances.

If Marshal Grechko did in fact perpetrate strategic frauds on his enemies, this does not mean that he was a fraudulent general. On his watch, the Soviet military did not surrender control of Eastern Europe. The price of his success, however, was a ruinously expensive arms race. If Grechko did carry out the greatest military deception of the Cold War,[43] his final achievement was to leave—as yet—no documentary record.

Conclusion

The spread of military technology and doctrine from the Soviet Union to its Warsaw Pact allies is not the typical case of interstate diffusion, be it among enemies or friends, examined throughout much of this volume, but rather a case of intraimperial imposition. The relationship between the Soviet Union and the Warsaw Pact militaries reveals that political and even economic concerns in the "leader state" drove the diffusion process. Soviet motivations differed markedly from their U.S. counterparts. International security reasons shaped Soviet behavior toward their Eastern European satellites, to be sure. But the most important explanation for their actions was "domestic politics"—safeguarding the po-

---

[43]The Soviet Ministry of Defense participated in several documented deceptions. These include (1) The attempt on 7 Nov. 1955 to make the Soviet air force appear larger than it was; (2) the attempt in the late 1950s to exaggerate the number of SS-4 and SS-5 missiles (these two cases parallel the creation of the NVA and Warsaw Pact chronologically); (3) the secret emplacement of missiles in Cuba during the early 1960s; (4) the construction of the Krasnoyarsk ABM facility in violation of the 1972 ABM treaty; and (5) the production of biological weapons in violation of a treaty prohibiting them. In the latter two cases, the Ministry of Defense evidently withheld this information from Gorbachev until it was brought to his attention by U.S. officials.

litical survival of the regime in Moscow, which meant safeguarding and legitimating the communist regimes in the satellite countries, too. This difference in political-structural circumstance explains the distinct patterns of diffusion we see in this case, including incomplete, uneven, or inefficient distribution of both equipment and information, and even intentional dissemination of inaccurate information.

CHAPTER SIX

The Diffusion of Nuclear Weapons

WILLIAM C. POTTER

One of the most influential studies of U.S. nuclear nonproliferation policy adopted as a guiding principle Florence Nightingale's admonition that "Whatever else hospitals do they should not spread the disease."[1] Retarding the spread of nuclear weapons (or disease) implies a sound knowledge of its causes. Although prescriptions for nonproliferation abound, the literature on nuclear proliferation presents a wide array of often speculative and contradictory insights on why nations embark or refrain from embarking on paths to acquire nuclear weapons. Even less well charted are the processes by which nuclear weapons diffuse. Also only dimly understood are the consequences of nuclear weapons diffusion at the national, regional, and international systems levels.[2]

This chapter distills from existing case studies of nuclear weapons decision-making the underlying factors responsible for national decisions to acquire, refrain from, or renounce nuclear weapons. Data from proliferation profiles of twenty-six states are used to explore a variety of propositions regarding technology diffusion and policy innovation, including the scope, rate, and pattern of

The author wishes to thank Filip Stabrowski, Mary Beth Nikitin, Jeff Steele, and Kahlil Thompson for their invaluable research assistance. Thanks are due also to Allene Thompson for her assistance with the manuscript preparation.

[1]Albert Wohlstetter et al., *Moving toward Life in a Nuclear Armed Crowd?* Report to the U.S. Arms Control and Disarmament Agency (Los Angeles: Pan Heuristics, 1976), p. 1. A revised version was published as *Swords from Plowshares* (Chicago: University of Chicago Press, 1979).

[2]Using diffusion of innovation terminology one can distinguish among functional-dysfunctional, direct-indirect, and manifest-latent consequences. See Everett Rogers with F. Floyd Shoemaker, *Communication of Innovations: A Cross-Cultural Approach*, 2d ed. (New York: Free Press, 1971), pp. 330–34.

diffusion, the relative importance of domestic and external factors, the constancy (or variation) in the explanatory power of alternative proliferation incentives and disincentives, and the relationship between the diffusion of technological know-how and political decisions to acquire nuclear weapons. Policy recommendations regarding proliferation management and control are derived from the analysis.[3]

Research Methodology

In the proliferation literature, the acid test for "going nuclear" or "proliferating" has traditionally been the detonation of a single atomic explosion. The emphasis has been on the divide between nuclear and non-nuclear status rather than on the disparity among proliferators (those who have crossed the divide) in terms of the number of subsequent detonations, the weaponization of the nuclear charge, the size of the nuclear weapons arsenal, the availability of invulnerable and reliable delivery systems, and the articulation of a strategic doctrine. This essay adopts the traditional test for the purpose of assessing proliferation determinants, although it acknowledges the analytical value of a more differentiated "ladder of nuclear weapons capability."[4]

Most attention in the past has been directed toward containing the spread of nuclear weapons to additional national governments. A growing and legitimate concern is proliferation by subnational or transnational groups, including terrorist organizations. This reorientation is prompted by the increased accessibility to nuclear weapons material and technical know-how following the collapse of the Soviet Union and the erosion of normative restraints against terrorist employment of large-scale violence, most evident after 11 September 2001.[5] Nevertheless, the focus of this chapter is on national proliferation decisions, since terrorist acquisition of nuclear weapons remains hypothetical.

Given the wide scope of this analysis, information gleaned from a comparative investigation of proliferation determinants in twenty-six states is presented in

[3]No attempt was made in this study to address rigorously the important but complex issues of diffusion processes or consequences.

[4]An exception to the traditional rule is employed for Israel and South Africa, neither of which unambiguously has tested nuclear weapons.

[5]See Graham T. Allison et al., *Avoiding Nuclear Anarchy: Containing the Threat of Loose Russian Nuclear Weapons and Fissile Material* (Cambridge, MA: MIT Press, 1996); William C. Potter, "Before the Deluge? Assessing Threat of Nuclear Leakage from the Post-Soviet States," *Arms Control Today* 25, no. 8 (October 1995): 9–16; Joseph F. Pilat, "Prospects for NBC Terrorism after Tokyo," in Brad Roberts, ed., *Terrorism with Chemical and Biological Weapons* (Alexandria, VA: Chemical and Biological Arms Control Institute, 1997), pp. 1–21; Jonathan Tucker, ed., *Toxic Terror: Assessing the Terrorist Use of Chemical and Biological Weapons* (Cambridge, MA: MIT Press, 2000). Many assumptions about the obstacles to nonstate actor acquisition and use of weapons of mass destruction must be reconsidered after the terrorist attacks of September 11.

TABLE 6.1

Hypothesized Proliferation Determinants

Determinants	Orientation	Illustrative sources	
I. National Prerequisites			
Economic wealth	Internal	Bull (1961)	Schwab (1969)
Scientific and technological expertise			Forland (1997)
II. Underlying Pressures			
Deterrence	External	Albright & Gay (1998)	OTA (1977)
		Beaton & Maddox (1962)	Quester (1973)
		Cohen (1998)	Rosecrance (1964)
		Dunn & Kahn (1976)	Schoettle (1976)
		Epstein (1977)	Singh (1999)
		Frankel (1993)	Thayer (1995)
		Greenwood (1977)	Waltz (1981)
		Holloway (1994)	
Warfare advantage and defense	External	(same as for Deterrence)	
Weapon of last resort	External	Cohen (1998)	Harkavy (1977)
		Dunn & Kahn (1976)	Hersh (1991)
		Haselkorn (1974)	OTA (1977)
Coercion	External		Dunn & Kahn (1976)
International status/ prestige	External	Ahmed (1999)	Jervis (1989)
		Chubin (1995)	Sagan (1996)
		Dunn & Kahn (1976)	Subrahmanyam (1998)
		Dunn (1998)	Suchman & Eyre (1992)
		Ghosh (1998)	
Revolutionaries/pariahs	External	Betts (1993)	Ghosh (1998)
		Chubin (1994)	Reiss (1995)
Assertion of autonomy and influence	External	Beaton & Maddox (1962)	Paul (1998)
		Epstein (1977)	Perkovich (1999)
		Kapur (1979)	Rosecrance (1964)
		Lewis & Xue (1988)	Schoettle (1976)
		OTA (1977)	Subrahmanyam (1998)
Economic spillover/ Compensation/ Cost effectiveness	Internal	Beaton & Maddox (1962)	Meyer (1984)
		Dunn & Kahn (1976)	OTA (1977)
		Eisenstadt (1999)	Potter (1995)
		Epstein (1977)	Quester (1973)
		Greenwood (1977)	Rosecrance (1964)
Domestic politics	Internal	Dunn & Kahn (1976)	OTA (1977)
		Flank (1993/94)	Perkovich (1996)
		Joeck (1986)	Perkovich (1998)
		Kampani (1998)	Perkovich (1999)
		Kapur (1979)	Sagan (1996)
		Lavoy (1993)	Wiseman & Treverton
		Meyer (1984)	(1998)
Technological momentum	Internal	Dunn & Kahn (1976)	Scheinman (1965)
		Rosecrance (1977)	

TABLE 6.1—*cont.*

Determinants	Orientation	Illustrative sources	
III. Underlying Constraints			
Military reaction by other states	External	Albright & Gay (1998) Dunn & Kahn (1976) Epstein (1977)	Greenwood (1972) OTA (1977) Quester (1973)
The strategic credibility gap/technical difficulties	External	Dunn & Kahn (1976) Dunn (1998) Epstein (1977) Greenwood (1977)	OTA (1977) Potter (1995) Quester (1973) Rosecrance (1964)
Absence of perceived threat	External	Carasales (1995) Quester (1973) Redick (1998)	Reiss (1995) Rosecrance (1964)
International norms	External	Dunn (1982) Epstein (1977) Greenwood (1977) Müller (1991, 1995)	Quester (1973) Solingen (1994) Spector (1990)
Economic and political sanctions	External	Ahmed (1999) Dunn & Kahn (1976) Epstein (1977)	Greenwood (1977) OTA (1977) Spector (1990)
Unauthorized seizure	Internal	Dunn & Kahn (1976) Dunn (1982) Greenwood (1977)	Meyer (1984) OTA (1977) Stumpf (1996)
Economic costs	Internal	Carasales (1995) Dunn & Kahn (1976) Greenwood (1977) Jonter (2001)	OTA (1977) Quester (1973) Schwartz (1998)
Public opinion	Internal	Dunn & Kahn (1976) Greenwood (1977)	OTA (1977) Quester (1973)
Bureaucratic politics	Internal	Barletta (1997) Betts (1980) Carasales (1995) Holloway (1994) Kapur (1979)	Redick (1998) Reiss (1988) Rosecrance (1964) Sagan (1996) Solingen (1994)
IV. Situational Variables			
International crisis	External	Chubin (1995)	Dunn & Kahn (1976)
Weakening of security guarantees	External	Dunn & Kahn (1976) Ganguly (1999) Goldstein (1993) Greenwood (1977)	Lefever (1979) OTA (1977) Rosecrance (1964) Willrich (1976)
Increased accessibility of nuclear materials	Internal/ External	Dunn & Kahn (1976) Ford/Mitre (1977)	Wohlstetter et al. (1979) Allison et al. (1996)
Vertical proliferation	External	Kapur (1979)	Schwab (1969)
Domestic crisis and leadership change	Internal	Dunn & Kahn (1976)	Kapur (1979)
Focusing events	Internal/ External	Ganguly (1999) Singh (1999)	Subrahmanyam (1998) Walsh (1997)

tabular form, with discussion limited to illustrative cases. The countries examined include all of the states that had demonstrated a nuclear explosive capability by November 2001, regional representatives from most lists of critical potential proliferators, those states that acquired nuclear weapons but chose to renounce them, and several states that have consistently pursued a policy of nuclear weapons abstinence although having long had the capability to join the nuclear weapons club. In some instances (Belarus, Kazakhstan, Ukraine, and Yugoslavia) the findings reported are based principally on my original case study research. Most of the findings are derived from a survey of secondary sources.

Proliferation Determinants

Table 6.1 provides a list of those factors often cited as proliferation determinants. This list distinguishes among hypothesized national prerequisites (i.e., necessary conditions), underlying pressures and constraints, and more transitory situational variables. It also indicates the internal or external orientation of the possible determinants.[6]

National Prerequisites

The question of capabilities versus intentions muddies the study of nuclear proliferation. One cannot infer intent from capability, although one might anticipate a rough correlation between the possession of certain national capabilities and a propensity toward nuclear armament. Most observers agree that a nuclear weapons option presupposes a certain level of economic wealth and technological know-how, although there is little consensus as to what constitutes prohibitive costs and requisite expertise.

The ability to predict national postures toward proliferation based on national economic wealth, scientific expertise, and technological skills has been eroded by the increased accessibility on a global scale of both nuclear technology and fissile material. Some analysts have abandoned the notion of indigenous prerequisites for weapons proliferation. One school of thought substitutes the notion of "technological imperative" for that of national technical prerequisites, arguing that

---

[6]Table 6.1 draws on the categorizations provided by Charles Kegley, Gregory Raymond, and Richard Skinner, "A Comparative Analysis of Nuclear Armament," in Patrick McGowan and Charles Kegley, eds., *Threats, Weapons, and Foreign Policy*, Sage International Yearbook on Foreign Policy Studies, vol. 5 (Beverly Hills, CA: 1980), pp. 231–55; and by Stephen M. Meyer, *The Dynamics of Nuclear Proliferation* (Chicago: University of Chicago Press, 1984), pp. 48–49. An earlier version of this analytical framework was proposed in William C. Potter, *Nuclear Power and Nonproliferation: An Interdisciplinary Perspective* (Cambridge: Oelgeschlager, Gunn and Hain, 1982), pp. 132–37; and was developed further in Potter, "The Politics of Nuclear Renunciation: The Cases of Belarus, Kazakhstan, and Ukraine," Stimson Center Occasional Paper No. 22 (April 1995).

once a nation acquires the physical ability to manufacture nuclear weapons it will inevitably do so.[7] A less deterministic conception of how nuclear technology diffusion impacts weapons proliferation emphasizes the "synergistic link" between civilian nuclear power and nuclear weapons. According to this perspective, national energy needs, international commerce in nuclear technology, and technology diffusion pressures are critical "contextual variables" that affect the calculus of the nuclear weapons decision.[8]

Underlying Pressures and Constraints

Sorting out the different pressures for and constraints on the decision to demonstrate a nuclear explosive capability according to the relative importance of internal or external considerations and military or political-economic objectives yields four clusters of proliferation incentives and disincentives: international security, international politics, domestic security, and domestic politics. (See Table 6.2.)

International Security Incentives

Deterrence of Adversaries. A desire to deter external threats is often cited as an underlying international security incentive for proliferation. One finds arguments that the acquisition of even rudimentary nuclear weapons can afford a measure of deterrence against nuclear attack or blackmail by a superpower, conventional attack by a regional adversary, and prospective acquisition of nuclear weapons by a regional rival. This is a perspective embraced by the neorealist school of international politics. It assumes that states in an anarchical environment will seek to acquire a nuclear deterrent—directly or through an alliance with a nuclear power—to enhance their security against external threats.[9] The assumption is that nuclear weapons would have the same stabilizing influence on regional balances as they allegedly have on U.S.-Soviet relations.[10]

---

[7]Meyer provides a good discussion of this school of thought and a sophisticated effort to test its explanatory power empirically.

[8]See Kegley et al., Nuclear Energy Policy Study Group, *Nuclear Power Issues and Choices* (Cambridge: Ballinger, 1977), pp. 38–40; and Joseph Nye, "Nonproliferation: A Long-Term Strategy," *Foreign Affairs* (April 1978): 601–23.

[9]The most influential articulation of this view is Kenneth Waltz, *Theory of International Politics* (New York: Random House, 1979). For a critique of the neorealist approach to nuclear decision-making, see Scott Sagan, "Why Do States Build Nuclear Weapons? Three Models in Search of a Bomb," *International Security* 1, no. 1 (winter 1996-97): 54–86.

[10]For a debate on the relevance of the regional adversary-superpower analogy, see Scott D. Sagan and Kenneth N. Waltz, *The Spread of Nuclear Weapons: A Debate* (New York: W. W. Norton, 1995). A useful primer on the hazards of making inferences about the nuclear policy of small powers on the basis of the experiences of the superpowers is Yehezkiel Dror, "Small Powers' Nuclear Policy: Research Methodology and Exploratory Analysis," *Jerusalem Journal of International Relations* (fall 1975): 29–49.

TABLE 6.2
Underlying Pressures and Constraints on Proliferation

	Domestic	External
Military	Domestic Security	International Security
Political-Economic	Domestic Political	International Political

Warfare Advantage. Possession of nuclear weapons may be sought as a means to achieve an advantage in warfare should deterrence fail. U.S. and British interest in the development of the atomic bomb can be attributed in large measure to their determination to wage war successfully against the Germans. Certain small and middle-range powers may seek tactical nuclear weapons in order to defend against nuclear or conventional attack by a superpower or regional adversary, particularly in the absence of credible security guarantees by a superpower.[11] A tactical nuclear force might be particularly attractive to a nation whose adversary possesses superiority in conventional force or to one where natural invasion corridors make defense by battlefield nuclear weapons feasible.[12] Some suggest that global proliferation of low-yield tactical nuclear weapons for border defense could be a stabilizing factor that would enable "any nation not now a nuclear power, and not harboring ambitions for territorial aggrandizement, to walk like a porcupine through the forest of international affairs: no threat to its neighbors, too prickly for predators to swallow."[13] International "pariah" states such as Iraq, Taiwan, and North Korea are most often thought of as interested in tactical nuclear arms, but national debates in Switzerland and Sweden have considered their role in support of armed neutrality.[14]

Weapon of Last Resort. Related to the "defense against invasion" incentive is the motivation to possess a weapon of last resort that would be used only if a nation were on the brink of total destruction. The rationale is psychological and punitive ("If we are going to go, we'll take someone with us"), as well as tactical,

---

[11]See Enid C. B. Schoettle, *Postures for Nonproliferation* (London: Taylor and Francis, 1979), p. 108, and Lewis Dunn and Herman Kahn, *Trends in Nuclear Proliferation: 1975–1995* (Croton-on-Hudson, NY: Hudson Institute, 1976), p. 96. See also Geoffrey Kemp, *Nuclear Forces for Medium Powers*, Adelphi Papers 106 and 107 (London: International Institute for Strategic Studies, 1974); and Thomas Jonter, *Sweden and the Bomb* (Uppsala: Swedish Nuclear Power Inspectorate, 2001).

[12]Dunn and Kahn, *Trends in Nuclear Proliferation*, 96.

[13]R. Robert Sandoval, "Consider the Case of the Porcupine: Another View of Nuclear Proliferation," *Bulletin of the Atomic Scientists* (May 1976): 19.

[14]Dunn and Kahn, *Trends in Nuclear Proliferation*, 3; and Jonter, *Sweden and the Bomb*.

in order to threaten convincingly that the possessor would escalate the conflict to such a level that the benefits achieved by the "victor" would be outweighed by the costs of achieving "total victory."

Coercion. Nuclear blackmail, intimidation of non-nuclear regional adversaries, and even use of nuclear weapons in "preventive first strikes" may be desirable policy options for the leaders of certain "crazy states" as well as those facing the prospect of a long-term deterioration of their security vis-à-vis non-nuclear opponents.[15] Nuclear weapons may be coveted as a security trump card whose possession would enable one to employ conventional forces for aggressive purposes with greater impunity.[16]

International Security Disincentives

Hostile Reactions of a Military Nature by Adversaries and Allies. The anticipation of a hostile response by both adversaries and allies may discourage proliferation. An adversary might threaten or undertake military action to destroy an incipient nuclear weapons force and production capability. A country weighing the decision to go nuclear may also be dissuaded by the fear that such action would provoke a neighboring adversary to follow suit, resulting in a costly escalation of the arms race without a commensurate increase in security. Undesirable allied responses of a military nature might take the form of a reduction or severance of established security guarantees and the disruption of the supply of important conventional armaments.[17]

The Strategic Credibility Gap/Technical Difficulties. The difficulty of obtaining the technical conditions associated with a credible nuclear deterrent (e.g., secure second-strike forces, effective systems of command and control, and reliable delivery vehicles) may diminish the attraction of nuclear weapons for potential proliferators. An embryonic and poorly defended nuclear force serves as an incentive for a preemptive strike and as a source of crisis instability.[18] Although compelling in the abstract, it remains to be demonstrated whether the logic of Western strategic analysis is relevant to nuclear decision-making in near-nuclear countries. Recent nuclear tests in South Asia suggest it may not be.

Absence of Perceived Security Threat. If acute external threats provide an ar-

[15]Yehezkel Dror, *Crazy States* (Lexington, MA: Lexington, 1971).
[16]Iraq is often cited in this connection.
[17]For a discussion of this point, see Dunn and Kahn, *Trends in Nuclear Proliferation*, p. 14, and U.S. Office of Technology Assessment, *Nuclear Proliferation and Safeguards*, vol. 1 (New York: Praeger, 1977), p. 97.
[18]This traditional perspective is challenged by Yehezkel Dror and Kenneth Waltz, "The Spread of Nuclear Weapons: More May Be Better," Adelphi Paper 171, International Institute for Strategic Studies, 1981.

gument for the acquisition of nuclear weapons, absence of a hostile international environment, or the perception of such a condition by a nation's leadership, may serve as a disincentive to acquiring nuclear arms. Even in a more threatening international milieu, confidence in the security guarantees of a militarily powerful ally might reduce the pressure to develop an independent nuclear deterrent.

International Political Incentives

Increased International Status. Nuclear weapons are a symbol of scientific expertise and technological development. They are almost synonymous with great power status and are viewed by many states as a source of international prestige and autonomy. Aside from bolstering a nation's self-confidence, an important consideration for many "have-not" nations and others who perceive a disparity between their actual and "rightful" place in the international pecking order, nuclear weapons capability may engender fear and respect.

A number of studies indicate that international prestige and influence are important incentives for potential proliferators from the developing world who despair over gross inequities in the global distribution of wealth and power.[19] Nuclear weapons appear as a useful lever in North-South politics and as a means to reshape the structure of the contemporary international order. Demonstration of a nuclear weapons capability or even the credible threat to proliferate might command the attention of the industrialized states and prompt greater economic assistance and political support. One study notes that "the developing states probably did not overlook the fact that India's economic aid from the Western industrial states was increased by some $200 million less than a month after its initial nuclear explosion."[20] Nuclear weapons also have a symbolic value for more developed states, a factor important in French nuclear decision-making.[21]

Increased Autonomy. Nuclear weapons may be sought to enhance intra-alliance influence and international freedom of action. Nuclear weapons might enable a state to exert greater influence in regional security arrangements and international political forums (e.g., the UN Security Council, General Assembly, and its specialized agencies).[22] French attempts to gain a greater voice in NATO

[19]See Ashkok Kapur, *International Nuclear Proliferation* (New York: Praeger, 1979); T. T. Poulose, "Nuclear Proliferation: A Third World Perspective," *Round Table* (April 1979); Amitav Ghosh, "Count Down," *New Yorker* (Oct. 26–Nov. 2, 1998): 186–97; and Samina Ahmed, "Pakistan's Nuclear Weapons Program," *International Security* 23, no. 4 (spring 1999): 178–204.
[20]See Ted Greenwood, Harold A. Feveson, and Theodore B. Taylor, *Nuclear Proliferation: Motivations, Capabilities, and Strategies for Control* (New York: McGraw-Hill, 1977), p. 51.
[21]See Sagan, "Why Do States Build Nuclear Weapons?" pp. 78–79.
[22]U.S. Office of Technology Assessment, *Nuclear Proliferation and Safeguards*, p. 94; and Greenwood et al., *Nuclear Proliferation*, pp. 48–49.

affairs and assert national autonomy influenced development of its *force de frappe.*[23]

Revolutionaries/Pariahs. International outcasts who face hostile regional adversaries (e.g., Israel, Taiwan, pre-1993 South Africa) or revolutionary regimes that reject prevailing international norms and institutions (e.g., North Korea, Iraq) are apt to have few qualms about violating international taboos. They may pursue nuclear weapons because of their revolutionary or pariah status.

International Political Disincentives

International Norms. The Non-Proliferation Treaty (NPT), subscribed to by 188 parties, more than any other international treaty, explicitly states that proliferation of nuclear weapons would seriously increase the danger of nuclear war. It was extended indefinitely on 11 May 1995 at the NPT Review and Extension Conference, without a vote by all 174 states parties present at the conference.[24] The nonproliferation norm also is embodied in the widely subscribed to safeguards statute of the International Atomic Energy Agency (IAEA), the growing number of regional nuclear-weapons-free zones, and in efforts taken historically by such politically diverse states as the Soviet Union and the United States to restrict the export of sensitive nuclear technology and fuel cycle components.[25] Although some countries may assume treaty membership and a public nonproliferation stance to conceal their real nuclear weapons ambitions, most international treaty commitments are not undertaken lightly or easily repudiated. It may even be the case that "the political commitments involved in the acceptance of the NPT and IAEA safeguards are as important as the accompanying physical constraints."[26]

Economic and Political Sanctions by Other States. The fear of political and economic reprisals may discourage potential proliferators. This concern is apt to be greatest among those nations that depend heavily on the major powers for economic assistance and technological aid. Would-be proliferators also run the risk

[23]Dunn and Kahn, *Trends in Nuclear Proliferation*, pp. 3–4.

[24]At the time of the conference, there were 178 states parties to the NPT. See Tariq Rauf and Rebecca Johnson, "After the NPT's Indefinite Extension: The Future of the Global Nonproliferation Regime," *Nonproliferation Review* 3, no. 1 (fall 1995): 28–42.

[25]On these issues, see David Fischer, *History of the International Atomic Energy Agency: The First Forty Years* (Vienna: IAEA, 1997); Jozef Goldblat, "Nuclear-Weapon-Free Zones: A History and Assessment," *Nonproliferation Review* 4, no. 3 (spring/summer 1997): 18–31; William C. Potter, "Nuclear Nonproliferation: U.S.-Soviet Cooperation," *Washington Quarterly* 8, no. 1 (winter 1985): 141–54; and Peter Lavoy, "Learning and the Evolution of Cooperation in U.S. and Soviet Nuclear Nonproliferation Activities," in George Breslauer and Philip Tetlock, eds., *Learning in U.S. and Soviet Foreign Policy* (Boulder, CO: Westview, 1991), pp. 735–83.

[26]Greenwood et al., *Nuclear Proliferation*, p. 52.

of censure and sanctions by international organizations, although no international agreement mandates such action.

Domestic Security Incentives and Disincentives

The extant literature does not indicate any domestic security incentives for acquiring a nuclear weapons capability.[27] Nuclear weapons differ from chemical weapons, which have been used in the past against domestic adversaries. The risk of unauthorized seizure of nuclear weapons may be a domestic security disincentive, especially for countries subject to frequent political upheavals and domestic turmoil.[28] Scenarios involving unauthorized acquisition of nuclear weapons include seizure of all or part of a nation's nuclear weapons stockpile by revolutionary groups or terrorists for political blackmail,[29] or a military "nuclear coup d'état."[30]

Domestic Political Incentives

Economic Spillover/Compensation/Cost Effectiveness. The economic potential of peaceful nuclear explosions (PNEs) was heralded by the United States in the 1950s as part of President Eisenhower's "Atoms for Peace" program. The U.S. government and industry spent more than $200 million to explore such economic applications of nuclear explosions in excavating canals and harbors, producing oil from shale, and gas and oil stimulation.[31] The Soviet Union pursued an even more extensive effort to exploit PNEs for peaceful economic purposes.[32]

Although most U.S. and Russian analysts became convinced in the 1970s, and late 1980s, respectively, that PNEs were not cost effective or posed significant en-

---

[27]One of the charges leveled against the head of Yugoslavia's secret police, Aleksandar Rankovic, when he was ousted from his post in 1966 was that he had sought to obtain nuclear weapons for his organization. Rankovic also had served as head of Yugoslavia's Federal Commission for Nuclear Energy. See William Potter, Djuro Miljanic, and Ivo Slaus, "Tito's Nuclear Legacy," *Bulletin of Atomic Scientists* 56, no. 2 (Mar./Apr. 2000): 63–70.

[28]One analyst has hypothesized that the "fear of losing civilian control over the military aspect of nuclear energy is a brake against nuclear weapons proliferation in third world countries where nuclear power struggles are unsettled and the authority structure is not completely legitimate and popular." See Ashok Kapur, "A Nuclearizing Pakistan: Some Hypotheses," *Asian Survey* 20, no. 5 (May 1980): 551.

[29]U.S. Office of Technology Assessment, *Nuclear Proliferation and Safeguards*, p. 98.

[30]Lewis Dunn, "Military Politics, Nuclear Proliferation, and the 'Nuclear Coup D'etat,'" *Journal of Strategic Studies* 1, no. 1 (May 1978): 31–50.

[31]A useful review of the economics of PNEs is provided by Henry Rowen, "The Economics of Peaceful Nuclear Explosions," report prepared for the International Energy Atomic Authority, 1976.

[32]See Anna Scherbakova and Wendy Wallace, "The Environmental Legacy of Soviet Peaceful Nuclear Explosion," *NIS Environmental Watch*, no. 5 (summer 1993): 33–50; and Milo Nordyke, "The Soviet Program for Peaceful Uses of Nuclear Explosions," Center for Global Security Research, Lawrence Livermore National Laboratory, October 1996.

vironmental hazards, many potential proliferators profess to retain the view that PNEs have substantial economic promise. The desire to obtain economic benefits is sometimes cited as an incentive to develop a nuclear explosive capability, despite the guarantee in the NPT that "potential benefits from any peaceful applications of nuclear explosions will be made available to non-nuclear weapon States party to the Treaty on a nondiscriminatory basis."[33]

Reference is occasionally made to the utility of a nuclear weapons program for retaining highly skilled scientists and technicians who might otherwise immigrate to nations with more sophisticated nuclear research establishments. The risk of a nuclear "brain drain" became especially pronounced with the collapse of the Soviet Union and the dramatic reduction in Russian expenditures at its nuclear weapons complex, leading to a number of U.S. and international initiatives in the 1990s to mitigate the brain drain threat.[34]

Until quite recently there was no indication that states might have contemplated the acquisition of nuclear weapons in order to trade that capability for financial dividends. The 1994 Agreed Framework to halt North Korean nuclear activities suggests that some states may view the pursuit of nuclear weapons as a means to induce financial compensation from proliferation opponents. One variant of this approach might be seen in Ukraine's nuclear diplomacy, which after mid-1992 was increasingly driven by efforts to secure the maximum financial dividend from denuclearization.[35]

Another economic argument advanced in support of nuclear weapons acquisition, greater investment in nuclear force modernization, and retention of nuclear weapons is their cost effectiveness in comparison to conventional arms.[36] Little empirical evidence supports the cost-effectiveness thesis, though the thesis continues to find expression among Russian and Indian defense planners and may explain the Iranian leadership's interest in pursuing a nuclear weapons option.[37]

[33]Article V, Treaty on the Non-Proliferation of Nuclear Weapons.

[34]These initiatives include the multinational-sponsored International Science and Technology Center in Moscow and sister organization in Kiev, and the U.S.-funded Initiative for Preventing Proliferation and Nuclear Cities Initiative.

[35]See Potter, "The Politics of Nuclear Renunciation," p. 46.

[36]Major General Volodymyr Tolubko, the most outspoken advocate of Ukraine's retention of an independent "nuclear defense shield," maintained that it would have been cost effective for Ukraine to retain a nuclear deterrent in lieu of a large standing army with conventional arms. Tolubko elaborated his views in a series of three articles in *Holus Ukrainy* in 1992.

[37]Michael Eisenstadt suggests that nuclear weapons may be "the only way for Iran to become a military power without destroying its economy; while a bomb could cost millions, rebuilding its conventional military would cost tens of billions of dollars." See Eisenstadt, "Instability in Central Asia and the Caucasus and Iranian Weapons Proliferation," paper prepared for the Conference on Energy, Weapons Proliferation and Conflict in Central Asia and the Caucasus, Washington, DC (20–21 April 1999), p. 4.

Bureaucratic and Domestic Politics. Despite an emphasis on the rational "security model" of nuclear decision-making,[38] domestically oriented pressures to go nuclear may be difficult to justify militarily or economically from a national perspective. These include pressures from various industrial, scientific, and military groups that would stand to benefit from an expensive nuclear program; broad-based public support for an independent nuclear force; and pressure from politicians anxious to divert attention from other policy failures.[39] According to a bureaucratic politics interpretation of nuclear decision-making, "bureaucratic actors are not seen as passive recipients of top-down political decisions; instead they create the conditions that favor weapons acquisition by encouraging extreme perceptions of foreign threats, promoting supportive politicians, and actively lobbying for increased defense spending."[40] A related perspective directs attention to the role of individual "myth makers." According to this idiosyncratic explanation for nuclear proliferation, "a government is likely to 'go nuclear' when proficient and well-positioned individuals who want their country to build nuclear bombs, exaggerate security threats to make a 'myth of nuclear security' more compelling."[41]

Technological Momentum. Development of nuclear weapons may result from technological momentum in which the technological feasibility of the project takes precedence over the military or political necessity of the task and in which there is no formal decision to go ahead. "Technological creep" occurs when significant progress toward a weapons capability is achieved by incremental advances in different fields of nuclear engineering without a formal decision being taken to develop a nuclear explosive. Lower-level French scientists and bureaucrats reportedly took major steps toward developing a nuclear weapon capability before they were specifically directed to do so by the national leadership.[42]

Domestic Political Disincentives

Cost. The economic cost of developing and maintaining a nuclear weapons program may be prohibitive for certain countries. The cost disincentive involves not only the absolute level of expenditures but also the opportunity costs of diverting monetary and manpower resources away from economic and social projects. Direct expenditures for nuclear weapons design and production in the

---

[38]Sagan, "Why Do States Build Nuclear Weapons?" p. 55.

[39]An early discussion of this determinant is provided by Dunn and Kahn, *Trends in Nuclear Proliferation*, p. 5.

[40]Sagan, "Why Do States Build Nuclear Weapons?" p. 64.

[41]Peter R. Lavoy, "Nuclear Myths and the Causes of Nuclear Proliferation," *Security Studies* (spring/summer 1993): 192–212.

[42]U.S. Office of Technology Assessment, *Nuclear Proliferation and Safeguards*, p. 100.

United States account for only a small portion of the overall cost of developing and maintaining the U.S. program—estimated at $5.5 trillion (in constant 1996 dollars). The indirect costs were related to deploying the bomb, maintaining operational control over it, and environmental remediation associated with its production.[43] Still, there remains a myth that one obtains "more bang for the buck" or "rubble from the ruble (or rupee)."[44] The costs of nuclear weapons also may be perceived to diminish with the growth of civilian nuclear power industries and the decline in the incremental cost associated with a weapons program."[45]

Public Opinion. Adverse domestic opinion may constrain acquisition of nuclear weapons in some nations. In Japan, Germany, Sweden, and Canada, public opposition had a decided effect on the decision to renounce nuclear weapons. This fear of adverse public opinion may be marginal for many nations lacking a strong democratic tradition.

Bureaucratic and Domestic Politics. Bureaucratic and domestic politics may work as a disincentive as well as an incentive to development of nuclear weapons. Competition for scarce resources could produce an alignment of bureaucratic actors who, for parochial organizational reasons, opposed the creation of new institutional actors and potential competitors. Certain branches of the military might oppose a nuclear weapons program if it were perceived to interfere with funding of preferred weapons systems or to shift the distribution of the fiscal pie. Domestic political disincentives to the acquisition of nuclear weapons may stem from key individuals whose stance is determined by personal philosophical convictions (Nehru is the most frequently cited example) or by calculations of self-interest. One should also note the intriguing thesis that links nuclear proliferation restraint to the growing impact of international economics on domestic politics. According to this thesis, the growing internationalization of markets, finance, and technology has reduced the ability of policy-makers to isolate nuclear policy from broader questions of economic development and domestic politics.[46] States increasingly are forced to acknowledge the "creeping linkages between nuclear policies and access to foreign capital and advanced technology."[47]

[43]See Stephen Schwartz, ed., *Atomic Audit: The Costs and Consequences of U.S. Nuclear Weapons since 1940* (Washington, DC: Brookings Institution, 1998).
[44]This perception is reflected today among most Russian defense planners who favor greater reliance upon nuclear as opposed to conventional weapons.
[45]On this point, see U.S. Office of Technology Assessment, *Nuclear Proliferation and Safeguards*, p. 96.
[46]See Etel Solingen, "The Political Economy of Nuclear Restraint," *International Security* 19, no. 2 (fall 1994): 126–69; and Solingen, "The Domestic Sources of Nuclear Postures: Influencing 'Fence-Sitters' in the Post-Cold War Era," IGCC Policy Paper, No. 8, University of California at San Diego (October 1994).
[47]Solingen, "The Domestic Sources of Nuclear Postures," pp. 4–5.

Situational Variables

Most analyses of nuclear weapons choice imply the operation of two sets of sufficient conditions: (1) the balance between underlying proliferation incentives and disincentives, and (2) the presence of situational factors that might precipitate a decision to go nuclear whenever incentives outweigh constraints.[48] The most widely cited potential "trigger events" are summarized below.

International Crises. International crises have been identified as precipitants of decisions to acquire nuclear weapons. The most commonly mentioned is the nuclearization of a neighboring state or regional rival. An action-reaction dynamic is assumed to operate in which one nation's acquisition of nuclear weapons intensifies an adversary's sense of insecurity and reduces some of the psychological and political barriers to proliferation. A widely shared nuclear taboo might be weakened, which could alter the balance in the domestic debate over the desirability of possessing nuclear weapons.[49] More generally, a foreign crisis may provide the opportunity for forging a new bureaucratic consensus in support of a decision to go nuclear.[50] As the spring 1998 events in South Asia demonstrated, one nation's nuclear testing combined with belligerent foreign policy statements made it virtually impossible in terms of domestic politics for a regional rival to refrain from following suit.

Weakening of Security Guarantees. During the Cold War, bipolarity and credible alliance guarantees by the two superpowers were often credited with reducing proliferation incentives.[51] Diminution of superpower alliance guarantees, or the perceived collapse of their nuclear umbrellas, was viewed as a destabilizing event that might lead to a revision of national security calculations and the decision to acquire a nuclear deterrent. A similar situation appears to apply in the post–Cold War era. Some analysts attribute the decision by India to test nuclear weapons in 1998 to the loss of a Soviet counterweight to a perceived threat from China.[52]

Increased Accessibility of Necessary Technology and Material. For some would-be proliferators, increased availability of nuclear technology and material might trigger a decision to initiate a nuclear weapons program. This danger often is associ-

---

[48]A similar formulation is suggested by Kegley et al., *Nuclear Power Issues and Choices,* p. 234.

[49]See Wohlstetter et al., *Swords from Ploughshares,* pp. 138–39; and Dunn and Kahn, *Trends in Nuclear Proliferation,* p. 8.

[50]Lewis A. Dunn and William H. Overholt, "The Next Phase in Nuclear Proliferation Research," in Overholt, ed., *Asia's Nuclear Future* (Boulder, CO: Westview, 1977), p. 11.

[51]See Richard K. Betts, "Incentives for Nuclear Weapons," in Joseph A. Yager, ed., *Nonproliferation and U.S. Foreign Policy* (Washington, DC: Brookings Institution, 1980), p. 117.

[52]The security guarantee was implied in the twenty-year treaty of "peace, friendship, and cooperation" with the Soviet Union signed in 1971. For a discussion of the collapse of the implied guarantee, see Samit Ganguly, "India's Pathway to Pokhran II: The Prospects and Sources of New Delhi's Nuclear Weapons Program," *International Security* 23, no. 4 (spring 1999): 167.

ated with the disintegration of the Soviet Union and the diminished security of its nuclear weapons and material stocks.[53]

Vertical Proliferation. The failure of the superpowers to implement their promise in Article VI of the NPT to undertake effective measures to halt the nuclear arms race (i.e., vertical proliferation) has long been a major complaint of non-nuclear parties to the NPT. The current stalemate in strategic arms control negotiations, reluctance of the United States and China to ratify the Comprehensive Test Ban Treaty, U.S. and Russian opposition to a no-first-use pledge, and inflexible positions assumed by all of the nuclear weapons states that are parties to the NPT during the strengthened NPT review process contribute to the gap perceived by many non-nuclear states to exist between the words and actions of the nuclear powers. This gulf bolsters the case of so-called Nth country advocates of an independent nuclear force option. As one Indian exponent of this perspective put it in 1979, "[U]ntil the existing five nuclear-weapons states are able to . . . reduce their nuclear arsenals in a manner that reflects a process of making real sacrifices, the in-house bureaucratic debates in Third World societies are not likely to be impressed with, say, President Carter's 'open-mouth' antiproliferation speechmaking."[54] Many representatives of the Non-Aligned Movement (NAM) repeat similar charges and are reluctant to condemn the Indian and Pakistani nuclear tests.

Domestic Crisis and Leadership Change. Domestic events might trigger a decision to go nuclear. In a political crisis, the leadership might attempt to capitalize on a nuclear weapons decision to divert domestic attention and restore popular confidence in the government. A change in political leadership might elevate to power individuals committed to a nuclear weapons program.

Focusing Events. "Focusing events" compel policy-makers to direct their attention to a problem that they otherwise might not have.[55] These events need not be of a crisis nature entailing surprise or an extremely short decision-making time for response. They do, however, imply a looming deadline, such as a pending election, forthcoming treaty negotiation, or treaty entry into force date. Examples of focusing events that may have served as nuclear decision-making catalysts are the opening for signature in 1968 of the NPT and the provision of the 1996 Comprehensive Test Ban Treaty (CTBT), which calls for a conference to be held three years after the opening for signature of the treaty to decide how to accelerate the ratification process.[56]

[53]See Allison et al., *Avoiding Nuclear Anarchy.*
[54]Kapur, "A Nuclearizing Pakistan," p. 167.
[55]See John Kingdon, *Agendas, Alternatives and Public Policies* (Boston: Little, Brown, 1984), pp. 99–105.
[56]On the relationship between the timing of the 1998 Indian nuclear tests and the CTBT, see K.

Proliferation Profiles

Table 6.2 summarizes the major findings from a survey of nuclear decision-making in twenty-six past and potential proliferators.[57] Although disparity in reliable information available across cases makes efforts at generalization hazardous, the summary data, together with the case studies they reflect, enable one to examine a number of propositions regarding underlying pressures for and constraints against nuclear weapons acquisition.

Scope and Rate of Diffusion

Immediately after Hiroshima, again in 1960 after the French test, after the Chinese nuclear explosion in 1964 and the Indian detonation in 1974, and most recently after the nuclear tests by India and Pakistan in 1998, an alarm has been sounded forecasting the rapid proliferation of nuclear weapons.[58] Underlying these prophecies was the assumption that a kind of domino theory or chain reaction was in effect, whereby the violation of the nuclear weapons taboo by one nation undermines the nuclear inhibitions for regional adversaries, leading ultimately to the global diffusion of nuclear weapons.

These proliferation prophecies have not come to pass. The number of states having unambiguously conducted a nuclear weapons test is seven: the United States (1945), the Soviet Union/Russia (1949), the United Kingdom (1952), France (1960), the People's Republic of China (1964), India (1974), and Pakistan (1998). In addition to these overt proliferators, one must include those states that have assembled nuclear weapons without unambiguously having conducted nuclear detonations and without proclaiming their status (Israel and South Africa), as well as those nations that inherited nuclear weapons that were on their territory at the time of the disintegration of a prior nuclear weapons state (Belarus, Kazakhstan, and Ukraine).[59] These twelve states compose the N of past proliferators.

Subrahmanyam, "Nuclear India in Global Politics," *World Affairs* 2, no. 3 (July–Sept., 1998): 23–26. See also Ganguly, "India's Pathway to Pokhran II," pp. 168–70.

[57]The one significant omission from the survey is Switzerland, which maintained a nuclear weapons program for many years. I have not been able to obtain sufficient information on the Swiss decision-making process to include the case in the survey. One relevant source, albeit an incomplete one, is Juerg Steussi-Lautenberg, "Historical Abstract on the Question of Swiss Nuclear Arms Program," Draft, 1995.

[58]One of the most widely cited forecasts was made by President Kennedy, who was haunted by the fear that by 1970 there might be ten nuclear powers unless a test ban treaty were concluded.

[59]It has been suggested, although not substantiated, that a nuclear bomb of French-Israeli design may have been tested by the French in the early 1960s. See Ernest W. Lefever, *Nuclear Arms in the Third World* (Washington, DC: Brookings Institution, 1979), p. 165, and "How Israel Got the Bomb," *Time* (12 April 1976): 39–40. See also Leonard S. Spector, *Nuclear Proliferation Today* (New York: Vintage, 1984), p. 120. There are conflicting reports about the possibility of a South African nuclear test in 1979. See ibid., p. 133.

TABLE 6.3

Summary of Proliferation Determinants

Country	Underlying pressures		Underlying constraints		Most likely precipitants
	Primary determinant	Secondary factor	Primary determinant	Secondary factor	
Argentina	5,6	9	9,11,13,15	14,16	19
Australia	5,9	1,2,5	9,13,15	14	25
Belarus		5,9	13,15	11,12,14,17,18	
Brazil	5,6	10	9,11,13,15	14,16	19
Canada	2		13,14	6,17,18	
Egypt	9	1,8	9,15,17		20,22,24
France		5,6,10		17	
India					
1974	5,6	1,9,10	19	11,12,14,15,17	19,20,24
1998	5,6,9	1,4,7	19	11	21,24,25
Iran	5,6,7	1,8,9	11	9,15,17	20,22,24
Iraq	1,4,6,7,9				22
Israel	1,3	2,6	11,15	17	
Italy		5,6	13,17	9,15	
Kazakhstan	1	5,6	15,17	9,11,12,14,21	
North Korea	1,2,7	8,9	11,12,17		
Norway	9,10	1,5,6	11,15,21	9,12,17,18	
PRC	1,5,6			9,11,17	
Pakistan	1,2,9	6	15	9,11	19,21,22
South Africa	1,7	6,9	9,13,14,15,16		20,24
South Korea	1,2	1,6	11,15	16	19,21
Sweden	2	1,10	9	12,14,15,17,18	25
Taiwan	1,2,3,6	5,7,9	11,15	12	19,21,24
UK					
WW II	2	8			20
postwar	5,6	10			20
USA	2	10			20
USSR	1		9 (for 1940 decision)		19
Ukraine	1,8,9	5,6	12,15,17,21, 24	9,11	
Yugoslavia					
1948	1,6	2,5	12,17	14	20
1974	5,6		12,17		19

KEY:

1. Deterrence
2. Warfare advantage/defense
3. Weapon of last resort
4. Coercion/compellence
5. Status/prestige
6. Autonomy/influence
7. Revolutionaries/pariahs
8. Economic spillover/compensation/cost effectiveness
9. Bureaucratic/domestic politics of other states
10. Technological momentum
11. Military reaction by other states
12. Strategic credibility gap/technical difficulties
13. Absence of perceived threats
14. International norms
15. Economic and political sanctions
16. Unauthorized seizure
17. Economic costs
18. Public opinion
19. Nuclearization or denuclearization
20. International crisis
21. Change in security guarantees
22. Increased access to material/technology/know-how
23. Vertical proliferation
24. Domestic crisis/leadership change
25. Focusing events

The number of states that have demonstrated both the inclination and technical ability to develop nuclear weapons is small, and the rate of weapons diffusion is slow (two in the 1940s, one in the 1950s, probably three in the 1960s, two in the 1970s, none in the 1980s, and one in the 1990s).[60] An increasing number of states have advanced to a nuclear "twilight zone" in which the absence of operational nuclear weapons is more a function of political will than of technical know-how. One of our most striking findings is the extent to which most countries with nuclear industry assets and ambitions kept open a nuclear weapons option even in the absence of perceived external threats. Australia, Italy, Norway, and Yugoslavia (post-1974) are little known but important examples of that phenomenon. Only postwar Canada and post-Soviet Belarus reached a quick and unequivocal decision to forswear nuclear weapons.

In a number of instances the retention of a nuclear weapons option was the consequence of technological momentum and bureaucratic inertia acting in conjunction with the difficulty of readily distinguishing between good "atoms for peace" and bad "atoms for war." Both U.S. and Soviet nuclear policy for several decades contributed to the diffusion of nuclear weapons technology by actively promoting the development abroad of civilian, and presumably safe, nuclear activities with little regard for their military or dangerous implications. Ironically, during the mid to late 1950s, there developed a U.S.-Soviet peaceful nuclear energy and prestige race in which both nuclear weapons states sought to expand their influence among scientific communities internationally by declassifying and disseminating a large volume of technical nuclear information. When in December 1954, following the initiation of the "Atoms for Peace" program, the U.S. Atomic Energy Commission decided to declassify information on the Argonne (Illinois) atomic pile, the Soviet Union responded in January 1955 by announcing its readiness to share technical data on the operation of its first atomic power plant.[61] Amazingly, unaware of or insensitive to the national security implications of their behavior, "the nuclear giants lifted their skirts of secrecy, each challenging the other to reveal more evidence of their dedication to the peaceful atom."[62]

[60]Israel probably had an operational nuclear weapons capability in the 1960s; South Africa in the 1970s. See Seymour Hersh, *The Sampson Option: Israel's Nuclear Arsenal and American Foreign Policy* (New York: Random House, 1991); and David Albright, "South Africa's Secret Nuclear Weapons," ISIS Report (May 1994). I do not include Belarus, Kazakhstan, and Ukraine in this grouping because they did not demonstrate an inclination to acquire nuclear weapons and in fact renounced those they inherited.

[61]See *Izvestiia* (18 January 1955). For a discussion of U.S. and Soviet nuclear export policy during this period, see William C. Potter, "Nuclear Export Policy: A Soviet-American Comparison," in Charles W. Kegley, Jr., and Pat McGowan, eds., *Foreign Policy: USA/USSR* (Beverly Hills: Sage, 1982): pp. 295–98.

[62]H. L. Neiburg, *Nuclear Secrecy and Foreign Policy* (Washington, DC: Public Affairs Press, 1960), pp. 113–14.

By the mid-1950s, this race had taken the form of exporting not only nuclear power information and technical experts but also research reactors.

According to conventional wisdom, nuclear technology largesse should not be confused with a proclivity to transfer nuclear weapons assistance. One analyst notes, however, that although "there was some concern about how India would behave after its 1974 nuclear test, the oft-encountered consensus within the non-proliferation community at the time was that no country that had acquired nuclear weapons would officially help another country to join the club."[63]

Our survey of nuclear decision-making in twenty-six states indicates that this assumption of restraint is not borne out at either the government-to-government level or at the level of firm behavior. India's apparent denials of requests for nuclear weapons assistance after 1974 contrast with direct U.S. government assistance to Great Britain, Soviet government aid to China, French government assistance to Israel, and Chinese government aid to Pakistan.[64] Great Britain was inclined to sell nuclear weapons to Australia in the 1960s, although the deal foundered on U.S. consent requirements and Australian bureaucratic politics.[65] Israel may have been engaged in a cooperative nuclear arrangement with South Africa, although the evidence is not clear.[66] Soviet assistance to China is described in vivid detail in Khrushchev's memoirs. Khrushchev recounts:

> Before the rupture in our relations, we'd given the Chinese almost everything they asked for. We kept no secrets from them. Our nuclear experts cooperated with their engineers and designers who were busy building an atomic bomb. The Soviet Union went so far as to promise the Chinese a prototype atomic bomb, whose shipment by train in 1958 was only halted at the last moment.[67]

[63]Lewis A. Dunn, "On Proliferation Watch: Some Reflections on the Past Quarter Century," *Nonproliferation Review* 5, no. 3 (spring/summer 1998): 63.

[64]See Margaret Gowing, *Britain and Atomic Energy* (New York: St. Martin's Press, 1964); Arthur Steiner, *Great Britain and France: Two Other Roads to the Atomic Bomb*, Monograph 7 (Los Angeles: Pan Heuristics, 1977); David Fischer, *Stopping the Spread of Nuclear Weapons: The Past and the Prospects* (London: Routledge, 1992); Roland M. Timerbaev, *Rossiys i yadernoe neposprostraneniye* [*Russia and Nuclear Nonproliferation*] (Moscow: Nauka, 1999); and John Wilson Lewis and Xue Litai, *China Builds the Bomb* (Stanford: Stanford University Press, 1988).

[65]See Jim Walsh, "Surprise Down Under: The Secret History of Australia's Nuclear Weapons," *Nonproliferation Review* 5, no. 1 (fall 1997): 1–20.

[66]For conflicting appraisals see Dunn, "On Proliferation Watch," 63; and Gerald Steinberg, "Israel: An Unlikely Nuclear Supplier," in William C. Potter, ed., *International Nuclear Trade and Nonproliferation* (Lexington, MA: Lexington Books, 1990), pp. 181–98.

[67]Nikita Khrushchev, *Khrushchev Remembers: The Last Testament* (Boston, 1974), p. 269, cited by Lewis and Xue, *China Builds the Bomb*, p. 60. See also Timerbaev, *Rossiya i yadernoye nerasprostraneniye*, pp. 127–35.

Diffusion Patterns

An examination of the underlying factors responsible for national decisions in twenty-six states to acquire, forgo, or renounce nuclear weapons indicates the absence of any typical proliferator profile. National nuclear decisions vary widely in terms of the specific mix of underlying pressures, constraints, and trigger events. That finding is significant and rather surprising given the presumption on the part of most U.S. policy-makers and security analysts that the source of proliferation decisions is obvious: "[S]tates will seek to develop nuclear weapons when they face a significant military threat to their security that cannot be met through alternative means; if they do not face such threats, they will willingly remain non-nuclear states."[68]

If conventional wisdom is mistaken regarding a simple, general explanation for why states go nuclear, its emphasis on the importance of international pressures on proliferation is well taken. International factors are part of the group of primary pressures in all but three of our twenty-six cases. They are present as primary constraints in all but two cases in which one can identify primary constraints. In contrast, domestic factors are evident as principal pressures in ten nuclear decisions and as primary constraints in nine.

The predominance of international over domestic pressures for, as well as constraints on, proliferation is more pronounced if one confines the sample to cases in which countries actually acquired nuclear weapons. For this set of states, international pressures are primary determinants for all cases, and domestic factors are present only for France (technological momentum), India 1998 (bureaucratic/domestic politics), and Pakistan (bureaucratic/domestic politics). The role of domestic factors as primary disincentives also is rare: India in 1974 before Nehru's death (his personal philosophical opposition) and the Soviet Union in 1940 (bureaucratic opposition by entrenched senior scientists to the nuclear research proposals of younger colleagues with less institutional power).[69]

No single international determinant stands out for the cases of actual weapons proliferation. The pursuit of deterrence, war-fighting advantage, status/prestige, and autonomy/influence are each present in three to five cases. What is notable is the tendency for the international political factors of status/prestige and autonomy/influence to cluster together. Both appear as primary determinants in the

[68]See Sagan, "Why Do States Build Nuclear Weapons?" p. 55, for a thoughtful critique of this perspective.
[69]The role of domestic/bureaucratic politics was important in South Africa's decision to renounce nuclear weapons but not to acquire them.

cases of France, India (1974 and 1998), China, and the United Kingdom (post-war).[70]

Disaggregation of international constraints does not yield any marked pattern for actual proliferators. If one expands the set to include all nuclear decision-making cases, three constraints predominate: concern about the military reaction by other states, absence of perceived threats, and fear of economic and political sanctions. Of the single factor proliferation pressures for the entire set of cases, deterrence and autonomy/influence considerations are the most frequent primary determinant (present in eleven cases each) followed by status/prestige (ten cases), bureaucratic/domestic politics (eight cases each), and warfare advantage (eight cases). None of the other single explanations were present in more than four cases.

Our proliferation profiles also indicate that there is no significant variation over time in the importance of deterrence and warfare advantage as primary incentives to acquiring nuclear weapons. More variable are the factors of autonomy/influence and status/prestige, neither of which played a role in the nuclear weapons decisions during World War II but became increasingly important for nuclear decision-making in the 1960s and persisted throughout the 1970s, 1980s, and 1990s. Although noticeable in the nuclear decision-making of Norway in the 1950s and Australia in the 1960s, bureaucratic and domestic politics emerged as major factors in proliferation calculations for numerous states in the 1970s and thereafter. Pursuit of nuclear weapons as a way of coping with one's pariah status or because of the revolutionary rejection of prevailing international norms is a relatively new development, evident in the decisions taken by North Korea, Iraq, Iran, and to some degree India and Taiwan.

Although proliferation disincentives have remained fairly constant—the most significant being the anticipated reaction of other states—political or security disincentives are absent for the first five nuclear weapons states. Stated differently, the major potential proliferators today and in the recent past, in contrast to the first members of the nuclear weapons club, appear to attach much more importance to the anticipated political and military reactions of other states. Their greater sensitivity to external factors is not surprising, given their lower ranking on most indices of international power relative to the first five nuclear weapons states—all of which were generally regarded as great powers or superpowers prior to their acquisition of nuclear weapons. Also contributing to this development is the growing internationalization of markets and the diminished ability of policy-makers to isolate nuclear policy from broader issues of economic development.[71]

---

[70]They also appear together for Argentina, Brazil, Iran, and Yugoslavia in the larger set of cases.

[71]See Solingen, "The Political Economy of Nuclear Restraint."

Were it not for the cases of France and South Africa and, to a much lesser degree India, one would be tempted to emphasize the existence of an acute security threat as the major factor that discriminates between necessary conditions (e.g., technical know-how and availability of fissile material and weapons fabrication facilities) and the determining or sufficient condition to "go nuclear." The perception of an acute security threat and the desire to achieve deterrence or warfighting advantages were the major underlying pressures for the United States, the Soviet Union, Great Britain (through 1945), the People's Republic of China, Israel, and Pakistan. The change in threat perception (from a situation of international crisis and severe military danger to one of relative security) largely accounts for the reversal of Canadian interest in nuclear weapons. The end of the Cold War facilitated the removal of nuclear weapons from three post-Soviet states that inherited them after the collapse of the Soviet Union. One can even argue that although the postwar British decision to develop nuclear weapons was not directly stimulated by international security concerns, the atomic explosion in 1952 was the fulfillment of a task whose decision and momentum were established during the war.[72]

One might infer from these historical cases that "near nuclear" states, which are relatively free from international security challenges, will be less inclined to opt for nuclear weapons than more security-conscious and threatened states. This logic would appear to be supported by the recent decision by Argentina and Brazil to abandon their nuclear weapons option and to join the NPT. The logic also calls attention to the latent proliferation risks posed by Taiwan, South Korea, and, to a lesser extent, Yugoslavia—not withstanding their formal support for or adherence to the NPT.

This security-oriented interpretation—although on balance correct—discounts France and post-1974 Yugoslavia, for which international security considerations were of secondary importance, and perhaps also India, depending on when one dates the Indian decision to develop a nuclear option. It also does not attach sufficient importance to psychological and nonrational bureaucratic political determinants—dimensions that our proliferation profiles and most other studies of proliferation decisions inadequately tap because of data limitations.[73]

[72]See Richard Rosecrance, ed., *The Dispersion of Nuclear Weapons* (New York: Columbia University Press, 1964), p. 300.

[73]It is probably not coincidental that the few studies that focus on the bureaucratic politics of nuclear decision-making also find considerations other than those of international security to be important. See, for example, Lawrence Scheinman. *Atomic Energy Policy in France under the Fourth Republic* (Princeton: Princeton University Press, 1964); Kapur, *International Nuclear Proliferation;* Lavoy, "Nuclear Myths and the Causes of Nuclear Proliferation"; Michael Barletta, "The Military Nuclear Program in Brazil," CISAC Working Paper, Stanford University, August 1997; and George Perkovich, *India's Nuclear Bomb* (Berkeley: University of California Press, 1999).

Although no typical Nth country exists in terms of the mix of underlying pressures, constraints, and precipitating factors, according to our comparative survey, a small number of variables appear to be of primary importance for the states examined. What varies most are not motivations or constraints; it is the sequence in which political decisions to "go nuclear" precede, follow, or coincide with technological developments. In the French, Indian, and South African cases, technological developments preceded and paved the way for political decisions to produce nuclear explosives. The reverse sequence occurred in the Soviet Union, the People's Republic of China, Israel, Iraq, and Pakistan, where political decisions to acquire nuclear arms were made in advance of major technological developments. Illustrative of a third pattern of nuclear decision-making, in which technology and politics go hand in hand, are the cases of the United States and Great Britain during World War II.

An alternative to conceptualizing "going nuclear" as a discrete event is to think in terms of a range of nuclear decisions related to increasing levels of nuclear capability. Such a ladder of nuclear capability, similar to one depicted in Figure 6.1, better captures the dynamic nature of the proliferation process and may direct attention to points along the vertical proliferation continuum that are most subject to external influence. A more adequate graphic representation of the process of going nuclear should also depict the manner in which technical capabilities intersect with military and political pressures. A crude effort to depict the juxtaposition of technical capabilities and military-political pressures with reference to our twenty-six cases is presented in Figure 6.2. An axis measuring the balance of proliferation incentives and constraints is superimposed on the nuclear capability axis. One can quarrel with the precise location of some of the states, but the graph calls attention to the two dimensions of going nuclear—technical capabilities (which do not end with the detonation of a single atomic explosion) and the balance of military, political, and economic pressures and constraints. By compiling similar plots for different points in time, one can portray the dynamic quality of proliferation and also discern some clues as to the impact of one nation's proliferation posture on the scope and pace of other nations' nuclear programs.

Policy Implications and Recommendations

Our comparative analysis highlights the multicausal nature of the diffusion of nuclear weapons. This finding suggests the need to tailor nonproliferation measures to specific cases. The comparative analysis also indicates the need to reconceptualize nuclear proliferation decisions and nonproliferation policy-making to take better account of the complex interplay of externally and internally oriented proliferation determinants.

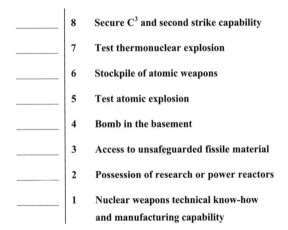

	8	Secure C^3 and second strike capability
	7	Test thermonuclear explosion
	6	Stockpile of atomic weapons
	5	Test atomic explosion
	4	Bomb in the basement
	3	Access to unsafeguarded fissile material
	2	Possession of research or power reactors
	1	Nuclear weapons technical know-how and manufacturing capability

FIG. 6.1. Ladder of nuclear weapons capability. Although not included in the ladder because of difficulty in locating its hierarchical position, possession of a nuclear weapons delivery system is a critical component of a state's nuclear capability.

Failure of Conception

The repeated failure by the U.S. government (and the nongovernmental non-proliferation community) to anticipate major nuclear proliferation developments is indicative of the problem of conceptualization. The U.S. intelligence lapse regarding the May 1998 Indian tests and the naive belief that after those detonations Pakistan could refrain from following suit are only the latest examples of a dismal forecasting record.[74] Although a number of factors have contributed to this situation, the 1998 events reflect a tendency to neglect the domestic political, bureaucratic, and cultural context in which proliferation decisions are made. Admiral David Jeremiah noted in a critique of the U.S. intelligence agencies' inability to foresee India's May 11 nuclear test that "senior US policy makers and intelligence officials had an 'underlying mind-set' that India would not test its nuclear weapons. . . . That fixed idea was unaffected by the fact that India's newly elected Hindu nationalist leaders openly and repeatedly vowed to deploy the bomb."[75] U.S. intelligence disregarded information that did not fit its preconceived view of what the world ought to look like. It conveniently saw a geostrategic world in which Chinese-Indian relations had improved, U.S.-Indian ties were on the mend, and thanks to the indefinite extension of the NPT, the nuclear prolifera-

[74]Earlier notable failures pertain to South Africa, Iraq, and North Korea.
[75]Tim Weiner, "Report Finds Basic Flaws in US Intelligence Operations," *New York Times,* 3 June 1998, p. 1.

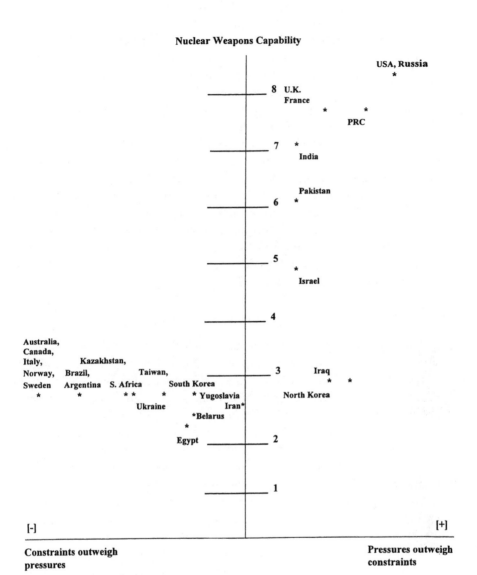

Nuclear Weapons Capability

USA, Russia
*

8 U.K.
   France
         *   *
              PRC

7 *
   India

   Pakistan
6 *

5 *
   Israel

4

                                                    3 Iraq
Australia,   *   *
Canada,
Italy,   Kazakhstan,   North Korea
Norway,   Brazil,   Taiwan,   South Korea
Sweden   Argentina S. Africa   * Yugoslavia
  *   *   * *   *   Iran*
                        Ukraine
                        *Belarus
                           *
              Egypt   2

                                   1

[-]   [+]

Constraints outweigh   **Pressures outweigh**
pressures   **constraints**

FIG. 6.2. Dual dimensions of proliferation for 2000

tion battle had been largely won. Despite clear indications of Indian preparations for a nuclear test two years earlier, it chose to discount the importance of national pride, Bharatiya Janata Party (BJP) vows to deploy the bomb, and "changes in the BJP party which advanced the careers of those favoring a test."[76]

Until we become more sensitive to these domestic motivations for "going nuclear," we will continue to miss both proliferation threats and nonproliferation opportunities. We also will be ill-prepared to devise a set of nonproliferation incentives and disincentives that have any prospect of impacting meaningfully on nuclear proliferation decision-making in at least some countries of concern.

The best corrective to this problem is not necessarily a larger nonproliferation intelligence budget or more classified information. Instead, it may be better use of open-source data and more analysts with advanced area studies training. As Senator Patrick Moynihan succinctly put it when asked, after the Indian and Pakistani tests, what needed to be done to improve the $27 billion/year U.S. intelligence effort: "Learn to read."

Susceptibility to Manipulation

A sound nonproliferation policy must be guided by the relative importance of alternative proliferation determinants and a recognition of their susceptibility to manipulation and influence. This latter calculation should moderate expectations about U.S. or international ability to influence—especially in the short term— proliferation decisions that are principally driven by domestic considerations. Although nonproliferation education and training to change mindsets, inculcate nonproliferation norms, and build communities of nonproliferation specialists may over time shift the balance of domestic political coalitions, a domestic-focused nonproliferation strategy will be difficult to sustain.[77]

Security incentives, on the other hand, are likely to be more susceptible to manipulation by an outside party.[78] Among the nonproliferation strategies that may be appropriate for reducing international security concerns are the extension of security guarantees by the nuclear-weapons states, negotiation and implementation of arms control and disarmament measures (e.g., nuclear-weapon-

---

[76]Mark Hibbs, "India Was Ready in 1996 to Test . . ." *Nucleonics Week* 39, no. 20 (14 May 1998): 1, 11.

[77]Additional measures consistent with a domestic politics approach to nonproliferation include the provision of information on the economic and environmental costs of pursuing a nuclear weapons option and the employment of scientists abroad who might otherwise become engaged in weapons activities. See Sagan, "Why Do States Build Nuclear Weapons?" p. 72 on this topic.

[78]This point is noted by Richard Betts, "Paranoids, Pygmies, Pariahs and Nonproliferation Revisited," in Zachary Davis and Benjamin Frankel, eds., *The Proliferation Puzzle: Why Nuclear Weapons Spread (and What Results)* (London: Frank Cass, 1993), p. 107.

free zones and a Comprehensive Test Ban Treaty), and provision of conventional arms transfers. These approaches try to affect the demand for nuclear weapons by reducing incentives for their acquisition.

Security assurances from the nuclear powers have been a prerequisite for the willingness of many states to adhere to the NPT and to justify their decisions in the face of domestic opposition. Security assurances have been especially important for such isolated and insecure parties as South Korea and Taiwan. They also proved significant in gaining the Ukrainian parliament's support for denuclearization and NPT accession.

Our review provides little encouragement for the utility of demand-side measures, such as sanctions, that seek to strengthen disincentives by raising the perceived costs of acquiring nuclear weapons. Their effectiveness depends very much on the motivations of the would-be proliferators, the state's economic and political vulnerability, and the degree of support from the international community for specific sanctions. While unilateral action by a great power has worked in selected cases where overwhelming leverage could be exerted (e.g., U.S. successes in inducing South Korea to rescind its order for a French reprocessing plant,[79] and Taiwan to reverse its reprocessing activities),[80] unilateral sanctions against other Nth countries are apt to be futile. For countries such as Iran and North Korea, even multilateral sanctions involving many of the principal nuclear supplier states could only raise the cost but not prevent the implementation of a decision to produce nuclear weapons. The difficulty of preventing nuclear proliferation by means of sanctions is well illustrated by South Africa's nuclear weapons program and by Iraq's post–Gulf War efforts to reconstitute its weapons of mass destruction.

The cases of India and Pakistan are indicative of another limitation of sanctions. Policies designed for discouraging proliferation may not be effective or appropriate for encouraging moderation once the nuclear threshold has been breached. This dilemma arose after the first Indian detonation in 1974 and again in 1998 after the new round of South Asian nuclear tests. In both instances U.S. policy has fluctuated between adopting punitive measures intended to induce a nuclear rollback and a more accommodating stance designed to curtail further weaponization efforts.

[79]For a discussion of U.S. sanctions in this case, see Ernest W. Lefever, *Nuclear Arms in the Third World* (Washington, DC: Brookings Institution, 1979), p. 130; and Leslie Gelb, "Arms Sales," *Foreign Policy* (1977): 11–13.

[80]See David Albright and Corey Gay, "Taiwan: Nuclear Nightmare Averted," *Bulletin of the Atomic Scientists* (July–Feb. 1998): 54–60.

Regional versus Global Approaches

The policy response that may be most appropriate in moderating the latest nuclear arms race in one region may have precisely the opposite effect in other regions. The logic that "the last culprit may have to be aided as much as punished" would be more compelling if one did not have to worry about how such assistance might alter the balance of perceived proliferation incentives and disincentives for other potential proliferators.[81] After the muted international response to the 1998 Indian and Pakistani tests, senior South African and Ukrainian diplomats argued forcefully that they could not accept a situation in which their states forswore nuclear weapons only to see additional states recognized as Nuclear Weapon States (NWS) by the international community. Also telling were the May 1998 remarks of a high-ranking Japanese official, that when Japan chose to adhere to the NPT, it did so with the clear understanding that the international community would not recognize any additional NWS. If that assumption was misplaced, Japan would have to rethink the role of the NPT in its security calculation.[82]

Future Trends

Recent years have not been the best of times for nuclear nonproliferation. Although the NPT remains the most widely subscribed treaty, with 188 states as parties, the long-term viability of the treaty is by no means ensured, as evidenced by the reopening of debates about the value of the NPT in a number of capitals during 1998. As Figure 6.2 suggests, the balance of proliferation incentives and disincentives for individual countries is not static, but subject to change in response to new international and domestic developments. Our examination of nuclear decision-making in twenty-six states reveals the recurrent tension in many countries between nonproliferation objectives and other foreign policy goals.

Significant differences also continue to exist among states about the importance of nonproliferation objectives relative to other domestic and foreign policy priorities. These differences have become increasingly noticeable in the post–Cold War policies of the United States and Russia, which have an unusual history of cooperation for nonproliferation. Regrettably, at a time when the global non-proliferation regime is under siege because of developments in South Asia, Iraq, and North Korea, a key cornerstone of the regime has begun to crumble as both

---

[81]The quotation is from George Quester, "Introduction: In Defense of Some Optimism," *International Organization* 35, no. 1 (winter 1981): 11.

[82]Remarks by participants at the Workshop on the Outcome and Implications of the Second Preparatory Committee Session for the 2000 NPT Review Conference, Southampton, England (29–31 May 1998).

Russia and the United States have increasingly emphasized short-term economic and political considerations over nonproliferation objectives.

It remains to be seen whether or not the U.S.-Russian nonproliferation partnership can be restored. It also is premature to predict either the near-term occurrence of defections from the NPT or the long-term integrity of the treaty. What can be said with confidence is that unless we improve greatly our understanding of nuclear decision-making, we are likely to be as surprised about the emergence of the next nuclear weapons state as we were when South Africa announced its nuclear weapons pedigree or when India and Pakistan tested.

References

Agrell, Wilhelm. "The Bomb That Never Was: The Rise and Fall of the Swedish Nuclear Weapons Programme." In *Arms Races: Technological and Political Dynamics*, ed. Els Petter Gleditsch and Olar Njolstad, 154–74. London: PRIO/Sage Publications, 1990.

Ahmed, Samina. "Pakistan's Nuclear Weapons Program: Turning Points and Nuclear Choices." *International Security* (Spring 1999): 178–204.

Albright, David. "Nuclear Rollback: Understanding South Africa's Denuclearization Decision." In *Pulling Back from the Nuclear Brink*, ed. Barry R. Schneider and William L. Dowdy, 67–79. Portland: Frank Cass Publishers, 1998.

Albright, David, and Corey Gay. "Taiwan: Nuclear Nightmare Averted." *The Bulletin of the Atomic Scientists* (Jan./Feb. 1998): 54–60.

Barletta, Michael. "The Military Nuclear Program in Brazil." CISAC Working Paper, Stanford University, Aug. 1997.

Barnaby, C. F. "The Development of Nuclear Energy Programs." In *Preventing the Spread of Nuclear Weapons*, ed. Barnaby, 16–35. London: Souvenir, 1969.

Beaton, Leonard, and John Maddox. *The Spread of Nuclear Weapons*. New York: Praeger, 1962.

Betts, Richard K. "Incentives for Nuclear Weapons." In *Nonproliferation and U.S. Foreign Policy*, ed. Joseph A. Yager, 116–44. Washington, D.C.: Brookings Institution, 1980.

Betts, Richard K. "Paranoids, Pygmies, Pariahs and Nonproliferation Revisited." In *The Proliferation Puzzle*, ed. Zachary S. Davis and Benjamin Frankel, 100–127. London: Frank Cass, 1993.

Bull, Hedley, *The Control of the Arms Race*. New York: Praeger, 1961.

Carasales, Julio C. "The Argentine-Brazilian Nuclear Rapprochement," *The Nonproliferation Review* (Spring–Summer 1995): 39–48.

Chari, P. R., et al. *Nuclear Non-Proliferation in India and Pakistan*. New Delhi: Manohar, 1996.

Chellaney, Braha. "India." In *Nuclear Proliferation after the Cold War*, ed. Mitchell Reiss and Robert Litwak, 165–205. Washington, DC: Woodrow Wilson Center Press, 1994.

Chubin, Shahram. "The Middle East." In *Nuclear Proliferation After the Cold War*, ed. Mitchell Reiss and Robert Litwak, 33–66. Washington, DC: Woodrow Wilson Center Press, 1994.

Chubin, Shahram. "Does Iran Want Nuclear Weapons?" *Survival* (Spring 1995): 86–104.

Cohen, Avner. *Israel and the Bomb*. New York: Columbia University Press, 1998.

Cole, Paul M. "Atomic Bombast: Nuclear Weapon Decision-Making in Sweden, 1946–72." *The Washington Quarterly* (Spring 1997): 233–51.

Dunn, Lewis A. *Controlling the Bomb*. New Haven: Yale University Press, 1982.

Dunn, Lewis A. "On Proliferation Watch: Some Reflections on the Past Quarter Century." *The Nonproliferation Review* (Spring/Summer 1998): 78–83.

Dunn, Lewis, and Herman Kahn. *Trends in Nuclear Proliferation, 1975–1995*. Crotonon-Hudson, New York: Hudson Institute, 1976.

Epstein, William. "Why States Go—and Don't Go—Nuclear." *The Annals of the American Academy of Political and Social Science* (Mar. 1977): 16–28.

Fischer, David. *Stopping the Spread of Nuclear Weapons*. London: Routledge, 1992.

Flank, Steven. "Exploding the Black Box: The Historical Sociology of Nuclear Proliferation." *Security Studies 3*, no. 2 (Winter 1993/94).

Ford/Mitre Study. *Nuclear Power Issues and Choices*. Cambridge: Ballinger, 1977.

Forland, Astrid. "Norway's Nuclear Odyssey: From Optimistic Proponent to Nonproliferator." *The Nonproliferation Review* (Winter 1997): 1–16.

Frankel, Benjamin. "The Brooding Shadow: Systemic Incentives and Nuclear Weapons Proliferation." In *The Proliferation Puzzle*, special issue of *Security Studies 2*, no. 3/4, ed. Zachary S. Davis and Benjamin Frankel (Spring/Summer 1993).

Ganguly, Sumit. "India's Pathway to Pokhran II: The Prospects and Sources of New Delhi's Nuclear Weapons Program." *International Security* (Spring 1999): 148–77.

Goldstein, Avery. "Understanding Nuclear Proliferation: Theoretical Explanation and China's National Experience." In *The Proliferation Puzzle*, special issue of *Security Studies 2*, no. 3/4 , ed. Zachary S. Davis and Benjamin Frankel (Spring/Summer 1993).

Ghosh, Amitav. "Countdown." *New Yorker* (Oct. 26–Nov. 2, 1998): 186–97.

Greenwood, Ted. "Discouraging Proliferation in the Next Decade and Beyond." In *Nuclear Proliferation: Motivations, Capabilities, and Strategies for Control*, Greenwood et al., 25–122. New York: McGraw-Hill, 1977.

Hamza, Khidhir (with Jeff Stein). *Sadam's Bombmaker*. New York: Scribner, 2000.

Harkavy, Robert. *Israel's Nuclear Weapons: Spectre of Holocaust in the Middle East*. Denver: University of Denver Press, 1977.

Haselkorn, Avigdor. "Israel—An Option to a Bomb in the Basement." In *Nuclear Proliferation Phase II*, ed. Robert Lawrence and Joel Larus. Lawrence: University Press of Kansas, 1974.

Hersh, Seymour M. *The Sampson Option: Israel's Nuclear Arsenal and American Foreign Policy*. New York: Random House, 1991.

Holloway, David. *Stalin and the Bomb*. New Haven: Yale University Press, 1994.

Jha, Prem Shankar. "Why India Went Nuclear." *World Affairs* (July–Sept. 1998): 80–97.

Jervis, Robert. "The Symbolic Nature of Nuclear Politics." In *The Meaning of the Nuclear Revolution*. Ithaca, N.Y., Cornell University Press, 1989, pp. 174–225.

Joeck, Neil. *Maintaining Nuclear Stability in South Asia*. Adelphi Paper No. 312, International Institute for Strategic Studies, Sept. 1997.

Joeck, Neil H. A. "Nuclear Proliferation and National Security in India and Pakistan," unpublished dissertation, UCLA, 1986.

Jonter, Thomas. *Sweden and the Bomb*. Uppsala: Swedish Nuclear Power Inspectorate, 2001.

Kampani, Gaurav. "From Existential to Minimal Deterrence." *The Nonproliferation Review* (Fall 1998): 12–24.

Kapur, Ashok. *International Nuclear Proliferation*. New York: Praeger, 1979.

Lavoy, Peter R. "Nuclear Myths and the Causes of Nuclear Proliferation." In *The Proliferation Puzzle*, ed. Zachary S. Davis and Benjamin Frankel, 192–212. London: Frank Cass, 1993.

Lefever, Ernest. *Nuclear Arms in the Third World*. Washington, D.C.: Brookings Institution, 1979.

Lewis, John Wilson, and Xue Litai. *China Builds the Bomb*. Stanford: Stanford University Press, 1988.

Müller, Harald. "Maintaining Non-Nuclear Weapon Status." In *Security With Nuclear Weapons?* ed. Regina Cowen Karp. New York: Oxford University Press, 1991.

Müller, Harald. "The Internationalization of Principles, Norms, and Rules by Governments: The Case of Security Regimes." In *Regime Theory and International Relations*, ed. Volker Rittberger. Oxford: Clarendon Press, 1995.

Nazir, Kamal. "Pakistani Perceptions and Prospects of Reducing the Nuclear Danger in South Asia." *CMC Occasional Paper No. 6*, Sandia National Laboratory. Jan. 1999.

Office of Technology Assessment. *Nuclear Proliferation and Safeguards*. New York: Praeger, 1977.

Pabian, Frank. "South Africa's Nuclear Weapon Program." *The Nonproliferation Review* (Fall 1995): 1–19.

Paul, T.V. "The Systemic Basis of India's Challenge to the Global Nuclear Order." *The Nonproliferation Review* (Fall 1998): 1–11.

Perkovich, George. *India's Nuclear Bomb*. Berkeley: University of California Press, 1999.

Perkovich, George. "Indian Nuclear Decision Making and the 1974 PNE." Unpublished manuscript, W. Alton Jones Foundation, Charlottesville, Va., 1996.

Perkovich, George. "Nuclear Proliferation." *Foreign Policy* (Fall 1998): 12–23.

Potter, William C. "Nuclear Export Policy: A Soviet-American Comparison." In *Foreign Policy: USA/USSR*, ed. Charles W. Kegley, Jr. and Pat McGowan, 291–313. Beverly Hills: Sage Publications, 1982.

Potter, William C. *Nuclear Power and Nonproliferation: An Interdisciplinary Perspective*. Cambridge: Oelgeschlager, Gunn, & Hain, 1982.

Potter, William C. "The Politics of Nuclear Renunciation: The Cases of Belarus, Kazakhstan, and Ukraine." Stimson Center Occasional Paper No. 22 (Apr. 1995).

Potter, William C., Djuro Miljanic and Ivo Slaus. "Tito's Nuclear Legacy." *The Bulletin of the Atomic Scientists* (Mar.–Apr. 2000): 63–70.

Quester, George. *The Politics of Nuclear Proliferation*. Baltimore: The Johns Hopkins Press, 1973.

Redick, John R. "Factors in the Decisions by Argentina and Brazil to Accept the Nonproliferation Regime." In *Pulling Back from the Nuclear Brink*, ed. Barry R. Schneider and William L. Dowdy, 67–79. Portland: Frank Cass Publishers, 1998.

Reiss, Mitchell. *Bridled Ambition: Why Countries Constrain Their Nuclear Capabilities*. Washington, D.C.: Woodrow Wilson Center Press, 1995.

Reiss, Mitchell. *Without the Bomb: The Politics of Nuclear Nonproliferation*. New York: Columbia University Press, 1988.

Rosecrance, Richard. "International Stability and Nuclear Diffusion." In *The Dispersion of Nuclear Weapons*, ed. Rosecrance. New York: Columbia University Press, 1964.

Sagan, Scott D. "Why Do States Build Nuclear Weapons? Three Models in Search of a Bomb." *International Security* (Winter 1996/97): 54–85.

Scheinman, Lawrence. *Atomic Energy Policy in France Under the Fourth Republic*. Princeton: Princeton University Press, 1964.

Schoettle, Enid C. B. "Arms Limitations and Security Policies Required to Minimize the Proliferation of Nuclear Weapons." In *Arms Control and Technological Innovation*, ed. David Carlton and Carlo Schaerf, 102–31. New York: Halsted Press, 1976.

Schwab, G. "Switzerland's Tactical Nuclear Weapon Policy." *Orbis*, Fall 1969: 900–914.

Schwartz, Stephen I., ed. *Atomic Audit*. Washington, D.C.: The Brookings Institution, 1998.

Singh, Jaswant. *Defending India*. New York: St. Martin's Press, 1999.

Solingen, Etel. "The Political Economy of Nuclear Restraint." *International Security* 19, no. 2 (Fall, 1994): 126–69.

Spector, Leonard (with Jaqueline R. Smith). *Nuclear Ambitions*. Boulder, Colo.: Westview Press, 1990.

Steiner, Arthur. *Canada: The Decision to Forego the Bomb*. Los Angeles: Pan Heuristics, 1977.

Steiner, Arthur. *Great Britain and France. Two Other Roads to the Atomic Bomb*. Los Angeles: Pan Heuristics, 1977.

Steussi-Lautenberg. *Historical Abstract on the Question of Swiss Nuclear Arms Program*. Draft, 1995.

Stumpf, Waldo. "South Africa's Nuclear Weapons Program: From Deterrence to Dismantlement." *Arms Control Today* 25, no. 10 (Dec. 1995/Jan. 1996): 4.

Subrahmanyam, K. "Nuclear India in Global Politics." *World Affairs* (July–Sept. 1998): 12–41.

Suchman, Marc C., and Dana P. Eyre. "Military Procurement as Rational Myth: Notes on the Social Construction of Weapons Proliferation," *Sociological Forum* 7, no. 1 (Mar. 1992): 137–61.

Thayer, Bradley A. "The Causes of Nuclear Proliferation and the Nonproliferation Regime." *Security Studies* 4, no. 3 (Spring 1995): 463–519.

Timerbaev, Roland. *Rossiya i yadernoye nerasprostraneniye, 1945–1968 (Russia and Nuclear Nonproliferation, 1945–1968)*. Moscow: Nauka, 1999: 127–35.

Van Dassen, Lars. "Sweden and the Making of Nuclear Non-Proliferation: From Indecision to Assertiveness." *SKI Report 98:16, Statens Kärnkraftinspektion*. Mar. 1998.

Vaziri, Haleh. "Iran's Nuclear Quest: Motivations and Consequences." In *The Nuclear Non-Proliferation Regime*, ed. Raju Thomas, 310–29. New York: St. Martin's Press, 1998.

Walsh, Jim. "Surprise Down Under: The Secret History of Australia's Nuclear Ambitions." *The Nonproliferation Review* (Fall 1997): 1–20.

Willrich, Mason. "Guarantees to Non-Nuclear Nations." *Foreign Affairs* (July 1966): 683–92.

Wiseman, Geoffrey, and Gregory Treverton. "Dealing with the Nuclear Dilemma in South Asia." Pacific Council Issue Brief, Pacific Council on International Policy (Nov. 1998).

Wohlstetter, Albert et al. *Swords from Plowshares*. Chicago: University of Chicago Press, 1979.

CHAPTER SEVEN

Revolution and Counter-Revolution

The Role of the Periphery in Technological and Conceptual Innovation

TIMOTHY D. HOYT

Most studies of military innovation focus on the impact of revolutionary changes in military capability resulting from great power conflict.[1] Military "revolution" among the "core states" tends to define critical periods in history, and global changes in economic and political dominance. The path of diffusion (innovator to adaptors; core to periphery) appears clear.[2] This chapter investigates four cases of military innovation by peripheral states in the Cold War. Peripheral innovation and diffusion challenges the prevailing theoretical model of military diffusion, which is based on neorealism. According to Kenneth Waltz, "Contending states imitate the military innovations contrived by the country of greatest capability and ingenuity. And so the weapons of the major contenders, and even their strategies, begin to look much the same over the world."[3] Lesser

[1] For example, see *The Military Revolution Debate*, ed. Clifford J. Rogers (Boulder, CO: Westview, 1995); Jeremy Black, *European Warfare 1660–1815* (New Haven: Yale University Press, 1994); Carlo M. Cipolla, *Guns, Sails, and Empires* (New York: Minerva, 1965); William H. McNeill, *The Pursuit of Power: Technology, Armed Forces, and Society* (Chicago: University of Chicago, 1982); and, in a slightly different perspective, Keith Krause, *Arms and the State: Patterns of Military Production and Trade* (Cambridge: Cambridge University Press, 1992).

[2] The concept of a world divided into core and peripheral states is widely accepted in the study of international relations, political economy, and in studies of the arms trade and militarism. See Immanuel Wallerstein, *The Modern World System*, vols. I and II (New York: Academic, 1974, 1980); Barry Buzan, *People, States, and Fear*, 2d ed. (Boulder, CO: Lynne Rienner, 1991); Mohammed Ayoob, *The Third World Security Predicament: State Making, Regional Conflict, and the International System* (Boulder, CO: Lynne Rienner, 1995); Mary Kaldor and Asbjorn Eide, *The World Military Order: The Impact of Military Technology on the Third World* (London: Macmillan, 1979).

[3] Kenneth N. Waltz, *Theory of International Politics* (New York: McGraw-Hill, 1979), p. 127. Waltz argues that the driving force in the diffusion of military technology and practice is the structure of the

states must imitate the greatest powers—failure to do so leads to insignificance or conquest.

Cases of peripheral innovation and diffusion challenge this neorealist view because the state that initiates the diffusion process is not a great power. In fact, the peripheral state rejects imitation of the greatest powers and seeks a different path. These cases expose the importance of nonsystemic, intrastate, or unit-level forces—leaders, industrialists, bureaucracies, and organizations—in the process of creating, demonstrating, and diffusing new military capabilities. The peripheral experience demonstrates that not all diffusion flows from the industrialized core to the developing periphery. Peripheral states develop capabilities, which elicit responses from other peripheral states both within and outside their immediate security environment, as well as from the major international powers.

Conflicts between peripheral states differ significantly from great power conflicts. The resources available to the combatants constrain the geographic scope of conflict, so these wars rarely spill into adjacent regions. Despite these limitations, conventional conflicts in the periphery have been sophisticated in terms of military operations and technology, and have also been surprisingly bloody.[4] Peripheral innovation and diffusion take place both as a response to developments within the region and as a response to import of concepts or capabilities from abroad, usually from core states. It provokes responses from immediately affected regional neighbors and, in some cases, from the major powers.

This chapter examines four recent cases of diffusion from the periphery: two each from Israel and Iraq. Both began the Cold War as classic "Third World" economies, heavily dependent on a single commodity for export and lacking industrial and economic infrastructure.[5] By 1990 each had carved a special niche for itself in the international community, based in part on the experience of costly and sophisticated regional conflict. These conflicts were fought with a mix of weapons and tactics diffused from the core and developed in response to local requirements.

Each case has had significant impact on periphery and core states. Diffusion has occurred primarily as a result of experience with or observation of military effectiveness in regional combat. In the Israeli case, development of the fast at-

---

international system, which is anarchical and forces states to rapidly copy "best practices" initiated by the leading powers.

[4] An excellent study of warfare in the periphery is Eliot A. Cohen, "Distant Battles," *International Security* 10 (spring 1986): 143–71.

[5] Israel's economy, in 1948, depended on export of citrus fruits and diamonds: virtually all manufactured and industrial goods were imported. Michael Barnett, *Confronting the Costs of War: Military Power, State, and Society in Egypt and Israel* (Princeton: Princeton University Press, 1992): pp. 14–17, 67–78. Iraq, obviously, continues to depend upon oil as the primary source of revenue. Both states also have national borders imposed by colonial powers.

tack craft/antiship missile combination and development of the remotely piloted vehicle feature the creation of "niche technologies" in response to specific regional military requirements. In the Iraqi case, use of chemical weapons and ballistic missiles for war and deterrence demonstrate conceptual innovation based on the creative use of technologies and capabilities already diffused from the core.[6]

The Israeli Navy, 1956–1973: Simultaneous Innovation in Core and Periphery

In 1973, Israel and the Arab navies fought the first naval engagements in which both sides used guided antiship missiles. The Arab navies were equipped with fast missile-armed attack craft (FACMs) developed by the Soviet Union in the late 1950s and early 1960s. Israel simultaneously designed a series of FACMs based on West German torpedo boat designs armed with indigenously developed and produced antiship missiles—the first deployed by a Western power. In addition to leading all major Western powers in development of this new weapon, Israel deployed it on a specialized vessel and developed new tactics, organizations, and doctrine for its naval force. These new capabilities proved decisive in the 1973 conflict and spurred the acquisition of similar capabilities by all major navies. The antiship missile currently constitutes one of the most important weapons in naval arsenals, and dozens of peripheral countries rely on FACM-based forces to defend their coasts and maritime interests.

The Israeli Navy (IDFN) has always been the smallest of Israel's three military services. The role envisaged for the IDFN by the architects of Israeli defense doctrine was small: "The navy was not to be used primarily for control of the open sea or to protect convoys, but was to be confined to protecting the coast, launching small landings on the enemy's beaches, and carrying out special operations against maritime targets."[7]

In the early 1950s trends in the regional naval balance were not positive for Israel. Egypt signed an arms agreement with Czechoslovakia (and, effectively, the USSR) in 1955, ensuring a steady flow of modern Soviet weapons and equipment for the future.[8] Moreover, the guns on Egypt's modern *Skory*-class destroyers out-

---

[6]By "conceptual innovation," I mean actions that result in a diffusion effect—imitation, countermeasure, or counterinnovation—based on existing or prediffused technologies. This can occur as a result of the development of new methods of using existing weapons or through a long time lag between diffusion and actual use. In the latter case, the "innovation" that begins the diffusion effect is actually the surprise caused by the use of a long-neglected capability or the violation of a widely accepted norm prohibiting its use.

[7]Yigal Allon, *The Making of Israel's Army* (New York: Universe, 1970), pp. 51–52.

[8]By 1967, for instance, the Egyptian Navy had seven destroyers, eleven submarines, eighteen missile boats, and other light craft. Yitzhak Rabin, *The Rabin Memoirs* (Boston: Little, Brown, 1979), p. 100.

ranged the weapons on Israel's obsolescent surface combatants.[9] In the eyes of some analysts the weakness of the IDFN imperiled Israel's ability to deter adversaries; in the event of an Arab naval blockade, Israel had no choice but to begin major land and air operations against the blockading states.[10]

IDFN force structure was based on two ex-British Z-class destroyers and a handful of motor torpedo boats.[11] There were insufficient numbers of ships to defend Israel's coastline. The torpedo boats were not effective against larger surface ships, and each destroyer had a crew of about 200 men. Loss of a destroyer would be a national catastrophe: in the entire 1956 Sinai campaign, for example, Israel had suffered 190 dead.[12] Two of Israel's three major population centers (Tel Aviv, Haifa) lay on the eastern edge of the Mediterranean Sea, vulnerable to attacks from Arab navies. As the Sinai Campaign demonstrated, even with the help of two of the largest and most capable fleets in the world (Britain and France), Israel's shores were not safe. On 31 October 1956 an Egyptian frigate bombarded Tel Aviv,[13] exhibiting the inability of the IDFN to defend Israel's coastline.

Financial constraints forced the IDFN to consider a variety of unorthodox options in an effort to find a role that the navy could perform effectively. These options included the establishment of an "all-commando" navy, limited to daring subsurface raids, or elimination of the surface force and concentration on submarines.[14] The latter option would have required the purchase of new vessels, in addition to significant expenditures for training crews and retraining existing personnel.

The IDFN reassessed its force structure, establishing requirements derived from Israel's short war doctrine and the IDFN's constrained role and resources in overall defense planning.[15] Israel required smaller naval units (to cut down on crew size, and allow procurement of more vessels) with high speed, large cruising radii (to enable ships to stay at sea for extended periods), and sufficient fire power to engage and defeat larger surface ships beyond the weapons range of

[9]Stewart Reiser, *The Israeli Arms Industry* (New York: Holmes and Meier, 1989), p. 62; Abraham Rabinovich, *The Missile Boats of Cherbourg* (New York: Henry Holt, 1988), p. 23.

[10]Edward Luttwak and Dan Horowitz, *The Israeli Army* (New York: Harper and Row, 1975), p. 133.

[11]Ibid.

[12]Eliot A. Cohen, Michael J. Eisenstadt, and Andrew J. Bacevich, *Knives, Tanks and Missiles: Israel's Security Revolution* (Washington, DC: Washington Institute for Near East Policy, 1998), p. 149.

[13]Chaim Herzog, *The Arab-Israeli Wars* (New York: Vintage, 1984), p. 138.

[14]Rabinovich, *The Missile Boats of Cherbourg*, p. 27.

[15]Writings on Israeli military doctrine include Yoav Ben-Horin and Barry Posen, *Israel's Strategic Doctrine*, RAND R-2845-NA (Santa Monica, CA: RAND Corporation, September 1981); Allon, *The Making of Israel's Army*, pp. 27–54, 62–71, 96–108; Michael Handel, "The Evolution of Israeli Strategy: The Psychology of Insecurity and the Quest for Absolute Security," in *The Making of Strategy*, ed. Williamson Murray, MacGregor Knox, and Alvin Bernstein (Cambridge: Cambridge University Press, 1994), pp. 534–78. Israeli strategy is driven by requirements to limit casualties, fight on enemy territory if possible, win quickly, and win decisively.

their adversaries.[16] It needed a design which, if necessary, could be manufactured in Israel, since supplies of foreign arms could not be assured. With such a force, the IDFN could protect the coast and carry out an aggressive campaign against an Arab coalition.

The IDFN therefore began what was then the most expensive, largest, and most complex undertaking in the history of Israeli military industry.[17] Financed in part by German reparations payments, Israel arranged for the manufacture of German *Jaguar*-class 250-ton patrol boats in France. These boats could, eventually, be manufactured at Israeli Shipyard Industries in Haifa. The *Jaguar*, originally designed to launch long-range torpedoes, was modified to Israeli specifications and was renamed the *Sa'ar* class.[18]

This procurement was coordinated with the Israel military's research and design bureau, Emet (later renamed RAFAEL), which was experimenting with new guided weapons throughout the 1950s. In 1952 a wooden glide-bomb named the Luz was developed for both air-to-sea and ship-to-ship use, as well as a series of radio-controlled sabotage boats called the G-11 series.[19] Absence of a reliable guidance system made the Luz an ineffective weapon. As late as 1962, tests of the ship-to-ship version of the Luz produced no hits, and some misses of up to a kilometer.[20]

These early experiments led to the development of the Gabriel ship-to-ship missile (ShShM), the first of its kind in the Western world. An independently developed inertial guidance system allowed the missile to fly close to the water to avoid detection.[21] In addition to the vessel and the weapon, new command procedures and training were required, and new electronic warfare suites were designed. According to Israeli reports, the combat information centers on the new *Sa'ar*-class fast attack craft were more complicated than those found in cruisers in the larger navies in the early 1960s. The IDFN also tripled the number of men passing through the officers training course and changed watch schedules from four to three hours in response to the complexity of these sophisticated new technologies.[22]

---

[16]Cmdr. Eli Rahav, IDFN (Ret'd), "Missile Boat Warfare: Israeli Style," *Naval Institute Proceedings* (March 1986): 108.

[17]Rabinovich, *The Missile Boats of Cherbourg*, p. 57.

[18]Rahav, "Missile Boat Warfare: Israeli Style," p. 110. Contemporary Soviet Type 183 patrol boats were inferior to the *Jaguar*. V. P. Kuzin and V. I. Nikol'skiy, *The Soviet Navy 1945–1991* (St. Petersburg: Historical Naval Society, 1996), p. 273. The Type 183 was later modified to become a FACM.

[19]Reiser, *The Israeli Arms Industry*, pp. 24, 61; Rabinovich, *The Missile Boats of Cherbourg*, p. 23.

[20]Reiser, *The Israeli Arms Industry*, pp. 61–62; Rabinovich, *The Missile Boats of Cherbourg*, p. 32.

[21]Gerald M. Steinberg, "Israel: Case Study for International Missile Trade and Non-Proliferation," in *The International Missile Bazaar*, ed. William C. Potter and Harlan W. Jencks (Boulder, CO: Westview, 1994), p. 235.

[22]Rabinovich, *The Missile Boats of Cherbourg*, pp. 57, 60, 179.

During the development of the *Sa'ar*-Gabriel combination, arms transfers to the Middle East region helped confirm the IDFN's assumption that missiles were the naval weapon of the future. The first *Komar*-class missile boats, equipped with Styx ShShMs, were delivered to Egypt in 1962, and the first *Osa*-class boat was delivered in 1966. The new Styx missile on these ships had more than twice the range of the Gabriel, still under development. A decision was made to proceed with the project and field an actual weapon, rather than scrap the existing research, start a new development cycle, and further postpone deployment.[23]

The sinking of the IDFN destroyer *Eilat* by an Egyptian-fired Styx in October 1967 (with forty-three dead and more than one hundred other casualties), and the later sinking of the fishing boat *Orit* off El Arish in May 1970 only confirmed the IDFN's commitment to ShShMs. Ironically, the IDFN, which had actively been preparing for missile war at sea for a decade, received the first of its new *Sa'ar* boats the day after the *Eilat* incident.[24] This was fortunate, as the successes of the 1967 war had nearly quadrupled Israel's coastline and vastly increased the IDFN's responsibilities.[25]

The entire project was closely coordinated with Israeli industry, to ensure that Israel could not only support but also repair, maintain, and, if necessary, replace the new vessels. The Six Day War, and the later raid on Beirut Airport, led to a French arms embargo on Israel: the last five *Sa'ar* boats were "stolen," with the tacit assistance of French authorities, from their internment in Cherbourg harbor on Christmas Day in 1969 and sailed to Israel. Future modifications of the *Sa'ar*, including the improved *Reshef-* and *Aliya*-class ships, would be in the hands of Israeli Shipbuilding Industries.

The unit cost of the *Sa'ar* was approximately $5 million: Israel could buy eight for the cost of a single destroyer. Development of the Gabriel missile, facilitated by contacts with Western guidance experts, took eight years and approximately $11 million. The Israeli Navy not only found an innovative solution for a serious military dilemma; it also found it at a cost that was affordable, and ensured that the capability could be manufactured and maintained locally.[26]

By 1973, Israel had integrated the new forces into the IDFN and developed appropriate doctrine and tactics to maximize their effectiveness. Despite the longer

---

[23]Reiser, *The Israeli Arms Industry*, p. 62. The range of the Styx was later increased to 80 km. The first deployments of FACMs in the Soviet fleet occurred in 1959 (Type 183, or *Komar*). *Osa*-class FACMs entered Soviet service the next year. Kuzin and Nikol'skiy, *The Soviet Navy 1945–1991*, pp. 277–78.

[24]Rahav, "Missile Boat Warfare: Israeli Style," p. 109.

[25]Luttwak and Horowitz, *The Israeli Army*, pp. 316, 395.

[26]Rabinovich, *The Missile Boats of Cherbourg*, pp. 67, 73. Funding for the *Sa'ar* boats came largely from German war reparations: more than 38 percent of all German payments came in the form of ships, machinery, and other heavy equipment. Reiser, *The Israeli Arms Industry*, p. 30; Barnett, *Confronting the Costs of War*, p. 64.

range of the Styx missile supplied to Arab forces, the IDFN sank fifteen Arab ships in the October War of 1973, at least eight through missile attack, and suffered no losses. Arab forces fired more than fifty Soviet Styx missiles, with a range twice that of the Gabriel, in encounters with the IDFN, but none hit. IDFN tactics emphasized closing the range with enemy ships as quickly as possible, and Israeli defense industries had developed a range of countermeasures to neutralize the capabilities of the Arab Styx.[27]

The results of the October War encouraged other developing countries to pursue coast-defense fleets of missile attack craft, and spurred development of new generations of antiship cruise missiles by both developing and developed arms producers. Some peripheral states imitated Israel's technology and tactics.[28] Others opted for less expensive (and usually inferior) Soviet equipment. A sea denial force based on missile-armed fast attack craft provided peripheral states with the ability to menace great power intervention forces cheaply and effectively.

Israeli innovation spurred countermeasures in the core countries. Faced with a significantly improved surface option for sea denial, these navies developed new equipment and tactics to minimize the threat posed by FACM-based coast defense forces. The use of sea-skimming missiles, in particular, posed a particular threat, prompting the development of automated point defense systems such as PHALANX (United States) and NULKA (Russia), and an upgrading of air defenses in general. Western and Soviet bloc navies accelerated development of their own longer-ranged, more sophisticated ShShMs, including the Harpoon (which has been adopted by the IDFN, thanks to alliance-related diffusion from the United States). Western navies also adopted new methods to counter the threat of FACMs, which proved their effectiveness in the Gulf War. These included Britain's use of Sea Skua missiles mounted on Lynx helicopters to engage FACMs before they came into missile range of the frigates that carried the helicopters.[29] Significantly, despite a fleet based on the FACM sea-denial concept and batteries of shore-launched antiship missiles, no Coalition vessel was hit by Iraqi antiship missiles during the Gulf War.

Innovation, in this case, was driven by security threats, and the innovation diffused to both core and periphery as a result of demonstrated effectiveness in

---

[27]According to Rabinovich, these included chaff, detection gear, deception apparatus, decoys, and an array of jammers. *The Missile Boats of Cherbourg*, p. 218.

[28]Exports of the Gabriel amounted to more than $1 billion, more than ten times the cost of Gabriel development and deployment. Gerald M. Steinberg, "Indigenous Arms Industries and Dependence: The Case of Israel," *Defence Analysis* 2 (December 1986): 304. Taiwan and South Africa, which suffered the same kinds of international diplomatic isolation that Israel did in the 1970s, were particularly eager to adopt Israeli equipment and methods. See Robert E. Harkavy, "The Pariah State System," *Orbis* 21 (fall 1977): 623–49, for a discussion of these links.

[29]Norman Friedman, *Desert Victory* (Annapolis, MD: Naval Institute Press, 1991), pp. 191, 209–10.

combat. Core states responded by selective emulation and adaptation (their own ShShMs) and a series of counterefforts to limit the effect of the new capability (both defensive technologies and new methods). Peripheral states responded by partial or total emulation from multiple suppliers.

Remotely Piloted Vehicles: Innovation in Technological Niches

Israel is one of the first states to develop and use remotely piloted vehicles (RPVs) and unmanned aerial vehicles (UAVs) to provide near–real time battle-field and operational intelligence. The use of RPVs and UAVs in the 1982 war in Lebanon demonstrated their utility as a force multiplier, providing many of the services of manned reconnaissance aircraft at much lower risk and cost. Israel demonstrated the ability to develop and field a new system more quickly and in-expensively than a superpower—the United States. Israeli firms remain the world leaders in this technology, licensing and selling their products to major peripheral powers (India) and to the armed forces of the United States.

The first major Israeli remotely piloted vehicle program was initiated by a private firm, Tadiran, after the successful use of a number of U.S. target drones to deceive Arab air defenses in the October War of 1973. Other states were experimenting with RPVs at the time. The United States looked on RPV technology as an area of considerable promise. Nevertheless, U.S. RPV projects were languishing by 1982: out of 986 RPVs built in the 1960s and 1970s, only 33 remained in U.S. inventory and all those were in storage.[30]

There are numerous advantages of RPVs over manned reconnaissance aircraft. There is no risk of pilot loss—a serious issue for Israel, which has occasionally exchanged dozens or hundreds of captured terrorists to gain the freedom of a single Israeli. RPVs and the more capable unmanned aerial vehicles are much less expensive than aircraft.[31] They are much smaller, decreasing the chance of radar detection. Finally, they are able to loiter in areas more easily than jet aircraft.

RPVs were used with enormous success in the 1982 conflict in Lebanon. Israel used Mastiffs and Scouts as part of a sophisticated combined arms attack against Syrian air defenses. In addition to providing real-time intelligence for both re-connaissance and battle damage assessment, they were used as airborne laser designators for aircraft-carried smart weapons, and may also have carried cluster

---

[30]Benjamin F. Schemmer, "Where Have All The RPVs Gone," *Armed Forces Journal International* (Feb.. 1982): 38.

[31]In 1986, for example, a complete system with five RPVs, spare parts, training, and auxiliary equipment cost less than $5 million. Michael Cieply, "Paper Planes," *Forbes* (26 Sept. 1983): 34, cited in Steven L. Spiegel, "U.S. Relations with Israel: The Military Benefits," *Orbis* (fall 1986): 486.

bombs. They may have been used in tandem with battlefield artillery rockets as well.[32] Their demonstrated effectiveness was undeniable, and within two years the U.S. Navy had purchased Tadiran's Mastiff.[33]

Compared with Israel's cheap and efficient Scout and Mastiff RPVs, the U.S. system then in development was gold-plated. The Aquila drone, canceled in the mid-1980s, was anticipated to cost $2.17 billion to develop. By contrast, Tadiran developed the Mastiff for only $15 million.[34] Within several years after the Lebanon conflict, the United States was purchasing and fielding Israeli-designed RPVs, and was involved in joint efforts to develop new systems and integrate existing systems into ground, naval, and amphibious units of the U.S. military. The Mazlat Pioneer was procured for the U.S. Marine Corps, Navy, and Army.[35] The new Hunter drone, a joint venture between IAI and the U.S. firm TRW, won a $200 million drone contract from the U.S. DoD in 1993.[36]

Israel's early RPVs have now evolved into more sophisticated Unmanned Aerial Vehicles, and their new capabilities suggest the potential for significant changes in the organization and force structure of future air forces. Some analysts, for example, have suggested that UAVs in Israeli service will replace manned aircraft for some types of combat roles, including long-range strike missions.[37] Israel exports RPVs to both the United States and developing countries (such as India). Official U.S. reports refer to this niche as an area of substantial Israeli comparative advantage: "Other countries (Israel is a good example) will compete in selected areas, such as remotely piloted vehicles and explosives technology. Indeed, if history is a guide, such countries may be able, because of their small size and focused efforts, to introduce certain technologies earlier than the wealthier but more ponderous superpowers."[38]

Again, innovation was prompted by a local military requirement, in this case spurred by the experience of the effectiveness of thick air defenses on piloted aircraft. As in the case of the ShShM, the technology itself was under development elsewhere, but Israel was able to tailor it to military requirements, and then integrate the new technology into its force structure and demonstrate its effectiveness in combat. The system diffused to both core and periphery because of the dem-

[32]W. Seth Carus, "The Bekaa Valley Campaign," *Washington Quarterly* 5 (autumn 1982): 39–41.
[33]"U.S. Buys Israeli Pilotless Planes," *New York Times*, 24 May 1984.
[34]Spiegel, "U.S. Relations with Israel: The Military Benefits," p. 486.
[35]"IAI: A Partner for the World," *Military Technology* (May 1992): 77.
[36]"IAI Wins $200 Million Drones Contract from US Defense Dep't," *Jerusalem Post*, 17 Feb. 1993.
[37]Cohen, Eisenstadt, and Bacevich, *Knives, Tanks and Missiles*, p. 126. The authors suggest that opposition is likely to be less adamant in the IAF than in other Western militaries.
[38]*The Future Security Environment*, Report of the Future Security Environment Working Group, submitted to the Commission on Integrated Long-Term Strategy (Washington, DC: Pentagon, Oct. 1988), p. 27.

onstration effect and the innovation's significant cost-effectiveness. Israeli RPVs and UAVs have been adopted by the U.S. armed forces, and integrated into several services. This innovation diffused, and continues to diffuse, to the periphery as well. India has directly emulated Israeli technology and practice. Other states, including Iraq, purchased RPVs from other suppliers but have not integrated them into their armed forces with great effectiveness.

Iraq's Use of Chemical Weapons: Adoption of Banned Weapons

During the Iran-Iraq War (1980–88), Iraq routinely used increasingly sophisticated chemical agents on the conventional battlefield, violating international norms and treaties that had held for almost sixty years. While the technologies and techniques that Iraq used were not innovative, the breach of international norms provoked a diffusion effect. Many states accelerated existing or planned chemical weapons (CW) programs, while core states responded by improving defenses and countermeasures and beginning a new series of agreements to strengthen and reinforce the norms and treaties that Iraq had violated.

The first recorded use of chemical weapons occurred in 1915, when German troops released large clouds of chlorine gas on French trench positions. By 1918, more than one million troops and civilians had been injured by chemical agents, and nearly 10 percent of those had died. After World War I, the Geneva Protocol of 1925 was drafted, to ban the use of poison gas in future conflicts. Nevertheless, chemical weapons were used in the 1930s by the Italians in their occupation of Ethiopia, and more recently in the 1960s by Egypt against rebels in Yemen.[39]

Iraq was a signatory to the Geneva Protocol of 1925, banning use of poison gas in wartime. Iraq, however, provided the first new case of verifiable and persistent mass CW use since World War I. Iraq was the first country to use nerve agents in combat.[40] Iraqi interest in chemical weapons dated back to the 1960s, and the basis of the CW industry was established in the Second Five Year Plan (1976–80). A "pesticide" plant began producing mustard gas at Samarra in September 1983.[41]

Estimates report that Iran suffered forty to fifty thousand casualties from chemical weapons during the Iran-Iraq War, including as many as five thousand fatalities. Many of these apparently occurred early in the conflict. Later in the war, Iranians were reportedly experiencing only 3 percent fatalities, despite Iraq's

---

[39]*The Biological and Chemical Warfare Threat* (copies in author's possession, n.d.), p. 23.

[40]*Conduct of the Persian Gulf War: Final Report to Congress* (Washington, DC: Pentagon, Apr. 1992), p. 16.

[41]Kenneth Timmerman, *The Death Lobby: How the West Armed Iraq* (Boston: Houghton Mifflin, 1991), pp. 134–35.

use of more sophisticated nerve agents. Chemical warfare casualties accounted for roughly 5 percent of the approximately one million Iranian casualties during the conflict.[42] Iraq initiated CW use in 1982, with limited amounts of CS gas (a nonlethal riot-control agent) fired against Iranian troop concentrations. Mustard gas, a lethal agent, was first used in 1983, and used in mass quantities during operations in 1984. Nerve agents were first used in 1984, and fired in mass quantities by 1985.[43] Each time Iraq introduced a new form of chemical agent into combat, it took time to integrate it fully into operational and tactical planning. By the end of the Iran-Iraq War, however, the Iraqi military had integrated chemical attacks into both offensive and defensive operations, using both aerial (aircraft and helicopter) and artillery/rocket delivery.

> By the beginning of 1988 the Iraqis had developed an effective offensive doctrine for the use of nerve agents, which fully integrated CW into fire support plans. Both nerve and blister agents were used successfully in the final offensives that defeated the Iranians in 1988. These weapons were targeted specifically against command and control facilities, artillery positions, and logistics areas.[44]

Iraq's chemical weapons manufacturing facilities were abundant and impressive. By 1990 production capability was between one and five thousand tons of all types per year, able to fill some 250,000 to 500,000 munitions annually.[45] By 1990, according to U.S. intelligence, Iraq's CW production capability was the largest in the developing world.[46] Iraq's stockpiles of chemical agents, deliverable via missile warheads, aerial bombs, artillery shells, rockets, and spray tanks, included mustard, Tabun (GA), Sarin (GB), Ricin (a nerve agent), and VX.[47] Other reports mention Sarin (GD) and "binary sarin" (a mixture of the nerve agents GB and GF).[48]

Iraqi chemical and, possibly, biological weapons posed a serious threat to

[42]Michael Eisenstadt, "'Sword of the Arabs': Iraq's Strategic Weapons," *Policy Paper* 21 (Washington, DC: Washington Institute for Near East Policy, 1990), p. 6; Anthony H. Cordesman and Abraham R. Wagner, *The Lessons of Modern War*, vol. 2 (Boulder, CO: Westview, 1990), p. 517.

[43]This account is derived from Stephen C. Pelletiere and Douglas V. Johnson, *Lessons Learned: The Iran-Iraq War* (Carlisle Barracks, PA: Strategic Studies Institute, U.S. Army War College, 1991), p. 97; Stephen C. Pelletiere, Douglas V. Johnson II, and Leif Rosenberger, *Iraqi Power and U.S. Security in the Middle East* (Carlisle Barracks, PA: Strategic Studies Institute, U.S. Army War College, 1990), p. 35; Cordesman and Wagner, *The Lessons of Modern War*, vol. 2, Table 13.4: "Claims of Uses of Chemical Warfare," pp. 508–9; Thomas L. McNaugher, "Ballistic Missiles and Chemical Weapons: The Legacy of the Iran-Iraq War," *International Security* 15 (fall 1990): 5–34.

[44]*Conduct of the Persian Gulf War*, p. 16.

[45]Eisenstadt, "Sword of the Arabs," pp. 5, 7; Anthony H. Cordesman, "Iraq and Weapons of Mass Destruction," *Congressional Record* (8 Apr. 1992): S 5066.

[46]*Conduct of the Persian Gulf War*, p. Q-3.

[47]Anthony H. Cordesman, *Weapons of Mass Destruction in the Middle East* (Washington, DC: Center for Strategic and International Studies, Sept. 1997), pp. 25–26.

[48]Timmerman, *The Death Lobby*, pp. 145–46; UN Press Release IK/27, 24 June 1991.

Coalition plans in the Gulf War. A manual published by the National Training Center at Fort Irwin just before Desert Storm stated:

> The Iraqi CW threat is serious. Iraqi forces are well-prepared for chemical warfare, even in the heat of the desert, and have demonstrated no compunction about using chemical weapons against either military or civilian targets.
>
> Iraqi forces would be highly likely, if not virtually certain, to use chemical weapons in any defensive situation where they were being pushed back. Iraqi forces would be highly likely to use chemical weapons as an integral part of any offensive. Once Iraqi forces begin using chemical weapons, they would probably be willing to use their entire chemical arsenal.[49]

In August 1990 it became apparent that Iraq was moving chemical weapons into southern Iraq and the Kuwaiti Theater of Operations.[50] In discussions of war preparations, U.S. intelligence flatly predicted that chemical weapons would be used.[51] The British advised against attempting to wait until summer to commence offensive operations, in part because of inclement weather and the difficulty of wearing protective clothing.[52]

Even during the war, the threat of Iraqi CW escalation caused Marine commanders to avoid use of nonlethal tear gas near Al Jaber on 25 February 1991.[53] Postwar U.S. analysis indicated several shortcomings in CW and BW preparedness for future conflicts, including the lack of reserve stocks of protective overgarments and shortcomings in existing protective suits for desert use. Collective protection systems were found to be inadequate, as were detection systems and overall BW defense capabilities.[54]

Iraq's use of CW was significant for several reasons. First, the world ignored it for political reasons, undermining a valuable international norm in the Geneva Protocol of 1925. This encouraged, and possibly accelerated, CW programs in other peripheral states for use as both a conventional force multiplier and a

[49]*The Iraqi Army: Organization and Tactics*, National Training Center Handbook 100–91 (Fort Irwin, CA: National Training Center, 3 Jan. 1991), p. 169.

[50]*Gulf War Air Power Survey*, vol. I: *Planning Report* (Washington, DC: Pentagon, 1993), p. 66; US News and World Report, *Triumph Without Victory: The Unreported History of the Persian Gulf War* (New York: Times Books, 1991), p. 124.

[51]Michael Gordon and Gen. Bernard E. Trainor, *The Generals' War: The Inside Story of the Conflict in the Gulf* (Boston: Little, Brown, 1995), p. 375; Rick Atkinson, *Crusade* (New York: Houghton Mifflin, 1993), pp. 86–87, 90. According to Atkinson, the U.S. military did not think CW use would present a major threat to the war plan, but the potential biological weapons threat was deemed serious enough to spur vaccination of 8,000 soldiers against botulinum toxin and 150,000 against anthrax.

[52]Gordon and Trainor, *The Generals' War*, p. 137. General Schwarzkopf addressed a similar concern when he told Secretary of Defense Cheney that the units likely to be deployed to the Gulf had CW training at the National Training Center in the California desert—conditions similar to those found in the Gulf in summer. Bob Woodward, *The Commanders* (New York: Simon and Schuster, 1991), p. 250.

[53]Gordon and Trainor, *The Generals' War*, p. 387.

[54]*Conduct of the Persian Gulf War*, pp. Q-9, Q-11.

"weapon capable of mass destruction" (WMD). The example of a nonindustrial-ized state developing such a massive capability sped diffusion to the periphery.[55]

Iraq's willingness to use CW in battle also prompted concern about Iraqi command and control of other nonconventional weapons, including biological weapons. According to the official DoD reports on the Persian Gulf War, the United States initially believed that authorization for CW use rested solely in the hands of Saddam Hussein.[56] There were concerns, however, about the circum-stances of predelegation to field commanders.[57] The fluid nature of the battle, in-tentionally forced on the Iraqis by the Coalition, increased the likelihood of CW use: "In a rapidly fluctuating tactical environment, the corps or division com-mander will probably have the authority to employ chemical weapons if they are in imminent danger of being overrun or defeated."[58]

Recent revelations to the UN Special Commission on Iraq indicate that Iraqi ballistic missiles (see below) were fielded with unconventional warheads (both chemical and biological), and that launch authority had been predelegated.[59] These revelations have apparently been confirmed by the memoirs of Staff Lt. Gen. Hazzim 'Abd-al-Razzaq al-Ayyubi, commander of the Iraqi surface-to-surface missile corps during the Gulf War.[60]

The relative ease with which Iraq amassed its chemical arsenal and production complex encouraged the West to respond with export controls and treaties. These culminated with the Chemical Weapons Convention (CWC), which opened for signature on 13 January 1993. The CWC entered into force on 29 April 1997, with 87 countries ratifying (and 165 signatures). The convention prohibits the development, production, acquisition, retention, stockpiling, transfer, and use of chemical weapons, and is enforced by relatively intrusive verification pro-

---

[55]These lessons were already being learned before the Gulf War. See Cordesman and Wagner, *The Lessons of Modern War*, vol. 2, p. 518: "These are lessons that many developing nations have already learned to heart, and further uses of chemical weapons, and possibly biological weapons, now seem all too likely."

[56]*Conduct of the Persian Gulf War*, p. Q-2.

[57]*The Iraqi Army*, p. 162, noted that initial release authority was in the hands of "the President of Iraq" (Saddam), but that release for some CW artillery and mortar munitions might be in the hands of corps commanders, with verbal approval from GHQ. Other sources note that Saddam delegated authority to three-star generals and corps commanders, and predelegated authority during the Gulf War. James Blackwell, *Thunder in the Desert* (New York: Bantam, 1991), p. 36.

[58]*The Iraqi Army*, p. 166.

[59]"Iraqi Nerve Gas Tests Confirmed," *Washington Post*, 25 June 1998, reporting that VX warheads had been loaded and deployed on ballistic missiles prior to the Gulf War; and "Tests Show Nerve Gas in Iraqi Warheads," *Washington Post*, 23 June 1998, which notes that as early as 1994, Wafiq Sammarrai, former chief of military intelligence in Iraq, reported that at least ten VX warheads and another ten filled with anthrax were available to Iraqi forces in 1991.

[60]"Forty Three Missiles on the Zionist Entity," *Al-'Arab al-Yawm* (in Arabic), FBIS-NES-98-326, 22 Nov. 1998. The memoirs were published over a period of days.

cedures.[61] It does not, however, ban research in defensive measures, and it can only be assumed that core states will continue to treat the CW threat very seriously, particularly in the Gulf region (where Iraq is not a signatory).

The pattern of innovation and diffusion differs significantly from the previous two examples. In this case, Iraq made use of weapons whose technology had been available for at least forty years (in the case of nerve agents, derived from German research before and during World War II), and whose effectiveness had been proven in combat in World War I (in the case of mustard and other agents). Iraq was willing, under significant threat to the existence of the ruling Ba'athist regime, to break international norms against the use of CW—norms so pervasive that they had only been violated covertly or with indignant public denials throughout the Cold War era. Iraq's innovation was conceptual: it broke with established practice and set off a diffusion effect.

The impact of Iraq's use of CW on the international system was heightened by several factors: its open and routine use of huge amounts of CW agents; the unwillingness of the international community to condemn its use, because of greater concerns about Iraq's Iranian adversary, which contributed (deliberately or inadvertently) to legitimizing CW use; the ability of a relatively underindustrialized and technologically unsophisticated Arab state to create a massive production line for CW agents in a short time; and, perhaps most important, the evident effectiveness of CW against Iran and, later, as a political weapon against Israel.

Diffusion patterns differ from the Israeli cases as well. Core states responded with efforts to prevent development and diffusion of CW capabilities, codified in the CWC. They also prioritized defensive countermeasures, including improved detection and protective gear for military expeditionary forces and increased concern for the effects of CW attacks on both combat troops and infrastructure. Peripheral states responded in a variety of ways, depending on the measure of the threat. Many states did nothing, because credible CW threats to them did not exist. Those that were most threatened (Iran, Israel) improved protective gear and other countermeasures, as well as establishing civil defenses and routines in the event of missile-borne chemical attack. Some states found Iraq's example compelling and attempted to emulate it. States with requirements for significant military capability and limited industrial potential found the CW option particularly cost effective—Syria achieved significant CW capability by the late 1980s, as did Iran, and even Burma/Myanmar (which may be focusing more on the internal security potential for CW demonstrated at Halabja) now has some CW capability.

[61] *The Biological and Chemical Weapons Threat*, p. 32.

Iraq and Ballistic Missiles: Conceptual Innovation

During the Iran-Iraq War, and later during the Gulf War, Iraq utilized inaccurate ballistic missiles derived from obsolete Soviet designs for attacking the cities of its enemies with conventional explosives. These missiles, dismissed as militarily ineffectual by most analysts, significantly affected the course of both wars, and had a disproportionate strategic and political impact on each conflict. The diffusion effect of Iraq's utilization and production of ballistic missiles extends on a global level, and responses include technology control regimes, missile defenses, development of missile forces as a deterrent, and preparation for preemptive strikes in times of tension.

Ballistic missiles have been an important element in the Middle East power balance since the 1960s, when the United States first expended enormous diplomatic energy (to little effect) trying to limit Egyptian and Israeli missile development programs.[62] By the early 1980s, a number of Arab states had purchased SCUDs and other surface-to-surface missiles (SSMs). Other developing states were undertaking indigenous production of long-range ballistic missiles.

Iraq was the first country, however, to operationalize the programs and use them in a major conflict. From February through April 1988, Iraq fired approximately two hundred ballistic missiles against Iranian cities, primarily at residential areas or economic targets.[63] Iraq averaged three SSM launches per day, and these attacks produced approximately ten fatalities per missile, roughly three times greater than German V-2 attacks on London and Antwerp in World War II. Ballistic missile attacks had an enormous impact on Iranian morale, spurring an evacuation of Tehran.[64] Combined with Iraq's successful spring offensives, the so-called War of the Cities persuaded Iran to end hostilities with Iraq by the summer of 1988.

Iraq's use of inaccurate, conventionally armed ballistic missiles for city bombardment came as a surprise to military analysts. Even more surprising was the fact that these missiles were largely the product of Iraqi industrial innovation and modification. Between 1980 and 1990, Iraq spent at least $3 billion on missile de-

---

[62]See, for example, *Memorandum for the President*, The White House, Subject: Second McCloy Mission on Near East Arms, dated 12 Aug. 1964 (declassified, formerly "Top Secret—Exdis"): "Mr. McCloy's objective is to let [Egyptian President Gamel] Nasser know we believe we can convince Israel to exercise nuclear and missile self-denial if Nasser will limit his acquisition of major offensive missiles either to the number he now has or to a low ceiling." The document further expresses concern over Israeli cooperation with France on surface to surface missiles.

[63]Cordesman and Wagner, *The Lessons of Modern War*, vol. 2, p. 366; *Conduct of the Persian Gulf War*, p. 15.

[64]Thomas G. Mahnken and Timothy D. Hoyt, "The Spread of Ballistic Missile Technology to the Third World," *Comparative Strategy* 9 (1990): 248; Cordesman and Wagner, *The Lessons of Modern War*, vol. 2, p. 500.

velopment, including an emergency investment of $400 million in 1987.[65] These efforts produced four variants of the Soviet SCUD-B tactical SSM, as well as other missile projects.[66]

The Al-Hussein, used in the War of the Cities, decreased the SCUD warhead from 800 kg to 190 kg and increased its range to 600 km (sufficient to hit Teheran).[67] The Al-Abbas, a longer-range derivative of the Al-Hussein, was originally tested at the end of the War of the Cities on 25 April 1988.[68] The Al-Hijarah, with a 750-km range, was tested during the run-up to the Gulf War. Finally, and most ominously, the Al-Abid space launch vehicle was tested on 5 December 1989, catching Western analysts totally by surprise. The military version, called the Tammuz-I, had a 2,000-km range. The missile was built in three stages, with the first stage consisting of a single SCUD or Iraqi SCUD variant, the second stage of a cluster of two similar rockets, and a third stage of five clustered rockets.[69]

Iraq was involved in other missile projects, the most important of which was the Condor-2/Badr 2000, a cooperative project with Egypt and Argentina to produce a two-stage, solid fuel missile with a range of approximately 1,000 km. The project sought a much more sophisticated and capable missile than Iraq's arsenal of SCUD derivatives. Loosely based on an Argentinean sounding rocket, the Condor-2 design was strongly influenced by designers from the German firm MBB, and by consultants from the independent firm of Consen. The same MBB employee who worked on the Pershing at the Pentagon also represented MBB in Iraq for the Condor, and thus was in a position to transfer missile technology to Baghdad.[70]

The phenomenal accuracy of the Pershing-2's terminally guided warhead, and the strong physical similarities between the Condor-2/Badr 2000 design and the Pershing, prompted a strong diplomatic response by the United States and other

[65]Timmerman, *The Death Lobby*, p. 255; Cordesman, "Iraq and Weapons of Mass Destruction," p. S5062.

[66]For an overview of the Iraqi missile programs, see Timothy D. Hoyt, "Rising Regional Powers: Case Studies New Perspectives on Indigenous Defense Industries and Military Capability in the Developing World." Ph.D. dissertation, Johns Hopkins University, 1996, pp. 386–96.

[67]W. Seth Carus and Joseph S. Bermudez, Jr., "Iraq's Al-Husayn Missile Program: Part One," *Jane's Soviet Intelligence Review* (May 1990): 204–9. According to Lt. Gen. Al-Ayyubi, the Al-Husayn missile had a range of 650 km. "Forty Three Missiles," entry dated 31 Jan. 1990 (published on 27 Oct. 1998).

[68]"New Missile with 900-km Range Launched 25 Apr." Baghdad INA in Arabic 1543 GMT 25 Apr. 1988 in *FBIS-NESA* (26 Apr. 1988), p. 24. The actual range of the Al-Abbas was closer to 800 km.

[69]"Satellite-carrier Rocket System Tested," Baghdad Voice of the Masses in Arabic 1136 GMT 7 Dec. 1989, in *FBIS-NESA* (8 Dec. 1989), p. 23; and "2,000-km Range Missiles Produced," Baghdad Voice of the Masses in Arabic 1230 GMT 7 Dec. 1989, in *FBIS-NESA* (8 Dec. 1989), p. 23.

[70]Gary Milhollin, "Building Saddam Hussein's Bomb," *New York Times Magazine* (6 Mar. 1992): 34. See also Herbert Krosney, *Deadly Business* (New York: Four Walls, Eight Windows, 1993), pp. 135–37. According to Krosney, Iraq's goals for the project included a 750–1,100 kg warhead and a Circular Error of Probability (CEP) of 0.001 (1 km at 1,000 km range).

Western countries. The Missile Technology Control Regime (MTCR), formally announced in 1987, helped stifle the international cooperation that threatened to make Condor-2 a reality. The project was not completed before the Gulf War, although United Nations Special Commission (UNSCOM) inspectors found information confirming that the project was not scrapped.[71]

Coalition estimates of Iraq's missile arsenal and tactics at the start of the Persian Gulf crisis were deeply flawed. Military planners did not expect Iraq's missile force to have much military utility. Prewar estimates of the number of SSMs available to Iraq ranged from three hundred to one thousand.[72] Little was known about the number of mobile launchers available to Iraq. The Coalition assumed that Iraq had approximately three dozen mobile launchers and twenty-eight fixed sites, but postwar assessments of the number of mobile launchers range from fifty to four hundred.[73] U.S. Air Force planners thought that the entire SCUD target set, including mobile launchers and production facilities, constituted less than one hundred "aim points."[74]

Coalition difficulties were compounded by Iraqi tactics. The United States assumed that the stationary launchers were an important target: Saddam used them as a diversion. Coalition intelligence relied on Soviet practice and doctrine as a model for Iraqi tactics. Soviet SCUDs take one hour to prepare for launching, one hour to fuel and position, and one hour to reload. They also make use of weather balloons to test atmospheric conditions. The SCUD's red fuming nitric acid and hydrazine fuel is volatile and corrosive: it cannot be mixed more than one day before launch.[75]

In fact, the Iraqis considerably cut down the refire time of the Al-Hussein,

[71]See Andrew Slade, "Condor Project in Disarray," *Jane's Defence Weekly* (17 Feb. 1990): 295; Michael Eisenstadt, "Like a Phoenix from the Ashes: The Future of Iraqi Military Power," *Policy Papers* 36 (Washington, DC: Washington Institute for Near East Policy, 1994), pp. 35–36; statement by Rep. Fascell of Florida, *Congressional Record* (2 Oct. 1992), Extension of Remarks E2884.

[72]According to a Defense Intelligence Agency memorandum classified "Secret" from Mar. 1991, estimates of the stockpile ranged from three hundred to seven hundred. Gordon and Trainor, *The Generals' War*, p. 241. The *Gulf War Air Power Survey*, vol. I: *Planning*, pp. 210–11, states that the Defense Intelligence Agency estimated that Iraq had bought six hundred SCUDs in 1976–79, and would have about four hundred available in Aug. 1990. "Iraq Demonstrates SSM Capability," *Janes' Defense Weekly* (15 Dec. 1990): 1212, estimates that Iraq bought fifteen hundred to two thousand SCUDs, and that four hundred to one thousand SSMs were therefore available for service.

[73]Prewar estimates can be found in *Conduct of the Persian Gulf War*, pp. 161, 225; and *Gulf War Air Power Survey*, vol. I, pp. 210–11. Postwar estimates of up to 225 launchers can be found in Cordesman, "Iraq and Weapons of Mass Destruction," p. S5064; and Jack Anderson and Martin Binstein, "Full Iraqi Capabilities Remain Unknown," *Washington Post*, 27 July 1992, reports that Iraq had fifty fixed sites and more than four hundred mobile launchers. The fact that Iraq produced launchers indigenously made accurate estimation virtually impossible.

[74]Atkinson, *Crusade*, p. 145.

[75]Cordesman, "Iraq's Weapons of Mass Destruction," pp. S5071–5072 n. 6; Blackwell, *Thunder in the Desert*, p. 35; Gordon and Trainor, *The Generals' War*, p. 241.

from approximately three hours to about sixty minutes.[76] Iraq's mobile launchers operated at night under radio silence, using East German decoy vehicles that were indistinguishable from a launcher at more than twenty-five yards. They launched SAMs simultaneously with SCUD launches as a deception measure.[77] Iraqi missile crews had more mobile launchers available than the Coalition anticipated, and they learned to erect and fire them quickly. Iraqi crews practiced "shooting and scooting," hid their launchers under highway culverts at night, and fired from presurveyed launch positions.[78]

Faced with a continuing political problem, as Israeli leaders considered retaliating against Iraq for SCUD strikes in January, the Coalition was forced to increase the priority of the "SCUD hunt," tasking additional air assets and shifting aircraft from other missions. By the evening of 20 January 1991, the number of aircraft hunting SCUDs in the western areas of Iraq had increased substantially.[79] Faced with continuing SCUD attacks, one option offered to Gen. Schwarzkopf was the diversion of all Coalition sorties to western Iraq for a three-day, massive SCUD hunt.[80] Active defenses, including the Patriot missile, were complicated by the fact that the Iraqi missiles flew 40 percent faster than normal SCUDs, had a smaller radar signature, and tended to break up when approaching the target.[81]

Despite the heightened priority given to the SCUD hunt, it is not clear which side won. The first two weeks saw the largest number of launches by the Iraqis, with the maximum salvo (number of missiles fired from different locations within three minutes) ranging from four to six. Counter-SCUD operations clearly had an impact in the third and fourth weeks of the war, with the total number of Iraqi missile firings reduced to four and five firings each week, respectively (and a maximum salvo of one). During the fifth and sixth weeks, however, Iraqi missile firings climbed back up to week two levels, with maximum salvoes of five (week five, a total of eleven launches) and three (week six, a total of seventeen launches). This suggests that the Iraqis adopted new tactics to counter the

[76]Cordesman, "Iraqi Weapons of Mass Destruction," p. S5064.
[77]Lawrence Freedman and Efraim Karsh, *The Gulf Conflict, 1990–1991: Diplomacy and War in the New World Order* (Princeton: Princeton University Press, 1993), p. 308.
[78]Friedman, *Desert Victory*, pp. 193–94; Gordon and Trainor, *The Generals' War*, p. 241. See also Lt. Gen. Al-Ayyubi's comments on training and improvements in "Forty Three Missiles," especially entries dated 16 and 17 Sept. 1990, which discuss the need to drastically cut the amount of time needed to prepare and fire missiles.
[79]US News and World Report, *Triumph Without Victory*, p. 261. According to Freedman and Karsh, the first phase of the air campaign originally allotted 15 percent of all strategic sorties to missile-related targets; after the first week of the war, this figure was increased to 40 percent. Freedman and Karsh, *The Gulf Conflict, 1990–1991*, p. 309.
[80]Gordon and Trainor, *The Generals' War*, p. 249.
[81]Atkinson, *Crusade*, p. 79.

Coalition reaction. These included firing from sites deeper inside Iraq and changing targets from population centers (Tel Aviv, Haifa, Riyadh, Dhahran) to military-related targets (Haf al-Batin, King Khalid military city, and Dimona).[82]

The conclusion of the Defense Intelligence Agency, in a secret memorandum released shortly after the war, was that:

> the Coalition's inability to permanently degrade SRBM command and control is also significant, despite determined efforts to incapacitate Iraqi military and civilian networks. Even during the last days of the war, Baghdad retained a sufficient capability to initiate firings from new launch areas and to retarget SRBMs from urban to military and high-value targets, such as the Dimona reactor.[83]

The "lessons learned" from the Gulf War make disturbing reading. Clearly, the "rogue" states (Iraq, Iran, Libya, Syria, North Korea) have all placed greater emphasis on mobile missiles and weapons of mass destruction. The general success of Iraqi cover, concealment, and deception tactics and the mobility of their missile force allowed them to avoid detection and destruction by allied air power.[84] The ability of Iraqi SCUDs to survive the U.S. air campaign was undoubtedly well received by the North Koreans, given their extensive missile protection schemes, which rely on terrain, underground facilities, and various cover, concealment, and deception practices.[85]

The Gulf War experience also demonstrated the utility of shooting first, especially at air bases and port facilities. In fact, an Iraqi missile landed just 130 yards from the USS *Tarawa* as it offloaded Harrier aircraft on 15 February 1991.[86] A hit would have been very serious, damaging or sinking a major naval vessel and possibly killing hundreds of sailors. A Condor-2, equipped with a terminally guided warhead, might have hit the ship. A missile attack on Al Khobar on 25 February 1991 killed twenty-eight soldiers and wounded ninety-eight more, and represented the highest single U.S. casualty total of the war. Salvoed missile fire at the Persian Gulf ports during the Desert Shield buildup might easily have caused significant casualties, or destroyed vital supplies.

Finally, Iraq's missile successes in the Gulf War reinforced the lesson that even a missile force derived from SCUD technology is a significant threat. Despite the MTCR, SCUD technology continues to be widely traded in the international

---

[82]Gordon and Trainor, *The Generals' War*, pp. 252, 259.

[83]Defense Intelligence Agency memorandum, Mar. 1991, classified "secret," cited in Gordon and Trainor, *The Generals' War*, p. 238.

[84]Patrick J. Garrity, *Why the Gulf War Still Matters*, Report No. 16 (Los Alamos National Laboratory: Center for National Security Studies, July 1993), pp. 32, 54, 66, 87.

[85]Ibid., p. 95.

[86]Gordon and Trainor, *The Generals' War*, p. 250.

market, primarily from North Korea. Iran, North Korea, and Syria have all deployed the SCUD-C, a North Korean variant with a range of 310 miles, since the
Gulf War.[87] North Korea has also designed three medium and intermediate
ranged ballistic missiles based on the SCUD (Nodong, Taepo Dong 1, Taepo
Dong 2), and contributed significantly to Pakistan's Ghauri SSM[88] and Iran's Sha-
hab-3.[89] India, while not relying on the SCUD or North Korean supplies, pursues
a variety of mobile missiles, including sea-launched versions of the Prithvi, the
new extended-range solid-fuel Prithvi-3, the Agni follow-on, and the subma-
rine-launched Sagarika.[90]

Iraq innovated based on older technologies (the SCUD) that had already diffused to the periphery. By modifying the design, producing increasing portions of
the missile indigenously, and using it as a substitute for long-range air strikes,
Iraq achieved substantial political results in two separate regional conflicts. Iraq's
evident success with its programs reinforced the desire of other peripheral states
for similar capabilities, and several states (including Egypt, Brazil, and Argentina)
engaged in cooperative missile development projects with Iraq. Other peripheral
states obtained their own missile production capabilities, in cooperation with
each other (North Korea, Pakistan, Iran) or with technology obtained illicitly or
covertly from major producers (Russia, China, Germany).

This peripheral diffusion and cooperation prompted a strong diplomatic response from the West in the form of the MTCR—an effort to deny the transfer of
militarily relevant rocket technology to suspect states. The demonstration effect
of the two Gulf Wars also prompted a range of defensive efforts, including upgrading SAMs (Patriot) and development of new forms of tactical antiballistic
missile defenses (Theater High Altitude Air Defense [THAAD], the U.S.-Israeli
Arrow project, and others). The potential combination of missile with CW or
BW warheads remains a specific area of concern today, prompting continued discussion of options including technology denial (through treaty and export controls), active and passive defenses, and new operational methods to reduce vulnerability to such strikes or to preemptively eliminate potential threats (counterproliferation).

[87]*Ballistic and Cruise Missile Threat* NAIC-1031-0985-98 (Wright Patterson AFB, OH: National Air
Intelligence Center, 1998): pp. 4–5, 7.
[88]*Ballistic and Cruise Missile Threat*, p. 9; "U.S. Says North Korea Helped Develop Pakistani Missile,"
New York Times, 11 Apr. 1998.
[89]"Iran Missile Test Shows Effort to Extend Range," *Washington Post*, 23 July 1998.
[90]*Ballistic and Cruise Missile Threat*, pp. 4–5, 9, 15; Chinatamani Mahapatra, "The U.S., China and
the Ghauri Missile," *Strategic Analysis* (June 1998) Internet version (http://ww.idsa-india.org/an-june8–
4.html), p. 1.

Conclusion

Diffusion from the periphery differs from diffusion from the core. Peripheral improvements in military capability are constrained by the regional focus of conflicts and resource limitations. Therefore, these improvements are not as broad in scope as major innovations from the core.[91] Diffusion results from observation of the effectiveness of new capabilities in regional combat. Peripheral diffusion prompts reaction at two levels. States in the periphery seek to imitate and adapt demonstrably effective practices, while states in the core attempt to imitate or counter them. Peripheral innovation can be "technological" or innovation with significant indigenous technological content (the Israeli cases), and "conceptual," based on previously diffused or mature technologies (the Iraqi cases).

Although the four cases represent examples of innovation, rather than emulation, the capabilities diffused (and continue to diffuse) to the periphery and core. The *Sa'ar*-Gabriel combination was emulated because of its effectiveness in combat, even if Israel's pariah status limited its ability to supply equipment or training. The United States emulated Israeli use of RPVs, to the extent of purchasing and employing them with U.S. military units, because of their proven effectiveness in 1982 and the inability of the United States to design and produce a viable alternative. Core states have also attempted to counter some of these capabilities, particularly the FACM/missile combination. Responses to conceptual innovation include broad technology-denial and export control efforts (MTCR), international treaty regimes (CWC), active defenses (missile defense: THAAD, improved Patriots, Israel's Arrow, and the recent surge in SA-10 and SA-12 exports from Russia), and passive measures (improvements in protective gear), as well as threats of unconventional retaliation (deterrence). Responses range from attempts to prevent diffusion to counter-innovation. Peripheral diffusion tends to take the form of total or partial emulation.

Neither state is the leader of a bloc, but it can be surmised that other pariahs (Taiwan, South Africa, Chile) imitated Israeli naval practices and force structure in the 1970s. Similarly, Iraq's use of both chemical weapons and ballistic missiles has reinforced the acquisition policies of other Arab states and other "rogue"

---

[91]New or improved military capabilities may be defined by significant changes in policies and plans, strategy and force employment, doctrine and operational/tactical concepts and techniques, force structure and organizational concepts, training, hardware systems, and/or technology. "Revolutionary" improvements will encompass some or most of these categories. Dr. David Andre, "Competitive Strategies as a Strategic Response to Proliferation," presentation for the Institute for National Security Studies and Nonproliferation Policy Education Center, 19 June 1998.

states. U.S. adoption of Israeli RPV capabilities was aided by the close relationship between the United States and Israel, which undoubtedly also increased U.S.
familiarity with the performance of the systems. Finally, the prestige and sym-
bolism of ballistic missiles and WMD clearly contribute to the ongoing proliferation of these systems in the international system. Iraq's effective use of these
technologies only heightens their attractiveness.

All four cases suggest that technological changes do diffuse more rapidly than
"software" changes (organization, doctrine, or macro-social change). Egypt and
Syria, for example, already had inferior Soviet attack craft/missile combinations,
but succumbed to superior Israeli organization, doctrine, and tactics. Iraq had a
FACM/missile-based navy, which was hammered by Coalition forces. Many
states, including Iraq, have RPVs, but few have used them as effectively as have
the Israelis. However, it appears that Iraqi conceptual innovation, both in the use
of CW and in their tactics and doctrine with mobile missiles, is spreading rapidly
to states that have access to those technologies. Innovations with higher costs
diffused more slowly than those with lower costs. Israel's innovations diffused
rapidly in part because of their very low costs and demonstrated effectiveness.
The high costs and resource requirements of CW and missile production programs provide a serious constraint to their diffusion, which also allows broad export controls, technology denial, and treaty regimes to be relatively more effective. Still, the example of Iraq demonstrates that even a relatively underindustrialized state can develop significant production capacity. In the final analysis, diffusion from the periphery, at least since the onset of the Cold War, appears to be
a natural offshoot of local security dilemmas and regional conflicts. Diffusion
from these peripheral conflicts to the rest of the periphery follows a predictable
pattern. Technologies and practices that prove effective in one regional conflict
will be imitated or adapted by other peripheral powers. The key factor stimulat-
ing diffusion is observation of demonstrated effectiveness.

Diffusion to the core states takes two forms. First, where niche technologies
are created that are useful, they may be absorbed or imitated by core powers. In
the case of Israeli RPVs, this was facilitated by a close U.S.-Israeli relationship. Israel produces electronic warfare technology, avionics, and missile systems at the
state-of-the-art level. These may diffuse to core or would-be core powers.[92] Other
states may also find competitive niches, based on the increasingly international-
ized electronics and computer software industries, which will prompt core power
imitation or adoption.

The second diffusion effect from periphery to core takes the form of core at-

---

[92]Israel has already engaged in technological cooperation with France (1955–68, including design of
the Mirage V aircraft), the U.K. (design of the Chieftain tank), the United States (Arrow ATBM and
other systems), and possibly China. See Hoyt, *Rising Regional Powers*, pp. 199–321.

tempts to prevent, counter, or limit the potential damage of peripheral technological and conceptual innovation. The Office of the Secretary of Defense defined
"issues of proliferation concern" on the basis of three criteria during the early
1990s: if (1) it enabled another state to inflict high-leverage strategic harm against
the United States or its friends; (2) the United States lacked effective defenses or
countermeasures against this capability; or (3) its mere acquisition could change
other states' perceptions as to who was the leading power in the region.[93] Three of
the cases fall under this definition: FACM-based sea denial and coast defense
forces, which threaten U.S. force projection efforts; Iraqi CW, which resulted in
re-evaluation of U.S. defenses and countermeasures, and threatened the perception of neighbors in the Persian Gulf; and Iraqi missile employment, which contributed to temporarily overthrowing Iran as the leading regional power, and
against which defenses and countermeasures were remarkably ineffective.

Strategies for prevention of diffusion include export controls, arms control
treaties, and other multilateral regimes and treaties. These have already been applied to limit certain types of diffusion (proliferation of missiles and weapons capable of mass destruction), and efforts have been made to expand them to other
weapons (limits on conventional arms sales, small arms, and land mines). These
measures are far from leakproof, especially where production capability has diffused. Examples of countermeasures include design of new naval forces and tactics for countering the FACM threat. Damage limitation and protection efforts
include active defenses (missile defense systems), passive defenses (protection
gear against CW), and civil defenses for civilian populations.

Two important trends will define the future international security regime: the
proliferation of commercial and military technologies, and the continuing wave
of savage regional conflict based in part on ethnic, cultural, and national antipa-
thies. Diffusion from the periphery will occur as a result of these conflicts, although in some cases (Rwanda, Zaire) it may have little strategic impact on the
core. Core states will continue to respond with a limited number of options:
imitation, prevention, countermeasures, and damage limitation. So far, these options have proven sufficient to avoid significant strategic harm. As new forms of
warfare arise, based on information and biological or genetic technologies, the
core may have to work much harder to anticipate and prevent threats, and develop new means to prevent their diffusion or counter their effects.[94]

[93]Henry Sokolski, "Fighting Proliferation with Intelligence," in Henry Sokolski, ed., *Fighting Proliferation: New Concerns for the Nineties* (Maxwell AFB, AL: Air University, 1996), p. 282.
[94]Sadly, the attacks of 11 September 2001 might be considered a case of conceptual innovation, as the
terrorists used commercial airliners in a shocking and effective new form of attack that has prompted
significant responses from the core states.

PART III

Diffusion and Military Transformation

CHAPTER EIGHT

Military Diffusion in Nineteenth-Century Europe

The Napoleonic and Prussian Military Systems

GEOFFREY L. HERRERA AND

THOMAS G. MAHNKEN

Between the French Revolution and World War I, two waves of military innovation swept through Europe. Building on tactical and organizational changes that
began during the second half of the eighteenth century, France brought about a
revolution in warfare by fielding a mass, national army organized into corps and
divisions and trained in a flexible tactical system. Under the command of Napoleon Bonaparte, the French army achieved a decade of nearly unbroken success
against adversaries who had failed to come to terms with the military revolution.
As France's power grew and Napoleon's victories mounted, his foes began to
adapt to this new style of warfare. Most countries were unwilling to adopt the
types of wholesale social and political reforms brought on by the French Revolution. Instead they attempted to copy Napoleonic tactics and organization with
measures requiring less jarring changes. Prussia took an innovative approach to
countering the Napoleonic threat, organizing a reserve system and a general staff.
By 1815, France's adversaries were able to master enough of the revolution to defeat Napoleon.

By midcentury, Prussia's response to Napoleonic warfare, coupled with the
military use of the railroad and telegraph and the adoption of rifled guns and artillery, transformed warfare once again. The success of Prussian military methods
in the Wars of German Unification led states both within and outside Europe to
imitate German combat techniques. The breechloading rifle, and the tactics
needed to employ it effectively, spread widely. Adapting to the railroad and the
telegraph was somewhat more difficult, since militaries were dependent upon the
civilian networks their countries developed. While many copied the Prussian re-

serve system and general staff in form, few adopted its substance. Yet by World War I, the diffusion of German military methods had produced armies that at least superficially resembled one another.

The Napoleonic revolution, drawing its power from the political changes triggered by the French Revolution, was confined largely to Europe. Even after the Napoleonic wars, European armies fought conflicts outside the continent with pre-Napoleonic methods. Prussian military techniques spread much more widely. The Wars of German Unification triggered efforts to adopt Prussian methods not only by European powers but also by states such as Japan and Chile. These, in turn, served as models for other states in Asia and South America.

No army was able to replicate the Napoleonic or Prussian military systems. Rather, they selectively imported new military techniques. Success in adopting military practices turned not only on military organization and culture but also on the fit between the new techniques and native social, political, economic, and military structures. Austria's multinational composition constrained its response to the threats posed by France and Prussia. Russia's feudal social structure limited its ability to adopt French and Prussian practices. Britain's relative insularity throughout the period reduced the perceived need for its army to modernize.

The Napoleonic Revolution in Warfare

The roots of the revolution in warfare that occurred during the Napoleonic wars lay in a series of tactical and organizational innovations. The French Revolution's transformation of French society yielded the *levée en masse* (universal conscription), an officer corps chosen on the basis of talent, and organizational innovation. The cumulative result of these changes was a new type of army capable of waging a new style of warfare.

The French republic inherited from the *ancien régime* Europe's most effective army. The French army's 1791 drill manual, the result of an intense debate over infantry tactics brought on by the Seven Years' War, outlined a highly flexible system of tactics. The manual envisioned forming battalions into columns to maneuver and assault, and lines to mass firepower.[1] It also called for dispersed groups of skirmishers (*tirailleurs*) to harass and distract the enemy's infantry and artillery units in preparation for an assault.[2] France's infantry was supported by the most mobile and effective artillery in Europe, the result of adopting the

---

[1] Peter Paret, "Napoleon and the Revolution in Warfare," in Peter Paret, ed., *Makers of Modern Strategy: From Machiavelli to the Nuclear Age* (Princeton: Princeton University Press, 1986), pp. 124–25.

[2] John A. Lynn, *The Bayonets of the Republic: Motivation and Tactics in the Army of Revolutionary France, 1791–1794* (Urbana: University of Illinois Press, 1984), p. 262.

Gribeauval system. The system standardized the caliber and construction of artillery pieces to ease manufacture and repair. Gribeauval's guns were shorter, lighter, and more mobile than previous pieces, and possessed superior range for a given powder charge.[3]

While the ancien régime refined the eighteenth-century paradigm of combat, the French Revolution transformed warfare. As Carl von Clausewitz wrote in his unfinished masterpiece *On War*: "[In] 1793 a force appeared that beggared all imagination. Suddenly war again became the business of the people—a people of thirty millions, all of whom considered themselves to be citizens." The political conditions that caused the French Revolution unleashed "new means and new forces, and . . . thus made possible a degree of energy in war that otherwise would have been inconceivable. It follows that the transformation of the art of war resulted from the transformation of politics. So far from suggesting that the two could be disassociated from each other, these changes are a strong proof of their indissoluble connection."[4]

The French Revolution's most important military legacy was the adoption of universal conscription, the levée en masse. Under the ancien régime, the army had been composed of professional soldiers, most of whom were peasants and many of whom served for life. Harsh drill and punishment were the rule, and desertion an endemic problem.[5] At first the French republic relied upon volunteers to fill the ranks of its army. On 21 February 1793, however, the government was forced to put all single men between the ages of eighteen and forty in a state of requisition. Six months later, the government issued the levée en masse decree, declaring all Frenchmen to be in a permanent state of requisition. As a result of this influx of manpower, in 1794 the army had a million men on its rolls and 750,000 present under arms, nearly three times the size of Louis XIV's largest army.[6]

The levée en masse transformed the republican army into a more accurate reflection of French society. The political changes wrought by the French Revolution produced an army of citizen soldiers, fighting to protect their rights and fulfill their patriotic duties.[7] Unlike the army of the ancien régime, the republican army was dependable and motivated. The result was a new style of warfare that

---

[3]John A. Lynn, "Nations in Arms, 1763–1815," in Geoffrey Parker, ed., *The Cambridge History of Warfare: The Triumph of the West* (Cambridge: Cambridge University Press, 1995), p. 192.

[4]Carl von Clausewitz, *On War*, ed. and trans. Michael Howard and Peter Paret (Princeton: Princeton University Press, 1976), pp. 591–92, 610.

[5]Steven T. Ross, *From Flintlock to Rifle: Infantry Tactics, 1740–1866*, 2d ed. (London: Frank Cass, 1996), p. 19.

[6]Lynn, "Nations in Arms," p. 193.

[7]Lynn, *Bayonets of the Republic*, pp. 43, 66.

broke the eighteenth-century mold. Instead of forming up into firing lines and delivering controlled musket volleys, French troops massed into assault columns supported by dense clouds of skirmishers, charging their enemies with bayonets.[8]

The French Revolution also transformed the army's officer corps. Under the ancien régime, officers had been drawn from the aristocracy. Moreover, military service was a part-time profession. The expansion of the republican army, combined with the resignation of aristocratic officers, created many vacancies. To fill the army's rolls, the republican government began selecting and promoting officers based on merit and seniority. By 1794 aristocrats constituted only 2 to 3 percent of the officer corps.[9] The new cadre of officers possessed a much closer relationship to their men. Officers were imbued with a spirit of initiative and leadership, permitting decentralized command and control impossible under the ancien régime.

As the army grew, it needed new tactical organizations to ensure effective command and control. Under the ancien régime, the largest fixed military unit had been the two-battalion regiment.[10] In 1792–93 France began to organize its army into divisions of two brigades, each containing two regiments. These divisions were combined-arms formations that included infantry, cavalry, and artillery units. Prior to the 1805 campaign, Napoleon combined divisions into corps of between seventeen thousand and thirty thousand men. While the composition of a corps varied, it included two to four infantry divisions, a cavalry brigade or division, thirty-six to forty field guns, and engineer and support units. The corps gave the French army the flexibility, resiliency, and cohesion that its adversaries lacked. It was capable of fighting independently for a day or more, militated against collapse, and permitted rapid regeneration.[11]

The organization of the army into corps expanded the size and scope of responsibilities of commanders and their staffs. Marshals and senior officers orchestrated foot, horse, and gun operations, while division, brigade, and battalion commanders directed tactical engagements. Staffs eased problems of supply and enhanced command and control. They also allowed Napoleon to divide his army so that it could advance along separate lines before converging on the foe.

The tremendous growth in the size of the army required new logistics. Rather

---

[8]John A. Lynn, "En Avant! The Origins of the Revolutionary Attack," in John A. Lynn, ed., *Tools of War: Instruments, Ideas, and Institutions of Warfare, 1445–1871* (Urbana: University of Illinois Press, 1990), p. 169.

[9]Lynn, *Bayonets of the Republic*, p. 67.

[10]Two regiments often combined on the battlefield to form a brigade. Two or more brigades sometimes served under a single commanding officer. Brigades were temporary formations lacking permanent staffs. See Ross, *From Flintlock to Rifle*, p. 30.

[11]Robert M. Epstein, "Patterns of Change and Continuity in Nineteenth-Century Warfare," *Journal of Military History* 56 (July 1992): 377.

than using a network of fixed magazines to support troops in the field, they lived off the land. French troops requisitioned food and clothing from the local populace. Instead of carrying tents and baggage with them, they slept wherever they could. Revolutionary soldiers could be trusted to return to their units. As a result, French troops were considerably more mobile than their adversaries.

Napoleon Bonaparte mastered these elements, and then this juggernaut of war, based on the strength of the entire people, began its pulverizing course through Europe. It moved with such confidence and certainty that whenever it was opposed by armies of the traditional type there could never be a moment's doubt as to the result.[12]

In 1805 the *Grande Armée* included 210,000 men and 396 guns. It relied upon national conscription and was backed by the full mobilization of the state, organized into seven army corps, an allied corps, a cavalry reserve, and the Imperial Guard. Each corps was a small army with its own staff, able to employ combined arms effectively and difficult to destroy.[13]

The Spread of the Napoleonic Military Revolution

New tactical and organizational methods brought France ten years of nearly unbroken victory against unreconstructed adversaries. To observers such as Clausewitz, Napoleon's success demonstrated that the "old methods of warfare had collapsed."[14] Between 1805 and 1807, the combination of modern fighting methods and Napoleon's operational genius gave France a series of quick, decisive battlefield victories. On 20 October 1805, Napoleon enveloped the Austrian army at Ulm, leading to the surrender of thirty-three thousand troops. On 2 December he destroyed a combined Austrian-Russian force at Austerlitz. Three weeks later he concluded the Peace of Pressburg, which detached Austria from the coalition opposing France, forced it to cede control of Venetia, and thereby made France the dominant power in Central Europe. The following year Napoleon shattered the Prussian army in less than two weeks at Jena and Auerstädt.

Napoleonic military methods spread through two distinct processes. For some states, French occupation or dominance produced military modernization. For others, reform came as the result of the need to devise new strategies to counter the French. Napoleon's adversaries were slow to react to the changes in warfare brought on by the French Revolution. Contemporary observers had difficulty determining what was new and different about revolutionary France's approach to warfare. Moreover, during the early revolutionary period the French army was

---

[12]Clausewitz, *On War*, p. 592.
[13]Epstein, "Patterns of Change and Continuity," p. 376.
[14]Quoted in Gunther E. Rothenberg, *The Art of Warfare in the Age of Napoleon* (Bloomington: Indiana University Press, 1978), p. 165.

anything but unstoppable. Napoleon's decisive victories of 1805 and 1806 forced his adversaries to undertake serious reforms. Even then, efforts focused on tactical innovations such as skirmishing and the attack column rather than fundamental political and military reform. Indeed, some practices, such as universal conscription and open access to commissions, were clearly incompatible with ancien régime values. Adopting Napoleonic methods in full would have required wholesale political and social change, something none of Europe's monarchies were willing to do.[15]

Napoleon's Allies

Napoleon's battlefield success gave France a growing number of allies, including Holland, Westphalia, the kingdoms of Italy and Naples, and the Grand Duchy of Warsaw. Allied armies, fighting side by side with the French, adopted elements of French tactics and organization. But they did not become replicas of the French army. While France's allies emulated elements of the Napoleonic military system, native institutions often blocked or channeled reform.

The Kingdom of Italy, ruled by Napoleon with his adopted stepson Eugene as his viceroy, imitated French military methods closely. The kingdom's army was organized on the French model, as was its reserve and *gendarmerie*. Indeed, fully two-thirds of its troops were French. Italian organization differed from the French model in one significant respect: instead of grouping divisions into corps, the Italians organized them—usually in pairs—into "lieutenancies" under the command of an acting lieutenant general.[16]

Harnessing Italian nationalism, troops were brave and highly motivated. But Italian social structure foiled adoption of modern military methods. Illiteracy made it difficult to train skilled noncommissioned officers, while the indifference of the aristocracy to the French-dominated kingdom limited officer recruitment.[17]

Napoleon's Polish allies also emulated French methods. In 1797, Napoleon formed a Polish Legion out of émigrés, prisoners, and deserters from France's adversaries; two years later he organized a second unit. These legions, operating as part of the *Grande Armée*, adopted French tactical, organizational, and administrative methods.[18] In 1806 the appearance of French units in the Prussian-con-

---

[15]Paret, "Napoleon and the Revolution in Warfare," p. 135.

[16]John R. Elting, *Swords around a Throne: Napoleon's Grande Armée* (New York: Free Press, 1988), p. 392.

[17]Ibid., pp. 392, 394–95.

[18]Jan Pachonski, "The Effects of Revolutionary and Napoleonic France on the Shaping of Polish National Military Forces, 1797–1814," in Bela K. Kiraly, ed., *War and Society in East Central Europe*, vol. IV: *East Central European Society in the Era of Revolutions, 1775–1856* (New York: Columbia University Press, 1984), p. 86.

trolled zone of Poland triggered a national uprising. By the end of the year, some thirty thousand Poles were under arms, and in January of the next year, Napoleon ordered the formation of a Polish army.[19] In July, Napoleon forced Prussia to give up its portion of Poland and announced the formation of the Grand Duchy of Warsaw.

The Grand Duchy's army used some Napoleonic military practices. In May 1808 its ruler, Prince Frederic August of Saxony, enacted conscription based on French methods. In 1810 it converted its three-brigade legions into divisions. It also adopted French tactical regulations, even when they were at variance with Polish weaponry.[20] The Poles did not, however, form light infantry units.[21] The duchy also retained a number of traditional practices, raising, for example, three small Cracus regiments, similar to Cossack units.[22]

In Poland, imitation was a two-way street. While the Poles emulated French practices widely, the French and others also adopted Polish methods. The Polish cavalry, unlike that of other European armies, was largely composed of lancers (*Uhlan*) whose combat effectiveness led other armies to reintroduce the lance into their cavalry arms. The French army added several Polish lancer regiments and later established six manned by Frenchmen. In 1813 the Bavarians and Prussians formed Uhlan units, followed two years later by the Westphalians.[23]

The Kingdom of Holland, ruled by Napoleon's younger brother Louis, adopted French tactics, organization, drill, weapons, and equipment.[24] However, the king refused to introduce conscription, and was thus unable to raise an army large enough to suit his brother. Moreover, the Dutch army was in a perpetual state of reorganization. In 1810, consequently, France annexed Holland, transforming Dutch army units into their French equivalents or incorporating Dutch troops directly into French battalions.[25]

The Kingdom of Westphalia, created at the Peace of Tilsit in July 1807 out of the states of Hesse-Kassel, Brunswick, Prussia's western provinces, and parts of Hanover, adopted Napoleonic political reforms and military methods as well. Much of the kingdom's army was composed of German and Polish units that had served in the French army. Westphalian soldiers were commanded by French of-

[19]Elting, *Swords around a Throne*, p. 405.
[20]Pachonski, "The Effects of Revolutionary and Napoleonic France," pp. 97–99; George Nafziger, Mariusz T. Wesolowski, and Tom Devoe, *The Poles and Saxons during the Napoleonic Wars* (Chicago: Emperor's Press, 1991), p. 44.
[21]Elting, *Swords around a Throne*, p. 405.
[22]Rothenberg, *The Art of Warfare*, p. 161.
[23]Brent Nosworthy, *With Musket, Cannon, and Sword: Battle Tactics of Napoleon and His Enemies* (New York: Sarpedon, 1996), p. 297.
[24]Otto von Pivka, *Dutch-Belgian Troops of the Napoleonic Wars* (London: Osprey, 1980), p. 4.
[25]Elting, *Swords around a Throne*, pp. 389–90.

ficers, wore French insignia, used French titles, and were taught French drill.[26] In 1809 the kingdom ruled by Napoleon's younger brother Jerome began to raise additional units through a French system of conscription. By 1812, Westphalia fielded an entire corps.

The Kingdom of Naples, ruled by Napoleon's older brother Joseph, was less successful in adopting French military methods. In 1807, Joseph introduced an attenuated system of conscription in an attempt to create a French-style army. In Naples, however, French dominance did not yield the political and social reforms it had in other parts of the continent. The troops of the new Neapolitan army, made up of the remnants of the former Bourbon army, brigands, and beggars, had low motivation and high desertion rates.[27] The officer corps was made up of former Bourbon officers and poor nobles.[28] Despite its trappings of modernity, the Neapolitan army was expensive and largely ineffective.

Napoleon's Adversaries

While Napoleon's allies had the French military system imposed upon them, his adversaries were compelled to imitate his methods to avoid defeat. Complete emulation of French practices would have required large-scale political and social reform. In the end, only Prussia was willing to risk upheaval to transform its military. Austria and Russia implemented limited tactical and organizational modernization without changing social institutions. Britain coupled limited tactical reforms with a strategy based on coalitions with continental states, a superior navy, and a strong economy.

The Prussian army that entered the Napoleonic wars was the legacy of Frederick the Great. The infantry employed three-rank linear formations and attempted to dominate adversaries by delivering controlled volleys while advancing steadily across the battlefield. As war with France approached, Prussia implemented limited tactical reforms to make its army, in the words of Frederick William III, "more like the French."[29] Such limited reforms did not, however, avert Prussia's crushing defeat at Jena and Auerstädt. Those disasters demonstrated beyond doubt the urgency of reform. The resulting Treaty of Tilsit deprived Prussia of its richest provinces and reduced its population to 4.5 million, while the Convention of Paris limited the Prussian army to forty-two thousand men.[30] Prussia's only hope of regaining its military strength lay in mobilizing the resources of the state.

[26]Otto von Pivka, *Napoleon's German Allies*, vol. 1: *Westfalia and Kleve-Berg* (London: Osprey, 1975), pp. 8, 20.
[27]Rothenberg, *The Art of Warfare*, p. 160.
[28]Elting, *Swords around a Throne*, p. 397.
[29]Rothenberg, *The Art of Warfare*, pp. 188–89.
[30]Ibid., p. 191.

The catastrophe convinced the King and his advisors that fundamental change was required to preserve Prussia. On 15 July 1807 the king established the Military Reorganization Commission. Dominated by liberals such as Major General Gerhard Johann von Scharnhorst, the commission members recognized that military reform had to proceed in parallel with political reform. The serfs and mercenaries that filled the rolls of the Prussian army lacked the ability and inclination to adopt the tactics of Napoleon's *Grande Armée*. Scharnhorst realized that Prussia had first to transform its subjects into citizens, and then train them in a more flexible and effective tactical system. On 9 October 1807 Prussia freed its serfs. The next year it enacted limited municipal self-government. No longer mercenary, the army was motivated by patriotism and hatred of the French.

Prussia countered Napoleonic military methods with an innovation of its own. Between 1803 and 1809, Prussia created a General Staff to anchor planning within the army. The General Staff allowed Prussia to harness the collective intelligence of its brightest officers as a counterweight to Napoleon's individual genius.[31] The chief of staff became a full partner in military decision-making, and general staff officers were seconded to army headquarters to plan future operations.

Prussia also transformed its officer corps. In September 1807, the Military Reorganization Committee recommended the abolition of the nobility's monopoly on entry into the officer corps. While the aristocracy continued to dominate, a portion of the corps was opened to merit appointments. Prussia also established a comprehensive and increasingly efficient system to educate its officers, culminating in the formation of the Prussian war college.

Frederick William III balked at proposals to institute universal military service. Under pressure from reformers, in 1813 he authorized the formation of volunteer *Jäger* battalions. By royal order he established a statewide militia (*Landwehr*), as well as a Landsturm consisting of all men between the ages of eighteen and sixty not in the Landwehr. The following year the Prussian army formalized its conscription system. Draftees spent three years in the regular army, two in reserve, thirteen years in the Landwehr, and ten in the Landsturm. This innovative approach formed the basis of Prussian army organization for the remainder of the nineteenth century and spread to other armies after the Wars of German Unification.

Prussian military reformers also introduced the brigade and corps as permanent higher tactical formations. Indeed, the Prussians went beyond the French by standardizing their organizations.[32] Prussia adopted the combined-arms brigade—a self-sufficient unit composed of infantry, cavalry, and artillery formations—as its

[31]Ibid., p. 192.
[32]Ibid., p. 193.

basic tactical unit.[33] Prussia also grouped four or more brigades and cavalry and artillery elements into a corps composed of some thirty to thirty-two thousand
men.[34]

The Prussian army that faced Napoleon at Waterloo was thus substantially
different from what it had been a decade earlier. Prussia had expanded, reorganized, and retrained its army. While Prussia had copied elements of the Napoleonic military system, it had also innovated by forming a General Staff and an
organized reserve system. In so doing it demonstrated that a stable, authoritarian,
yet reform-minded regime could institute an effective program of military modernization.

Austria possessed the largest army continually engaged against France. Prior to
1805, the Habsburg army was a classical eighteenth-century force manned primarily by long-service professionals employing linear tactics and organized into
regiments.[35] Early efforts to reform the Austrian army in response to the French
threat were stillborn. Archduke Charles, Austria's most talented commander,
made several attempts to modernize the bureaucracy and military, but the emperor rejected his proposals. The regime's reluctance to reform flowed from the
very nature of the Habsburg empire with its diverse political institutions and na-
tionalities.[36] For Austria, becoming a nation-in-arms bore the risk of anarchy.

Humiliating defeats at Ulm and Austerlitz convinced the military leadership of
the need for change. Easiest to implement were tactical reforms. In January 1806,
Austria adopted a new cavalry regulation, followed in March 1807 by one for the
infantry.[37] Austria also began employing the division "mass," composed of six
companies positioned in two columns, and the battalion "mass," a single column
one company wide and six deep.[38] But these changes were incremental at best.

After Ulm and Austerlitz, the Austrian army began organizing its units into
divisions and corps. Austrian corps, like the French, varied in size and composition but were generally smaller than their French counterparts. In 1808–9, for example, each Austrian corps was composed of twenty to thirty infantry battalions,
sixteen to twenty-four cavalry squadrons, and seventy to ninety guns.[39]

While the Austrian army embraced tactical and organizational reform, the empire's social composition blocked recruitment modifications. The polyglot make-

[33]Nosworthy, *With Musket, Cannon, and Sword*, p. 167.

[34]Ross, *From Flintlock to Rifle*, p. 147.

[35]Robert M. Epstein, *Napoleon's Last Victory and the Emergence of Modern War* (Lawrence: University Press of Kansas, 1994), pp. 22–23; Scott Bowden, *Napoleon and Austerlitz* (Chicago: Emperor's Press,
1997), ch. 3.

[36]Rothenberg, *The Art of Warfare*, p. 166.

[37]Nosworthy, *With Musket, Cannon, and Sword*, pp. 159–60.

[38]Ibid., p. 160.

[39]Rothenberg, *The Art of Warfare*, p. 170.

up of the Habsburg empire prevented a levée en masse. The Hungarians refused to accept conscription, despite attempts to make service more attractive by shortening the term of enlistment. Moreover, the Austrian nobility feared that discharged soldiers might fuel popular revolt. Austria would not open its officer corps to promotion based on talent. Instead the government made incremental adjustments to the existing system by dismissing generals whose battlefield performance was lacking.[40]

Archduke Charles's attempt to create a militia was a temporary success. Harnessing German nationalism within the empire, in June 1808 the emperor established a German-speaking militia, the Landwehr, enrolling all men not already serving in the army.[41] While the strength of the Landwehr reached 240,000 troops, the emperor disbanded it the following year, fearing an armed populace and national revolt.[42]

Before 1805 the Russian army also remained wedded to eighteenth-century military methods. Its tactics were conservative, relying on deep, dense columns of bayonet-armed infantry. Cooperation between combat arms was lacking. The army suffered from a primitive staff system and possessed no fixed formations above the regiment.[43]

In Russia, as in Austria, the defeat at Austerlitz awakened interest in tactical reform. In the years that followed, Russia adopted elements of Napoleonic tactics including the battalion column, increased use of light infantry, and employment of skirmishers in close coordination with line and column formations.[44] It also strengthened its artillery arm.[45]

After 1807, Russia began to expand its army. By June 1812 it had 409,000 men under arms.[46] Shortly before the French invasion, the Russian minister of war, Barclay de Tolly, began organizing the army into corps of two infantry divisions, a regiment or brigade of cavalry, and one or more artillery companies.[47] While the reorganization was clearly patterned on the French model, the low level of enlisted and officer training and rivalries within the command prevented Russian troops from achieving the flexibility or speed of the French.[48]

Like Europe's other monarchies, the Russian government was concerned

[40]Ibid., pp. 166, 169–70.

[41]Epstein, "Patterns of Change and Continuity," p. 378.

[42]Rothenberg, *The Art of Warfare*, p. 172.

[43]Epstein, *Napoleon's Last Victory*, p. 23; Bowden, *Napoleon and Austerlitz*, ch. 2.

[44]Ross, *From Flintlock to Rifle*, p. 140.

[45]Rothenberg, *The Art of Warfare*, pp. 201–2.

[46]Ross, *From Flintlock to Rifle*, p. 141.

[47]Philip Haythornthwaite, *The Russian Army of the Napoleonic Wars*, vol. 1: *Infantry, 1799–1814* (London: Osprey, 1987), p. 9.

[48]Rothenberg, *The Art of Warfare*, p. 203.

about the threat that revolutionary political and social ideas posed to the empire's stability. Like Austria, Russia attempted to adopt French tactics and organizations without fundamentally changing its political institutions. Throughout the war, the enlisted ranks of the army were filled by conscripted serfs selected by local authorities to fill quotas. Officers were recruited from the Russian upper class and foreigners.[49] Despite their best efforts, however, the Russian leadership did not completely avoid the contagion of revolutionary thought. Officers infected with French political ideas launched the 1825 Decembrist uprising against the czar.

The British army remained a pre-Napoleonic force throughout the Napoleonic wars. Because of its small size and limited use, the British army never faced the need to reform that Continental armies did. Indeed, as Hew Strachan has observed, "Waterloo was won by an army that had made minimal concessions to Napoleon or to the nineteenth century."[50]

Rather than abandoning the linear system of combat, the British refined it to increase its effectiveness. The army abandoned the three-rank firing line in favor of a two-rank line, a move that allowed troops to achieve fire superiority by bringing more muskets to bear on a broader front than the French. The infantry supplemented its firepower with guns loaded with canister rounds placed in direct support of the troops.[51] While the French tried to work their men up before an attack, the British attempted to maintain calm. On the battlefield, British troops would advance slowly at a deliberate pace, delivering concentrated volleys at decreasing ranges.[52]

The British also formed light infantry units. While some had advocated the move as early as 1797, not until Napoleon organized his "Army of England" at Boulogne did the army decide to transform entire line battalions into light infantry. Between 1803 and 1804, Sir John Moore trained three light infantry regiments; others followed later in the wars.[53] By 1809 about 20 percent of the British army could operate as skirmishers, and British light troops were every bit as good as their French counterparts.[54]

While the British introduced some limited tactical reforms, they did not change their army's composition. The size of the army in the field rarely reached forty thousand. Officers were generally drawn from the British upper class and

[49]Ibid., pp. 196–97.

[50]Hew Strachan, *From Waterloo to Balaclava: Tactics, Technology, and the British Army, 1815–1854* (Cambridge: Cambridge University Press, 1985), p. 1.

[51]Rothenberg, *The Art of Warfare*, p. 183.

[52]Nosworthy, *With Musket, Cannon, and Sword*, p. 223.

[53]Richard Glover, *Peninsular Preparation: The Reform of the British Army, 1795–1809* (Cambridge: Cambridge University Press, 1963), pp. 126, 129.

[54]Ross, *From Flintlock to Rifle*, p. 128.

purchased their commissions and promotions.[55] Throughout the Napoleonic wars, Parliament shrank from conscription, relying instead upon voluntary enlistment. As a result, the ranks were filled with long-service volunteers, most of whom were uneducated and many of whom were outcasts.[56]

The basic administrative unit of the British army remained the infantry regiment. While the army formed temporary brigades to execute specific missions, Wellington first organized divisions in 1809 on the Iberian Peninsula. Divisions remained the highest tactical formation until 1815, when he organized the combined British and Allied force into three infantry and one cavalry corps.[57]

The British army did not embrace Napoleonic warfare because, unlike its Continental allies, it did not have to. The British army, unlike its Prussian, Austrian, and Russian counterparts, never suffered a traumatic defeat at the hands of the French. The Royal Navy, backed by the British treasury, bore the brunt of the struggle against France. Britain was Napoleon's only adversary to retain its traditional approach to combat.

Industrialization and Military Change

If the Napoleonic wars forced states to adapt to the rise of nationalism, the Wars of German Unification marked the first successful attempt to cope with the military challenges posed by the economic revolution that accompanied industrialization. Landes defines industrialization as the substitution of machines for human effort, and the substitution of inanimate for animate sources of power.[58] Industrialization altered the nature and production of weapons, the transport of troops and supplies, and methods of communication. It increased the complexity, scope, scale, and lethality of warfare. Mass-produced rifles, iron railways, steel battleships, and eventually trucks, tanks, and airplanes became de rigueur by the early twentieth century as major armed forces adapted to the industrial revolution. If the Napoleonic revolution was about new principles for employing men in combat, the industrial revolution was about new principles for managing men and materiel under increasingly mechanized and complex circumstances.

Prussia adapted to the industrial era sooner and more effectively than its rivals. The rest of the world was slower to respond; some states did not fully adapt to the industrialization of war until well into the twentieth century. Some states at-

---

[55]Officers of Royal Artillery and Engineers, by contrast, were professionals. In these branches, technical training was a prerequisite for commissioning, and promotion was by seniority. See Rothenberg, *The Art of Warfare*, p. 176.

[56]Ibid., p. 174.

[57]Ibid., p. 182.

[58]David Landes, *The Unbound Prometheus: Technological Change and Industrial Development in Western Europe from 1750 to the Present* (New York: Cambridge University Press, 1969), p. 41.

tempted to emulate Prussian military methods but failed; others managed to adopt aspects of Prussian practices; still others chose an alternative pattern of adaptation even while admitting the superiority of the Prussian example.[59]

Industrialization changed the technological, tactical, operational, and administrative (including educational and training) dimensions of warfare. The technological dimension (or *techne*) embraces the thing(s) only. Adopting new techne is rather simple, but how it is put to use can vary considerably.

Three technologies are central to military developments in this period: the breechloading rifle, the railroad, and the telegraph. At the tactical level, new technologies included the rifle, lightweight rifled, steel-barreled artillery, smokeless powder, and, in the early twentieth century, machine guns, TNT, and other high explosives. In each case, the new materials and mass-production techniques made the tools of war cheaper and faster to mass manufacture, easier to use, more powerful, longer ranged, and deadlier.

At the operational level, innovations in transportation and communications technologies—the railway and telegraph revolution—changed the relationship of time and space to military objectives. Most obviously, railways increased the speed at which troops could move, and the telegraph increased the speed of communication between headquarters and front. In so doing, the twin technologies changed the nature of mobilization, deployment, and even movement during battle.

The combination of these changes made warfare a much larger and more complex undertaking. The increased size of armies made possible by a combination of technological change and the rise of nationalism increased the cost and complications associated with recruitment and training. Increased complexity placed a greater premium on management and organization—not just of armies in the field but increasingly also of prewar preparation and planning. The industrial revolution thus had a profound effect on the administrative and training side

---

[59]The Prussian system may not have functioned as well (or in the same manner) as it was reputed. The same military and political leadership that constructed the Prussian military machine led the army into the disaster of World War I. Herwig argues that because the political objectives of the Prussian and German militaries far exceeded their resources, they must be considered a failure. German autocracy was incompatible with mass politics: "Germany proved incapable (or unwilling) fully to comprehend and to adjust to the new industrial war of the masses." Holger H. Herwig, "The Dynamics of Necessity: German Military Policy during the First World War," in Allan Reed Millett and Williamson Murray, eds., *Military Effectiveness*, vol. 1: *The First World War* (Boston: Allen and Unwin, 1987), p. 107. See also Holger H. Herwig, *The First World War: Germany and Austria-Hungary, 1914–1918* (New York: St. Martin's, 1997). Cox argues that the French created a more "modern" system than the Germans. The French industrialized military was under more civilian scrutiny than the German, and achieved a more stable relationship with democratic politics. The Germans created a highly unstable coupling of an atavistic political system with a modern industrialized economy and military. Gary P. Cox, *The Halt in the Mud: French Strategic Planning from Waterloo to Sedan* (Boulder, CO: Westview, 1994), p. ix.

of military practice as well as on the battlefield. These observations provide the four-part analytic framework for this section of our study: the technological, the tactical, the operational, and the administrative. The following sections detail the Prussian and German developments in each of the four areas, and the adaptive efforts of six militaries: the Austrian, French, Russian, British, Japanese, and Chilean.

The Prussian Adaptation

The Prussian approach successfully applied industrial technology to the tactical, operational, and administrative dimensions of war. No other state excelled in all three. The reform of the Prussian army after Jena and Auerstädt bore full fruit in the second half of the century. Helmuth von Moltke, the head of the Prussian General Staff, honed the application of industrialization to warfare over the course of the Danish War (1864), the Austro-Prussian War (1866), and the Franco-Prussian War (1870–71).

At the tactical level, the breechloading rifle posed both problems and opportunities. Until the 1850s the smooth-bore muzzle-loading musket was standard issue in European armies. The accuracy of the muzzle-loader dropped off precipitously after about 160 yards; the weapon was time consuming to load, and it required that the infantryman stand to do so.[60] European armies had adapted their tactics to the limitations of their guns. Infantrymen were massed tightly together and trained to fire and load in sequence to ensure a continuous stream of fire and maximize its deadliness.

Gunmakers had long known that rifling the barrel would increase the range and accuracy of the weapon and that breechloading would allow marksmen to reload much faster and while kneeling or even prone.[61] But a rifled barrel was much harder to load at the muzzle, and existing breechloaders, limited by available metals and production techniques, were too fragile for military use. The Prussian craftsman Nikolaus von Dreyse had worked out a solution to the problem of the breechloader earlier in the century, but it was not until the 1850s that developments in breechloading and rifling received formal military attention. In

---

[60]In 1813, Scharnhorst conducted a series of tests on accuracy and rates of fire of six different muskets and one rifle using two kinds of bullets. The average rate of fire for the muskets was two to two and a half rounds per minute. With small targets, the rifles were twice as accurate at 160 yards and four times more accurate at 240 yards. Yet loading, aiming, and firing a rifle took five times as long at 240 yards as a musket. Scharnhorst concluded: "Rifle and musket must have about the same effect in the same period of time; but the musket needs three to four times as much ammunition as the rifle." Cited in Peter Paret, *Yorck and the Era of the Prussian Reform, 1807–1815* (Princeton: Princeton University Press, 1966), p. 271. See also William H. McNeill, *The Pursuit of Power: Technology, Armed Force, and Society since A.D. 1000* (Chicago: University of Chicago Press, 1982), p. 231.

[61]The rifles of the 1860s had an effective range of about 1,000 yards. Hew Strachan, *European Armies and the Conduct of War* (London: Allen and Unwin, 1983), p. 112.

1858 the regent Prince William decided, after a lengthy debate (including consideration of a French rifling system designed in 1849 by the French army captain Claude Etienne Minié) to have the entire Prussian army equipped with the Dreyse needle-gun.[62]

Many European armies adopted rifles. For example, the Minié was standard issue in the French army by 1857, and the British troops in Crimea carried the same weapon.[63] In 1860, however, Prussia alone possessed large numbers of breechloaders. The needle gun allowed troops to fire four times faster than they could with muzzle-loading rifles. But crude construction, a fragile firing pin, stiff bolt action, and a leaky breech gave the weapon the disconcerting tendency to explode.

Each European army adapted to the advent of the rifle in its own way. The British introduced the "line of battalion columns," a group of columns joined behind a skirmish line of light infantry and reinforced by artillery. The Russians, lacking money and expertise, sought to counter the rifle with fast-moving "steamroller" columns so deep that they could take heavy casualties and still overrun an enemy. The French planned to use battalion masses alternatively deployed in column and line to deliver salvo fire, then charge with the bayonet.[64]

The Prussian army wavered until 1864, when it adopted fire tactics to exploit the superior rate of fire of the needle gun. Instead of close-order ranks, the needle gun–armed Prussians dispersed the infantry in open order formations, stretching thin the ranks of men. The critical issues were fire control and marksmanship, and therefore discipline. In open ranks, infantry are farther from their commanding officer and harder to control. With the breechloader, officers were concerned that infantry would fire all their ammunition at the opening stage of the engagement and effectively disarm themselves. The Prussians compensated with extensive training. Infantry were well drilled in marksmanship and the judicious use of their ammunition (for example, before the Austro-Prussian War, the Prussian army was allowed five times the number of practice shots during training as their Austrian counterparts).[65] The superior range, accuracy, and loading speed more than compensated for the decrease in troop density, in effect making the same number of men much deadlier all the while increasing the amount of territory they could effectively defend or attack.

The results were quick, disciplined, and powerful army units. The Prussians

---

[62]Dennis E. Showalter, *Railroads and Rifles: Soldiers, Technology, and the Unification of Germany* (Hamden, CT: Archon, 1975), p. 73.

[63]McNeill, *Pursuit of Power*, p. 231.

[64]Geoffrey Wawro, "An 'Army of Pigs': The Technical, Social, and Political Bases of Austrian Shock Tactics, 1859–1866," *Journal of Military History* 59, no. 3 (July 1995): 412–13.

[65]Geoffrey Wawro, *The Austro-Prussian War: Austria's War with Prussia and Italy in 1866* (Cambridge: Cambridge University Press, 1996), p. 134.

displayed their new military prowess first in the 1864 war with Denmark, but their most effective demonstration came in the 1866 war against Austria. Prussian infantry faced the Austrian shock tactics with open ranks and fired at will to crushing effect. Foreign observers credited the needle gun with Prussia's stunning victory. While the incompetent Austrian command and its outdated tactics and strategy were also to blame, the lopsided casualty figures underscore the superiority of Prussian methods. Units facing anywhere from two to ten times their number routinely beat back Austrian advances or successfully attacked. Losses like those at Vysokov on 27 June—where Austrian casualties were five times those of the Prussians—were routine. During the decisive battle of Königgrätz, the Prussians lost 8,800 men to 41,500 for the Austrians.[66]

Prussian adaptations to the new operational environment were superior to those of their European counterparts. The advent of dense railway and telegraph networks on the Continent increased the speed with which armies could mobilize, deploy, and coordinate their movements. The field telegraph allowed for faster and surer command and control of armies in the field. Typically, even with the rather slow speeds of the early rail networks, movement by train was six times faster than marching. Trains could also move large quantities of troops efficiently, and, perhaps most important, armies arrived at their destination rested and (in principle) well fed. Three- to four-day marches, often carrying troops through the night with little or no sleep or food, would do as much to sap the fighting strength of units as poor training, weapons, or tactics. Railways and telegraph networks could not eliminate all such movements, but planned maneuvers such as initial mobilizations could move more troops to the intended battlefield faster—thereby increasing the overall power of armies and giving the strategic advantage to the faster and more efficient mobilizer.

Dense rail networks had to be constructed, and, unlike rifled shoulder weapons, were beyond the fiscal reach of European governments and private firms acting alone.[67] A lengthy railway line—with its attendant locomotives, rolling stock, switching and communications equipment, operations and maintenance staff, and so forth—was the most expensive capital project an economy could undertake in the mid-nineteenth century. Public-private cooperation was necessary. All governments, even the laissez-faire British, had to fund railway construction to some extent—either through loan supports or outright financing of certain strategic lines.

Despite government aid, the European rail net was mostly laid by private en-

[66]Gordon A. Craig, *The Battle of Königgrätz: Prussia's Victory over Austria, 1866* (New York: J. B. Lippencott, 1964); Wawro, *Austro-Prussian War*. Casualty figures from Wawro, p. 143; and Craig, p. 166.
[67]See Geoffrey L. Herrera, "The Mobility of Power: Technology, Diffusion, and International System Change," Ph.D. dissertation: Princeton University, 1995.

trepreneurs to serve the interests of private industry and trade. With few exceptions—the most dramatic being the Berlin-Königsberg Railway in East Prussia and the Trans-Siberian Railway in Asiatic Russia—states did not build railways for military purposes but, in the event of war, were left with the rail net their private economy had produced.[68] Thus the more developed the economy, the denser the national rail net; and the denser the rail net, the greater the military possibilities and the greater the strategic advantage.

The German states were late industrializers, but robust ones. When Prussia confronted Austria in 1866, the Prussian army had five rail lines with which to direct their mobilization into Austria. Austria had one. In the 1870–71 war, the French were as well supplied with railway lines as the German states. But France's disastrous rail mobilization in 1870 shows that a dense rail net was only a necessary condition for the successful exploitation of the military potential of railways.[69] Railways are complex technological systems. For militaries to use railways effectively, they must spend time planning train schedules, figuring speed and carrying capacity, arranging for loading and unloading, and coordinating with civilian railway authorities. The Prussian army was the first to form a railway section within its General Staff—a move that acknowledged the importance of planning when it came to the strategic use of railways.[70] The Prussian rail net was no different from the French, or any other. It was mostly privately owned and administered. The Prussian difference was in the way the army coordinated with civilian companies and planned the mobilizations in advance.

Prussia's wars against Austria and France demonstrated Berlin's mastery of the strategic use of railway mobilization. Napoleonic doctrine, spread by his success and proselytizing strategists like Jomini, stressed the importance of internal lines and the concentration of forces before the battle to maximize the power the commander could bring to bear against the enemy. Moltke realized the inherent limits to the doctrine, and that the railway-telegraph combination offered new options. Armies continued to grow in size throughout the nineteenth century, as did the amount, weight, and complexity of material they brought to the battlefield. Concentration before battle in the Napoleonic mode became logistically unwieldy—too many soldiers, horses, wagons, and guns competing for the same space. Meanwhile, the rapidity and ease of mobilization by rail allowed for sepa-

---

[68]William O. Henderson, *The State and the Industrial Revolution in Prussia, 1740–1870* (Liverpool: Liverpool University Press, 1958), p. 167; Steven G. Marks, *Road to Power: The Trans-Siberian Railroad and the Colonization of Asian Russia, 1850–1917* (Ithaca: Cornell University Press, 1991).

[69]Thomas J. Adriance, *The Last Gaiter Button: A Study of the Mobilization and Concentration of the French Army in the War of 1870* (New York: Greenwood, 1987); Michael Howard, *The Franco-Prussian War: The German Invasion of France, 1870–1871* (New York: Macmillan, 1961), pp. 68–71; Edwin A. Pratt, *The Rise of Rail Power in War and Conquest, 1833–1914* (London: P. S. King and Son, 1916), pp. 138–48.

[70]Pratt, *Rise of Rail Power*, p. 104.

ration before battle. Troops would still have to make the trip from the railhead to the field of battle on foot—as Prussian forces did in all three midcentury wars. But precise use of rail could bring troops closer to the battlefield much faster and with more carefully controlled timing—thereby reducing the uncertainty of long marches. If timed right, the separated components of the army could meet on the field of battle, avoid the delay of a similarly sized concentrated force, and so bring more force to bear faster. The army in effect operated on external, not internal lines. Moltke realized, as Showalter puts it, that "[v]ictory in war . . . was not merely a function of numbers; it depended at least as much on the time needed to bring these numbers into action."[71]

Railways altered the relationship between time and space in military campaigns, and Moltke was the first to grasp this. Using the railroad for mobilization and deployment also created new vulnerabilities. If one component arrived late (or early), the entire battle plan was jeopardized. Prussian commanders made both mistakes at the battle of Königgrätz in July 1866, much to Moltke's chagrin, but the Austrians were unable to capitalize on it. Nevertheless, Moltke's strategic intent was to encircle the Austrians not by out-maneuvering them on the battlefield but by outflanking them before the battle was even engaged. In the end, the plan worked; movement by rail set the three Prussian armies around a broad perimeter. Their (somewhat imprecise) marching completed the encirclement. The Second Army arrived late in the afternoon of 3 July, but in time to complete the envelopment and routing of the Austrians.[72] The Prussian army commanders had more faith in Moltke's system in 1870, and the mobilization went more smoothly. In the end, the results were identical—a decisive victory for the Prussians via strategic encirclement.

The Prussians successfully adapted their strategy to the new technological environment in both the Austro-Prussian and Franco-Prussian wars. The industrial strength of the Prussian economy was crucial, but equally important was the use they made of their industrial raw materials. Railway use required precise timing, which in turn increased the importance of careful peacetime planning. None of the European armies could match the Prussian General Staff's training, technical expertise, diligence, and attention to detail. As Howard put it, "The Prussian General Staff acted as a nervous system animating the lumbering body of the army, making possible that articulation and flexibility which alone rendered it an effective military force."[73] The superiority of the Prussian General Staff points to the third facet of the changing military environment—the administrative.

The railway, telegraph, and rifled breechloader shared a common feature: their

[71]Showalter, *Railroads and Rifles*, p. 57.
[72]Craig, *Battle of Königgrätz*; Wawro, *Austro-Prussian War*.
[73]Howard, *The Franco-Prussian War*, p. 24.

effective use placed increased importance on planning, preparation, and administration. They required a managerial or bureaucratic revolution within the army. The Prussian administrative achievement had two dimensions: establishing a permanent, politically insulated technical and planning institution (the Great General Staff) and creating a nation-in-arms—a system of manpower recruitment based on compulsory service combining active-duty forces and sizable, well-trained reserves.

The centerpiece of Prussian adaptation to industrial warfare was the General Staff. Other European armies had central planning institutions. But the Prussians embraced technical and scientific principles and practices and extensive prewar planning with an unmatched degree of enthusiasm and attention to detail. The Prussians were the first to establish a railroad department within their General Staff, and technical and scientific training for officers exceeded anything mandated elsewhere.[74] The focus on training and technical education was in part cause, in part consequence, of the embourgeoisement of the Prussian officer corps.[75] While the army reforms of the nineteenth century—especially promotion by performance at the Kriegsakademie—had radically changed the class composition of officers, bourgeois officers were overwhelmingly concentrated in technical fields such as railways, mapping, and artillery. The infusion of bourgeois attitudes helps explain the greater technical orientation of the Prussian military; their concentration in technical fields (rather than more prestigious areas such as cavalry) is evidence of lingering aristocratic privilege and bias.[76]

The Prussian General Staff also differed markedly from other European general staffs in its political role. Over the course of the nineteenth century, it moved from advisor to the Prussian monarch to play a key role in the formulation of strategy and even, in the end, policy. This transformation was due in large measure to the singular figure of Moltke. He demanded autonomy, and his unparalleled military successes combined with a certain lack of political ambition led the king to place a great deal of trust (and power) in him.[77] Yet the political insulation was also a product of the institutional design and political culture of the Prussian military.[78] Overall, the technical expertise and political empowerment of the General Staff created a command system that combined technical mastery

---

[74]Pratt, *Rise of Rail Power*, p. 104; Strachan, *European Armies*, p. 126.

[75]Walter Görlitz, *History of the German General Staff, 1657–1945* (New York: Praeger, 1953), p. 96.

[76]Gordon A. Craig, *The Politics of the Prussian Army, 1640–1945* (New York: Oxford University Press, 1964).

[77]Ibid., pp. 180–216, 226–30; Larry H. Addington, *The Patterns of War since the Eighteenth Century* (Bloomington: Indiana University Press, 1994), p. 50.

[78]Craig, *Politics of the Prussian Army*. Hajo Holborn, "The Prusso-German School: Moltke and the Rise of the General Staff," in Peter Paret, ed., *Makers of Modern Strategy: From Machiavelli to the Nuclear Age* (Princeton: Princeton University Press, 1986), p. 284.

with war-planning autonomy, direct access to the sovereign, and insulation from the political system.[79]

The last piece of the German achievement was the army recruitment system that allowed Germany to put more troops on the field than all European armies save the Russian—and, more important, better-trained ones.[80] The system brought the entire population of young and middle-aged men into military service. All Prussian men were required to give three years of active service followed by eight in the reserves. In both they received extensive training, and few exemptions were permitted. The reserve system was an economizing measure—the Prussian army could have a massive, well-trained army without having to pay for all of it during peacetime.[81]

The Prussian reformers in the two decades between 1810 and 1830, and Moltke in the 1850s and 1860s, realized that war had become a complex administrative undertaking. As Howard has put it, the year "1870 was as much a victory for Prussian bureaucratic methods as it was for Prussian arms. . . . The romantic heroism of the Napoleonic era . . . was steam-rollered into oblivion by a system which made war a matter of scientific calculation, administrative planning, and professional expertise."[82]

The Spread of the Prussian Model

The Prussian army adapted its manpower recruitment, tactics, strategy, and administration to better fit the mechanization and increased complexity of industrial-era warfare. The system was less than perfect—the General Staff struggled with the problem of supplying troops after mobilization into World War I. The army was, as the war demonstrated, too dependent on a willful, intelligent monarch. The system was an aberration of Germany's peculiar autocratic political system. But because of its spectacular battlefield successes against Austria and France, it was widely imitated.[83]

After 1870, France, Austria, Italy, Russia, and Japan attempted to copy the German General Staff. Many others—including Greece, Turkey, China, Chile, and Argentina—either sent observers to Germany or received a German military mission.[84] The German conscript system, while controversial in nations where the aristocracy was accustomed to getting a free pass from military service if they

[79]Wawro, *Austro-Prussian War*, p. 284.

[80]Addington, *Patterns of War*, p. 103.

[81]Ibid., p. 52.

[82]Michael Howard, *War in European History* (New York: Oxford University Press, 1976), p. 101.

[83]Martin Van Creveld, *Supplying War: Logistics from Wallenstein to Patton* (New York: Cambridge University Press, 1977); Herwig, "Dynamics of Necessity"; Wawro, *The Austro-Prussian War*, p. 284; Howard, *War in European History*, p. 101.

[84]Görlitz, *History of the German General Staff*, p. 97.

wished, was also widely imitated. Many states also rushed to accommodate the newest weapons technology. Indeed, from 1870 to 1914 rapid technological change was often driven forward by the militaries themselves.[85]

Yet factors such as geostrategic position, social structure, economic development, and national wealth, along with administrative disorganization and ineffective command, complicated emulation of Prussian military techniques. Although historians note that European armies successfully emulated Prussian practices—for example, Addington writes: "In the years after 1870 France, Italy, Austria-Hungary, and Russia followed Germany in adopting the *Nation-in-Arms* and the German-style war college and general staff"[86]—upon closer examination, imitation of Prussian innovations was partial at best.

Adoption of breechloading rifles (and later machines guns) was easy enough for those militaries that could afford the new weapons, and most militaries learned not to attack using Napoleonic-style shock tactics. Yet none learned the central tactical lesson of Moltke's successes: increased firepower made the tactical defensive a necessity. The Germans even managed to unlearn the lesson.

Most learned that the key to mobilizing via rail was extensive prewar preparation. But no amount of preparation could compensate for economic backwardness and an undeveloped railway net. The length of Austria's rail network increased by nearly six times between 1866 and 1914, but it was still about one-third the length of Germany's. Russia's backwardness severely hampered its efforts in the Russo-Japanese War, and on the brink of World War I, loans from ally France were needed to boost the efficacy of the empire's western strategic railways.[87]

In army administration, most adopted the form of the Prussian General Staff without the substance. The emphasis on education, training, and technical expertise cut against the grain of the dashing and brave officer that was the heart of the corps's self-image. The technical (and command) incompetence of the French and Russian officer corps was evident even into World War I.[88] In addition, none of the imitators achieved the level of separation from day-to-day poli-

---

[85]David G. Herrmann, *The Arming of Europe and the Making of the First World War* (Princeton: Princeton University Press, 1996); Strachan, *European Armies*, pp. 108–28.

[86]Addington, *Patterns of War*, p. 102.

[87]Figures from B. R. Mitchell, *European Historical Statistics, 1750–1970* (New York: Columbia University Press, 1975). Bruce Menning, *Bayonets before Bullets: The Imperial Russian Army, 1861–1914* (Bloomington: Indiana University Press, 1992); L. C. F. Turner, "The Russian Mobilization in 1914," in Paul M. Kennedy, ed., *War Plans of the Great Powers, 1880–1914* (London: Allen and Unwin, 1979), p. 257.

[88]Allan Mitchell, "'A Situation of Inferiority': French Military Reorganization after the Defeat of 1870," *American Historical Review* 86, no. 1, Supplement (Feb. 1981): 49–62; Douglas Porch, *The March to the Marne: The French Army, 1871–1914* (Cambridge: Cambridge University Press, 1981); Menning, *Bayonets before Bullets*; Herrmann, *Arming of Europe*, p. 92.

tics of the German War Ministry. Finally, although all undertook conscription reform in earnest, none were able to replicate completely the German nation-in-arms. In most instances, lingering aristocratic privilege (and the advantages of wealth) prevented full and complete implementation of universal service laws. Reforms undertaken in the Austrian, French, Russian, British, Japanese, and Chilean armies in the wake of the Prussian achievement attest to the incomplete diffusion of the German innovations.

Although an oft-stated intention, Austro-Hungarian emulation of the Prussian army was thwarted by political divisiveness and an intractable faction of the officer class. The persistent economic and educational backwardness of the empire prevented the army from wholeheartedly taking on the range of reforms that required money and conviction.

Austria had an opportunity to adapt to industrial warfare well before the decisive 1866 encounter with Prussia. The rifle had been available to European armies since the Crimean War, but the Austrians, aside from some minor changes to linear formations, retained Napoleonic shock tactics. The army could not afford to train its uneducated troops to use rifles effectively, and the officer corps and emperor were ideologically committed to the older tactics.[89] In the 1864 Danish War, the Austrians and Prussians together trekked north to bring the German states of Schleswig and Holstein back into the German Confederation. The Austrians employed shock tactics and the Prussians line and enfilade, or raking gunfire, almost side-by-side. The effectiveness of the rifle was clear to Moltke and his staff, who even checked Prussian troops for evidence of hand-to-hand combat after a Danish bayonet charge (there was little, indicating that the Prussian casualties had been able to stop most of the Danes with rifle fire before they reached the Prussian line). But the Austrians evaluated and then rejected the breechloaders after the war.[90]

Two years later, in the war with Prussia, not only did the Austrians fail to adapt their shock tactics to Prussian firepower but, in addition, every aspect of the campaign showed them to be substantially behind their European counterparts. Along with Austria's deficient railway net, the incompetence of the Austrian command was severely exposed. The most elementary of staff functions, such as the provision of accurate and up-to-date maps, proved beyond the capability of the Austrian general staff. Orders were delivered slowly, incompletely, or not at all. General Ludwig Benedek (the commander in the Bohemian theater) and his officers refused to use the field telegraph provided for them. The troops

---

[89] Given its long range, firing a rifle required a bit of calculation—first of distance to the target and then of the parabolic offset—something the Austrian officer corps felt their peasant conscripts could not do. Wawro, "An 'Army of Pigs,'" p. 419.

[90] Ibid., pp. 424–31.

themselves were uneducated conscripts frequently unable to understand their German-speaking officers. The officer corps, in turn, did not think their infantry capable of following orders of even moderate complexity, so tactics were kept simple.[91]

After the war, the Austrian military undertook a series of reforms based loosely on the Prussian model. In 1868 the army set the term of service for conscripts at three years in the regular army and eight more in the reserves, a direct emulation of Prussian practice. In 1881 the modernizer Friedrich Beck was appointed chief of the General Staff. Beck worked to centralize war-planning in the General Staff, reform the reserve system, expand the railway network, and streamline mobilization. In a significant modernizing step, he converted the national-based home defense units (the Austrian Landwehr and Hungarian *Honvéd*) into a reserve army. And by 1889 the Austrian army adopted the Mannlicher, a rifle taking advantage of the latest technological developments, including smokeless powder and magazine loading.[92]

But as one of the most economically backward of the great powers (more than half of Austria's population was engaged in agriculture in 1910, two-thirds in Hungary),[93] the empire could scarcely afford a modern military. Budgets were routinely shortchanged with clear effects on manpower and equipment. Austrian artillery in 1914 was obsolete—without more recent developments such as pneumatic recoil mechanisms and gunner's shields. Their Serbian opponents in 1914, most of whom carried the German Mauser, were armed with a superior rifle. Problems with manpower persisted: even with the three-year term of service, in 1910 the army called up only 126,000 conscripts out of a population of fifty million, and in 1914, forty-eight Austrian divisions faced ninety-three Russian and forty-six Italian.[94]

Inadequate industrial development is only part of the story, however. Political intrigue in court and nationalist tensions in the parliaments severely constrained military reforms—and the big budgets that needed to accompany them. After the 1866 defeat, Hungarians demanded and obtained political reform that gave them increased autonomy within the imperial structure. The Dual Monarchy made the two states (Austria and Hungary) sovereign equals and created a unique governmental structure with only three shared ministries (foreign affairs, war, and finance), two coequal parliaments, two capitals (Budapest and Vienna), and three

[91]Craig, *Battle of Königgrätz*; Wawro, *Austro-Prussian War*.

[92]Strachan, *European Armies*, pp. 127, 113; Scott W. Lackey, *The Rebirth of the Habsburg Army: Friedrich Beck and the Rise of the General Staff* (Westport, CT: Greenwood, 1995), pp. 6–7.

[93]John W. Mason, *The Dissolution of the Austro-Hungarian Empire, 1867–1918* (London: Longman, 1997), p. 24.

[94]Strachan, *European Armies*, p. 109; Gunther E. Rothenberg, "The Austro-Hungarian Campaign against Serbia in 1914," *Journal of Military History* 53, no. 2 (Apr. 1989): 128–29, 134.

armies—the Landwehr, the Honvéd, and the imperial army.[95] The Compromise of 1867 made the empire ungovernable, inevitably affecting military matters. The Hungarians were convinced that the imperial army was a tool of domestic subjugation and successfully pressed to keep military budgets as small as possible. Imperial military budgets were always the smallest of the great powers and often barely a quarter of German or Russian expenditures. Even when spending was relatively free—for example, during post-1866 railway construction—the political situation intruded. The two halves of the monarchy built their own systems but (at least for the Hungarians) deliberately limited the connections between them. In 1914 the only route by rail from Austria to Bosnia was through Budapest.[96]

Within the military itself, commitment to reform was half-hearted. Beck and the other chiefs of staff never really attempted to separate their office from the War Ministry as Moltke had done. The staff remained trapped in the webs of ministerial political intrigue and pressures from the legislature.[97] Thus Austria went to war in 1914 undermanned, underequipped, and poorly led. Economic backwardness and the Dual Monarchy made meaningful military reform impossible. The Austrian army looked like a modern, German-style military, but its accomplishments were superficial. By war's end, the army was shattered and control over imperial military policy had passed to Berlin.[98]

The experience of the French army is remarkably similar to that of the Austro-Hungarian, with one significant difference. As a wealthy country, France did not suffer as much from the economic constraints faced by the Austrians. The obstacles were all political. Responding to industrialization's opportunities, France became a leading innovator before the Wars of German Unification (e.g., the Minié rifle). The French army conducted the world's first railway mobilization during the War of 1859, in which Piedmontese and French troops fought the Austrians in Italy. The mobilization was something of a disaster, but Moltke dispatched several officers to carefully study France's experience.[99]

While Prussia learned from the French experience, Paris learned from Berlin. The key to the Prussian victory in 1866, the French army decided, was the superiority of the needle gun (ignoring the strategic and tactical blunders of the Austrians). In August 1866, the French army adopted the *chassepot*, a rifled breech-

---

[95]Mason, *The Dissolution of the Austro-Hungarian Empire*, pp. 6–8; Rothenberg, "The Austro-Hungarian Campaign," pp. 127–28.

[96]Rothenberg, "The Austro-Hungarian Campaign," p. 128; N. Stone, "Moltke and Conrad: Relations between the Austro-Hungarian and German General Staffs, 1909–1914," in Paul M. Kennedy, ed., *The War Plans of the Great Powers, 1880–1914* (London: Allen and Unwin, 1979), pp. 243–44.

[97]Geoffrey Wawro, "The Rebirth of the Habsburg Army [Book Review]," *Journal of Military History* 60 (July 1996): 563.

[98]Stone, "Moltke and Conrad."

[99]Showalter, *Railroads and Rifles*, pp. 41–42.

loader with longer range, higher rate of fire, and a more reliable firing mechanism than the Dreyse needle gun. The French army thus entered the 1870–71 war with nearly a million shoulder arms superior to those of the Germans.[100] In 1866, France also began acquiring the *mitrailleuse*, the world's first machine gun, with a range of nearly two thousand yards and a rate of fire of 150 rounds per minute. While a revolutionary technological development, the Ministry of War kept the weapon so secret that few commanders knew that it existed. The French did not extend their technical achievements into the realms of policy and organization, however. After the Franco-Prussian War, they lagged behind the Germans, and in military and political crisis. Second-hand experience is never as compelling to militaries as that gained directly.[101]

During the Franco-Prussian War, the French army was deficient in all three functional areas. Despite the chassepot, French tactical doctrine was confused. Some officers had recognized the value of the tactical defense and became obsessed with finding and holding good defensive positions (often at the expense of flexibility), but others still relied on Napoleonic bayonet charges by mobile columns. French troops received inadequate training with the new rifles to use them effectively.[102] The railway mobilization was "incompetent" and "deplorable," thanks to the total inattention by the General Staff to the need for coordination.[103] In the end, individual commanders and departments within the War Ministry were allowed to arrange their rail transport as they pleased, troops and ammunition arrived in different places, and units arrived in separate pieces—often separated by several days.[104] The general staff and the indecision of Napoleon III rendered the French army without a clear plan for the war. The chaos at the railheads made this absence all the worse.[105]

Defeat shook the French military establishment to its foundations. There was a shared recognition that the French military should copy the institutions thought responsible for the Prussian victory—the general staff, the nation-in-arms, and an organized railway mobilization. The ignominious departure of the Emperor Napoleon III to the Netherlands and the establishment of the Third Republic created the requisite political atmosphere for change. Newly unified Germany was (correctly) thought likely to be France's future military opponent. The rambunc-

---

[100]Richard Holmes, *The Road to Sedan: The French Army 1866–70* (London: Royal Historical Society, 1984), p. 200; Howard, *Franco-Prussian War*, p. 35.

[101]Mitchell, "A Situation of Inferiority."

[102]Holmes, *Road to Sedan*, pp. 211, 204–5; Wawro, "An 'Army of Pigs,'" pp. 412–13.

[103]Howard, *Franco-Prussian War*, pp. 26, 68.

[104]Holmes, *Road to Sedan*, p. 177; Adriance, *The Last Gaiter Button*, p. 107; Howard, *Franco-Prussian War*, p. 68.

[105]Adriance, *The Last Gaiter Button*, pp. 55–62; Holmes, *Road to Sedan*, pp. 165–79.

tious politics of the Third Republic and the resistance of the military bureaucracy to change inhibited the French from fully emulating the Prussian example. It too took on the form, but not the substance, of the Prussian example.

In purely technological terms, France remained a great innovator. The period between 1870 and 1914 was one of rapid technological change—with the introduction of the machine gun, quick-firing artillery, the field telephone and telegraph, TNT, dirigibles, and eventually airplanes. France regained its reputation as the nation most likely to pioneer new technology.[106] There is even evidence to suggest that France, through the works of the Commission Supérieure des Chemins de Fer, closed some of the gap between the performance of the German and French railway mobilization systems.[107] In these areas and others, Germany was the conscious model.[108] But in two critical areas—conscription policy and the general staff—the chaos of French politics and the conflicting interests of factions within the military prevented complete emulation.

Prussia had enjoyed the advantage during the war of better trained recruits, especially among the reserves. They had accomplished this, all of Europe was sure, through effective conscription and training systems. All Prussian men gave the mandated three years of active service and eight in the reserves. They received extensive training in both services, and few exemptions were permitted.

The French relied on a seven-year term of service, making the army closer to a professional army than a conscript army or "army of the people." Conservative forces in France favored a longer term of service, while the Left, nostalgic for the revolutionary nation-in-arms and enraptured by the people's armies that sprang up after the surrender of the emperor in 1870, favored a shorter term. An 1872 reform law split the difference between the existing French system and the Prussian, settling for a five-year term, the worst of both worlds. Five years meant that a standing army made up of each year's eligible population was too large for the military budgets to absorb, so the same corrupting system of exemptions for the wealthy or well-connected was re-established. A large standing army meant that little money was available for adequately training the reserves. The law was

---

[106]Addington, *Patterns of War*, pp. 103–4; Herrmann, *Arming of Europe*, pp. 67–78.

[107]Allan Mitchell, surveying the French railroad reforms of the 1880s and 1890s, concluded: "At the very least, the evidence indicates that the strategic importance of railways was an instruction acquired at great cost by the French from the Germans in the years after 1870." French efforts included organization (the creation of a Commission Supérieure des Chemins de Fer in 1872), construction (eighteen thousand kilometers of new track, double-tracking existing lines, and expanding train platforms), and quasi-nationalization of key lines (191 lines designated in the "general interest"). Allan Mitchell, *Victors and Vanquished: The German Influence on Army and Church in France after 1870* (Chapel Hill: University of North Carolina Press, 1984), pp. 60–64. See also Pratt, *Rise of Rail Power*, pp. 151, 170.

[108]David B. Ralston, *The Army of the Republic: The Place of the Military in the Political Evolution of France, 1871–1914* (Cambridge: MIT Press, 1967), p. 139.

amended to three years in 1889 and again to two in 1905, but German population growth had already begun to outstrip the French, so it was too little too late. The French army would be unable to match the Germans numerically in 1914.[109]

Nor did the French general staff follow the Prussian pattern. Neither the left-leaning civilians nor the more conservative army generals were interested in having a chief of the general staff with the same powers, autonomy, and influence as a Moltke. Despite feverish reform efforts, few attempts were made to create a strong chief of staff.[110] If the German General Staff was a unique product of the autocratic Prussian political system, the French general staff reflected the severe political divisions between the army and civilian leadership. Between 1871 and 1880, the French explicitly rejected the Prussian model. The debate turned on two issues: how much autonomy from the war minister the chief of the État-Major should have; and whether the staff should be open, rotating officers between standard field and staff assignments, or closed, where the staff is separate, elite, and trained especially and exclusively for staff work. In the Prussian system, the chief had considerable autonomy from the War Ministry, a feature that kept the army far removed from day-to-day politics; and staff officers rotated between staff and nonstaff assignments to prevent the staff from becoming a sclerotic bureaucracy. The French chose subordination to the War Ministry—a solution that pleased both the left-leaning civilian politicians worried about an independent army and conservative officers concerned about a new power center—and a closed system mirroring the École de État-Major's monopoly on staff careers. Subordination guaranteed continued politicization, and the closed system kept the staff from direct contact with combat matters and encouraged routine administration and bureaucratization rather than initiative and innovation.[111]

A reform law passed in 1880 replaced the École de État-Major with the École Supérieure de Guerre, modeled on the Prussian Kriegsakademie. The high command was restructured again under War Minister Freycinet (1888–93). The Conseil Supérieur de la Guerre (CSG) was transformed from an infrequently meeting advisory board of retired and nearly retired generals to a de facto general staff with regular meetings and war-planning oversight. Freycinet sought to insulate the CSG and the military leadership generally from political control—with once again the German military as the explicit model.[112] But the war minister retained authority over the staff. The political volatility of that position—there were forty-one ministers between 1871 and 1914—prevented the establishment of a general

[109]Strachan, *European Armies*, p. 109; Mitchell, "A Situation of Inferiority," pp. 50–51.
[110]Ralston, *Army of the Republic*, pp. 138–60.
[111]Mitchell, "A Situation of Inferiority," pp. 58–60; Mitchell, *Victors and Vanquished*, p. 83.
[112]Mitchell, *Victors and Vanquished*, pp. 106–7.

staff insulated from politics and clear in its role.[113] Political and bureaucratic skills were valued more highly than military ones, and the staff was routinely ignored by other ministry and army officials.[114] Joffre's appointment as the new chief of the general staff in 1911, even though he frankly acknowledged that he knew little of staff work, was indicative of the French style of army command.[115]

On the outskirts of the European great power club, Russia had a complicated reaction to changes in the European military environment. Russia was unique among the great powers in the depth of its battlefield experience with industrial-era warfare. Between 1850 and World War I, it faced the threat of rising Germany, two catastrophic military defeats (in the Crimea and against Japan), and a war they had won, but at great cost (against Turkey). The Russian army was in a constant state of "reform" from the mid-nineteenth century on. Yet despite losses in Crimea and Manchuria and a long, costly war with Turkey, reform energies were often misdirected. Opinion and policy vacillated between the Prussian style, emphasizing an autonomous general staff, a nation-in-arms, and open-order fire tactics; the (Napoleonic) French style stressing shock tactics and a politically subservient general staff; and a reactionary, uniquely Russian response—an authentically Russian way of war—that rejected Western methods and depended on a huge army and the vast stretches of Russian territory to provide security.[116]

Emulation of Prussian military methods began during the Napoleonic wars,[117] but Russia's first military crisis of the nineteenth century came with the Crimean defeat. The French and British had used rifled shoulder weapons against the Russian muskets, an advantage that proved decisive at Alma, Balaclava, and Inkerman. An overall sense of societal inferiority generated the most anxiety. Under Alexander II, the 1860s and 1870s were the period of the Great Reforms: serfdom was abolished, the legal system reformed, censorship relaxed, and universal military service introduced—all for the purpose of transforming Russia into a modern European nation-state.[118] Reformers in the military, led by War Minister D. A. Miliu-

---

[113]Porch's breakdown of the terms of service makes the point even more dramatically. Between 1871 and 1914, twelve ministers had one-year tenures; thirteen had two-year tenures; three had three-year tenures; one had a six-year tenure; and two had seven-year tenures. Porch, *March to the Marne*, pp. 47, 53, 255; Mitchell, "A Situation of Inferiority," p. 60.

[114]For example, the 1899 reforms following the Dreyfus affair set the general staff back. Regular meetings of the CSG were suspended and future meetings came only at the request of the minister. Porch, *March to the Marne*, p. 67.

[115]Porch concludes: "It is simply ludicrous to suggest . . . that Joffre was more powerful than Moltke. The government may have become reconciled in the face of the growing German threat to name a chief of the general staff. But Joffre's authority was hedged by so many safeguards . . . that the army continued to function, or not to function, largely as before. War ministers and ministry officials often simply ignored him." Porch, *March to the Marne*, pp. 172, 253; quotation from p. 190.

[116]Menning, *Bayonets before Bullets*, p. 207.

[117]Görlitz, *History of the German General Staff*, p. 41.

[118]Perceived inferiority to Europe had previously ushered in reform in Russia. In the late seven-

tin, took steps to secure the latest technology, where they could afford it. They were equally interested in social reform. They argued that, in the event of war, a European-style army could expand to two or even three times its peacetime size through the judicious use of a reserve system and save money besides. But the absence of free labor prevented the state from having a fully functioning reserve army.[119]

Despite freeing the serfs, entrenched gentry interests made extraction of military service difficult. Integration of the peasantry into a modern nation-at-arms became a lengthy ordeal. In 1864 the term of military service was shortened to fifteen years, but this change kept the army closer to the eighteenth-century professional ideal than that of the modern reserve army. By 1874, nine of the fifteen years were served in the reserves; by 1906 regular service was reduced to three years—making the Russian army on par with the rest of Europe.[120] The educational levels and training of the conscripts remained a problem, but the army, especially after 1905, made steady progress. By World War I, Russia's conscript army was the biggest in Europe and reasonably well trained.[121]

In the area of technology and infantry tactics, the Russians also made slow but steady progress. The army switched from smooth-bores to rifles in 1857, to breechloaders from muzzle-loaders in 1867, and to the metallic cartridge, smokeless-powder-firing Mossine in 1891. But shifts in tactics were partial. Some Russian officers preferred Prussian open-order fire tactics, but the traditionalists—advocating the bayonet and massed attack—generally prevailed. The costly victory over Turkey in 1878 led to changes in army regulations in favor of fire tactics. But practice was another matter. The Russians suffered heavy losses in the battles of Inkou and Sandepu during the Russo-Japanese War because army commanders insisted on massed attacks with bayonets against dug-in Japanese forces with modern rifles and machine guns. By the twentieth century, Napoleonic and French tactics were out; Moltke and the Germans were in. Like militaries across Europe, the Russian army leadership had yet to fully grasp the defensive power of modern infantry weapons on the eve of World War I.[122]

The record of reform in the strategic and operational realm is even less im-

teenth and early eighteenth centuries, concern about relative backwardness and security vulnerability (triggered by a series of military defeats to Sweden) produced a period of intense reform followed by an equally vigorous counter-reformist period after Peter's death. See David B. Ralston, *Importing the European Army: The Introduction of European Military Techniques and Institutions into the Extra-European World, 1600–1914* (Chicago: University of Chicago Press, 1990), pp. 12–39.

[119]Menning, *Bayonets before Bullets*, pp. 7–13.

[120]Ibid., p. 13; Strachan, *European Armies*, p. 110.

[121]David R. Jones, "Imperial Russia's Forces at War," in Allan Reed Millett and Williamson Murray, eds., *Military Effectiveness*, vol. 1: *The First World War* (Boston: Allen and Unwin, 1987), p. 309.

[122]Menning, *Bayonets before Bullets*, pp. 30, 41–44, 136–42, 186; Strachan, *European Armies*, p. 113; Jones, "Imperial Russia's Forces at War," p. 310.

pressive. Russian geography, inadequate economic development, and a casual approach to war planning conspired to turn Russian use of strategic railways, and mobilization preparations generally, into a farce. The lightning mobilization of the Prussians at the outset of the Franco-Prussian War startled the Russians. The underdeveloped Russian railway net became an obsession for many military planners. But military concern made little difference. The limited investment capital went first to economic projects, which fit poorly with the army's strategic requirements.[123] Russia did build one great strategic railway—the famous Trans-Siberian Railway—but its history is a record of Russian incompetence and sloppy planning. Russian planners saw that a railway linking Moscow with their expanding Asian colonial empire would serve as a check against Chinese and British incursions (they were little concerned with the Japanese at first). Construction began in 1893, but poor technical standards and dubious cost-saving measures caused one Russian railway engineer to refer to it as "a monument to Russian official bungling," while a minister thought it a "toy railroad."[124] Passage around Lake Baikal was not completed prior to the outbreak of the Russo-Japanese War, a failure that cost the army dearly as it struggled to move troops east. General A. N. Kuropatkin, who led the Russian effort against the Japanese, would later blame the railway for the loss. Whether true or not, the line did prove incapable of performing the task its planners had intended.[125]

Problems of strategic infrastructure were as much a function of obsolete doctrine and poor administration as of economic backwardness. Russia's experience in the Napoleonic wars exerted an enduring influence on the army.[126] Reformers like Miliutin vacillated between the Napoleonic and Prussian camps. He undertook to reform the War Ministry in 1868, including the creation of a main staff. It functioned somewhat like the Prussian Great General Staff. But it was not autonomous from the ministry, a conscious decision by Miliutin to avoid dilution of his personal power. It did little war planning (given Russia's vast size, that was thought unnecessary) or technology assessment. The Franco-Prussian War gave some urgency to expand those duties. The head of the Nicholas Academy—the Russian equivalent of the Kriegsakademie—wrote a detailed memo to Miliutin in 1873 outlining Russia's mobilization shortcomings and advocating more strategic railways and peacetime planning. These arguments circulated, were taken seriously, but did not spur fundamental change. Many officers felt that

[123]Menning, *Bayonets before Bullets*, pp. 16, 19–20, 116–17

[124]Quoted in Marks, *Road to Power*, pp. 191, 200.

[125]Ian Nish, *The Origins of the Russo-Japanese War* (London: Longman, 1985), p. 139; Marks, *Road to Power*, p. 205. It took the army more than two years to return all of its troops from Manchuria. Herrmann, *Arming of Europe*, p. 62.

[126]John Shy, "Jomini," in Peter Paret, ed., *Makers of Modern Strategy: From Machiavelli to the Nuclear Age* (Princeton: Princeton University Press, 1986), p. 156.

Russia's relative backwardness made such a rationalized general staff unnecessary, and that same backwardness was also a source of strength.[127]

In 1881 there was another push to construct a Prussian-style general staff—one with more autonomy from the War Ministry and closer contact with the czar. That attempt failed because of internal politics within the War Ministry.[128] A final and more serious reform effort came in the wake of the loss to the Japanese. In 1905 a new institution, the Main Directorate of the General Staff, was created and made independent of both the main staff and the War Ministry. It was subsumed by the ministry in 1909, but this move actually had the effect of centralizing some planning functions. As in Austria and France, the politics of the War Ministry took their toll—the Main Directorate had five chiefs in the five years from 1909 to 1914.[129]

Throughout the period, even when modernizing forces had the upper hand, the uneven and often poor quality of the officers persisted.[130] A 1904 survey showed that only 45 percent of staff officers had the training and education required for their position. The most egregious example was General N. P. Linevich, the commander of a field army in Manchuria, who could neither read military maps nor interpret a railway mobilization schedule.[131] Russian performance in World War I confirmed this picture of halting, incomplete reform along Prussian lines.

Britain, as in the Napoleonic example, is the aberrant European case. With the exception of small arms, the British did not respond to German innovations, primarily because they did not have to. As Britain was not a Continental power, the British army did not have to prepare to meet the Germans in battle, and so had little need for the changes to army organization and tactics the Germans had developed. As an imperial power, the British were much more concerned with naval technology, and with fighting rebellion in the overseas colonies. Only as World War I approached did the British take greater interest in Continental military affairs, and then only partially. The British army created a general staff in 1906 but stuck with a professional army and eschewed conscription up through the war. Their only concession to Continental practices was a 1908 reform of the reserves that created the Territorial Army.[132]

The organizational and administrative adaptations may have been unnecessary for the British strategic position, but as in other cases, the British army readily

[127]Menning, *Bayonets before Bullets*, pp. 17–20.
[128]Ibid., pp. 96–97.
[129]Ibid., pp. 200–221; Strachan, *European Armies*, p. 127.
[130]Jones, "Imperial Russia's Forces at War," p. 309.
[131]Menning, *Bayonets before Bullets*, pp. 100–101.
[132]Strachan, *European Armies*, pp. 128, 110.

took to new military technology. Their troops in the Crimea were equipped with a licensed version of the Minié rifle, and in 1889 the army adopted a new rifle, the Lee-Metford, using the metallic cartridges and smokeless powder of the day. They kept up with other developments as well—the machine gun, the airship and airplane, rifled artillery, and so forth—though they lagged somewhat behind the earliest adopters (usually the French).[133]

Prussian military practices spread beyond Europe. In the wake of the Wars of German Unification, Japan and Chile emulated the Prussian system because it was thought the most effective of the available choices. Both then proceeded to emulate only partially the Prussian example—an outcome best explained by the nature of the fit between the import and the domestic social and economic structure and military organization, and not so much by the nature of the external threat they faced.

The Opium War of 1841–42 first demonstrated to the Japanese the power of Western military technology.[134] But the problem threatened Japanese shores when Admiral Perry anchored unimpeded in Edo Bay for ten days in 1853.[135] In 1863, the *bakufu*, or shogunate, ordered all foreigners expelled. Europeans responded by casually destroying a series of fortified coastal defenses, so in 1866 the Japanese turned to the French for help.[136] This decision came at an unfortunate moment for the bakufu. Some of the more enterprising *han*, or clans, had already begun to import Western technologies and train Western-style armies to topple the shogun, and within months he was dead and the Meiji Restoration began.[137]

While Japanese military officers were impressed by Prussia's performance during the Franco-Prussian War, the 1877 Satsuma Rebellion, an uprising of disenchanted and disenfranchised samurai, highlighted the inadequacies of French military practices. Under French tutelage, the Japanese army had no mobilization plans, no reserve army to speak of, and no plans for the organization of supply or transport. The mission was gone by 1880 and replaced by a German one.[138]

German military culture proved more compatible with the Japanese than had the French (German monarchism versus French republicanism), and the accomplishments of the mission were impressive.[139] The Japanese army formed an

[133]McNeill, *Pursuit of Power*, p. 231; Strachan, *European Armies*, pp. 113–14; Herrmann, *Arming of Europe*, p. 68.

[134]D. Eleanor Westney, "The Military," in Marius B. Jansen and Gilbert Rozman, eds., *Japan in Transition, from Tokugawa to Meiji* (Princeton: Princeton University Press, 1986), p. 170.

[135]Ralston, *Importing the European Army*, p. 142.

[136]Ibid., p. 150; Ernst Leopold Presseisen, *Before Aggression: Europeans Prepare the Japanese Army* (Tucson: Published for the Association for Asian Studies by the University of Arizona Press, 1965), p. 3.

[137]Ralston, *Importing the European Army*, p. 154.

[138]Presseisen, *Before Aggression*, p. 55; Ralston, *Importing the European Army*, p. 169; Westney, "The Military," p. 184.

[139]Westney, "The Military," p. 190; Presseisen, *Before Aggression*, p. 136.

autonomous general staff and increased its planning, preparation, and army exercises. The army reorganized itself into divisions on the German model and created a nation-in-arms based upon a short-service army and a reserve system.[140] By 1890 the army's size had increased substantially, troops were outfitted with modern weapons (including a domestically produced rapid-firing rifle), and the general staff corps had developed plans for mobilization and operations on the islands and even in China.[141]

In two areas, the Japanese emulation was less than complete. The head of the German mission, Major Jacob Meckel, recommended a system of internal railways to provide for defense against invasion. He reasoned that the current system of coastal railways would be vulnerable to naval bombardment. But the plan was expensive with no economic merit, so his suggestion was ignored.[142] Meckel was also an opponent of open-order fire tactics. Dispersed formations and fire control problems led to indiscipline and hindered offensive maneuvers. Modern weapons made massed assaults difficult, but he taught the Japanese to use sustained artillery barrages and terrain for cover as preparation for close-order tactical offensives. These tactics called for extremes of duty and sacrifice and fit well with Japanese military culture. The bloody frontal assaults favored by the Japanese command in the Russo-Japanese War (fully one-third of the besieging forces at Port Arthur were lost during infantry offensives) had Meckel's imprint.[143]

Japan's adaptation of the German model spread to China. Impressed by Prussia's performance in the Franco-Prussian War, in the late 1800s several provincial leaders established military schools manned by German instructors. German officers also served as brigade, battalion, and company commanders in Chinese army units. Beginning in 1901, however, Japanese officers began replacing German instructors in Chinese military schools. Provincial officials also sent large numbers of Chinese officers to study in Japan. In 1906 there were 691 Chinese officers and cadets in Japanese military schools. The Japanese had already patterned their army on the successful Prussian model; the Japanese claimed they had developed military practices uniquely suited to Asian armies. Japanese instructors were also willing to learn to speak Chinese and were less expensive than the Germans.[144]

In Chile, no European (or North American) threat spurred innovation. Instead a dangerous and shifting regional security situation did. Several smaller wars in the Southern Cone in the nineteenth century and two large ones—the

[140]Presseisen, *Before Aggression*, pp. 116–17, 135–36; Addington, *Patterns of War*, p. 128.

[141]Kozo Yamamura, "Success Ill-Gotten? The Role of Meiji Militarism in Japan's Technological Progress," *Journal of Economic History* 37, no. 1 (1977): 117; Presseisen, *Before Aggression*, p. 136.

[142]Presseisen, *Before Aggression*, p. 122.

[143]Ibid., pp. 83, 136, 145–46.

[144]Ralph L. Powell, *The Rise of Chinese Military Power, 1895–1912* (Princeton: Princeton University Press, 1955), pp. 41, 62, 161–62, 236.

Paraguayan War (1865–70) and the War of the Pacific (1879–83)—were fought with massive casualties on all sides and permanently removed two states, Paraguay and Peru, from their status as major regional powers.[145] Chile was the ostensible winner of the War of the Pacific, but its victory was Pyrrhic. The geostrategic situation remained tenuous. The other two regional powers, Brazil and Argentina, allied against Chile, fueling a severe arms race with Argentina in 1895.[146]

The Chileans turned to the Germans for instruction and technology—contracting for a mission in 1885. In a situation parallel to that of the Japanese, a single charismatic individual, Emil Korner, became a dominant reform figure.[147] He supervised the rebuilding of the Chilean army along German lines in five areas: instruction and training, organization via a general staff, a conscription and reserve system, armaments, and doctrine. Just as Japan subsequently re-exported the Prussian military system to China, so too Chile, the "Prussia of South America," spread German practices to Ecuador, El Salvador, Colombia, and Venezuela.[148]

Emulation was not total. The Chileans favored a more specialized style of officer training, like the French, rather than the generalist approach favored by Moltke (and Korner). The role of the general staff was never quite fixed. At first, army command was highly centralized in the general staff, a practice that ran against the German tendency to give the individual corps and divisions considerable flexibility. A set of reforms launched in 1906 pushed the army in the other direction, and the power of the general staff was severely diluted. Finally, Chile's level of economic development placed clear constraints on military modernization.[149]

Conclusion

Both the Napoleonic and Prussian military revolutions consisted of military innovations embedded within broader social and economic transformations. They were adaptations to novel circumstances. The age of nationalism and the industrial era were Rubicons that human history will not cross twice. We could conclude that all periods of rapid change in military practices are in fact symp-

[145]Joao Resende-Santos, "Anarchy and the Emulation of Military Systems: Military Organization and Technology in South America, 1870–1914," in Benjamin Frankel, ed., *Realism: Restatements and Renewal* (London: Frank Cass, 1996), pp. 223–24.
[146]Ibid., pp. 229, 231.
[147]Frederick M. Nunn, "Emil Korner and the Prussianization of the Chilean Army: Origins, Processes, and Consequences, 1885–1920," *Hispanic American Historical Review* 50, no. 2 (May 1970): 300–322.
[148]Resende-Santos, "Anarchy and the Emulation of Military Systems," pp. 240–41.
[149]Ibid., pp. 228, 235–38.

toms of broader and deeper socioeconomic transformations, and build that assumption into our theories of diffusion and emulation. Such an assumption throws us out of the theoretical universe of international relations theory—where the international political realm has considerable autonomy—into a messier, more historical-sociological realm. Thus all of the states considered in our two examples were swept up in the broader context of social transformation and so were, in a sense, really unable to choose to emulate or not. Emulation was thrust upon them—and not just by the competitive, anarchical nature of the international political system, but also by underlying currents of socioeconomic, political, and even cultural change. The rise of mass politics and nationalist movements were coming, whether the militaries wished for them or not. Staving off the industrial revolution was not really a choice militaries could make. The Austrian army, which fought so hard against change, was in the end literally undone by its inability to cope with both revolutions. Austria's fate was determined by defeat in war only in a proximate sense. The greater failure lay in Austrian society's inability (or unwillingness) to adapt to the forces of nationalism and industrialization.

Similar claims could be made for other periods of change—the rise of air power, for example, required a response. But an additional observation on the comprehensive social nature of these two transformations points to a final difference: nationalization and industrialization were simultaneously domestic and international and in fact transcend the domestic-international divide. So the extent to which the emulation of military practices flowing from these two transformations can be analyzed in isolation from the broader social context is limited. A large portion of our analysis of the spread of Napoleonic and Prussian military practices in the nineteenth century rests on the fact that each of the responding societies was passing through a decisive period of socioeconomic transformation.

Nevertheless, our findings do loosely fall within the explanatory rubric of emulation and diffusion studies. We include a table summarizing these findings below (see Table 8.1). The appearance of some great military threat or a catastrophic defeat in war provided a strong impetus for military reform. The vanquished sought to emulate whatever was currently acknowledged as the best military practice—whether practiced by the victor or some other state. The Prussians after Jena and Auerstädt adapted French army tactics and organization to their resources and purposes. The French after 1870 sought to copy the Prussian Great General Staff. The Russians after 1905 tried to abandon Napoleonic tactics and organization for German ones.

However, as we found no perfect imitations, the straightforward adoption of best available practices was always strongly conditioned by local custom and tra-

TABLE 8.1
Summary of Findings

Nationalism	Conscription	Officer Corps Composition	Organization	Tactics
Napoleonic Innovations	Universal conscription, levée en masse, a republican army	Full-time occupation, active command in battle, careers open to talent, promotion by merit	Permanent combined-arms divisions, autonomous corps	Shock tactics, skirmishing, combined-arms tactics
Kingdom of Italy	Yes	Partial	Partial	Yes
Poland	Yes	Yes	Yes	Yes
Kingdom of Holland	No	Yes	Yes	Yes
Westphalia	Yes	Yes	Yes	Yes
Kingdom of Naples	Partial	No	Yes	Yes
Prussia	Eventually	Partial	Partial	Yes
Austria	No	No	Yes	Yes
Russia	No	No	Yes	Yes
Great Britain	No	No	Eventually	Partial

Industrialism	Technology	Tactics	Operations and Strategy	Administration
Prussian Innovations	(1) Breech-loading rifle and rifled artillery (2) Railroad (3) Telegraph	(1) Open order battle lines (2) Marksmanship (3) Fire control	(1) Economic development (2) Mobilization planning	(1) Mass, reserve army (2) Autonomous general staff with technical proficiency
Austria-Hungary	Partial	Eventually	No	No
France	Yes	Yes	Partial	Partial
Russia	Eventually	No/Offset	Partial	No
Great Britain	Yes	No/Offset	N/A	No
Japan	Yes	No	No	Partial
Chile	Yes	N/A	No	No

dition, local patterns of social power, and day-to-day politics. The Prussians managed to create a French-style nation-in-arms but without the messy democratic politics that were supposed to go along with it. As Hajo Holborn puts it, "National service, the logical outcome of national and liberal thought in America and France, became in Prussia a device for strengthening the power of an absolutist state."[150] Intrigue in the imperial court prevented the Russian general staff from ever becoming an autonomous and objective war-planning body. The national problem in Austria, symbolized by the unwieldy Dual Monarchy, under-

[150]Holborn, "The Prusso-German School," p. 282.

mined attempts to create an efficient, well-trained infantry. The Japanese settled on the Prussian military system because it fit the quasi-feudal structure of Japanese society and provided the best solution to domestic political and military unrest, but not before a serious flirtation with French methods and training. So in these instances we find not just that local conditions and attitudes refracted imported military practices in odd ways but also that the refraction was often intentionally directed at a domestic political problem or conflict. The international threats the practices were ostensibly imported to fend off are of secondary importance. The external threat environment was only a partial guide to the success or failure of emulation. Britain—far removed from the Continental eruptions of the Napoleonic and Prussian military revolutions—provides the best case for the external threat argument, but only in the negative. With little pressure to adapt to the changing military environment, the British did not, or only very slowly. Meanwhile, the best example of the failure of the external environment to prompt meaningful change is the French army after 1870—despite fierce internal political and external military pressure to adapt. The tremendous reform-oriented energy was absorbed by and directed toward battles with radical Third Republic legislatures and intramilitary political intrigue.

A final conclusion concerns the role of technology in these transformations. In all cases, and among all recipients within cases, the acquisition of new technology was always the first and easiest step. This observation is tempered somewhat by the history of the railway, where the expense and dependence on civilian investment prevented the more economically backward states from building a sufficient net. Even so, successful importation of the tactics and the administrative and training apparatus that made the military practice effective was the harder step. Where the needed innovation required substantial social and political transformation, comprehensive emulation was even harder still.

CHAPTER NINE

Beyond Blitzkrieg

Allied Responses to Combined-Arms
Armored Warfare during World War II

THOMAS G. MAHNKEN

Tanks first appeared on the battlefield during World War I, shielding the infantry's advance across no-man's-land by crushing obstacles and destroying machine-gun nests. While tanks made their debut during the Battle of the Somme in 1916, it was the British army's massed use of armor at Cambrai in November 1917 that captured the imagination of many military observers.[1] During the battle, a force of nearly five hundred British tanks achieved a spectacular breakthrough in German lines but lacked the ability to exploit the breach. By the end of the war, a handful of British officers, such as J. F. C. Fuller and Giffard LeQ. Martel, had begun arguing that tanks held the potential to transform warfare.

Armies followed divergent approaches to armored warfare in the years that followed the war. While Britain led the world in tank technology and concepts throughout the early 1920s, it did not begin fielding armored divisions until 1939. The Soviet army, by contrast, fielded an impressive mechanized force in the 1930s, only to dismantle it on the eve of World War II because of faulty lessons drawn from the Spanish Civil War, Stalin's purges of tank enthusiasts, and the emergence of a powerful cadre of cavalry officers within the army's leadership. The U.S. Army experimented with tank formations but did not field large armored units prior to World War II.

Ironically it was the German army, which had been forbidden from possessing

---

[1]Brian Holden Reid, "Fuller and the Revolution in British Military Thought," in Reid, *Studies in British Military Thought: Debates with Fuller and Liddell Hart* (Lincoln: University of Nebraska Press, 1998), p. 54.

tanks until 1935, that developed the most effective approach to armored warfare. Instead of using tanks exclusively to support traditional combat arms, the army fielded versatile, combined-arms Panzer divisions. The German army combined the tank with innovative operational concepts and organizations in a way that transformed warfare.

Germany's development of combined-arms armored warfare has become the canonical case of peacetime military innovation. Indeed, many have argued that Germany's use of armor during the early campaigns of World War II triggered a revolution in military affairs.[2] Germany's decisive defeat of France and the Low Countries in May–June 1940 offered a vivid demonstration of the deadly effectiveness of combined-arms armored warfare. Mainstream international relations theory hypothesizes that a competitive strategic environment should stimulate the spread of military innovations as states seek to ensure their security.[3] Thus Germany's adversaries should have been expected to adopt combined-arms armored warfare out of fear for their own survival. In fact, the historical record reveals a more complex picture. Following the fall of France, the British, Soviet, and U.S. armies expanded their armored forces considerably and adopted variants of combined-arms doctrine and organization. The process of adaptation was, however, neither smooth nor painless: the difficulty of reforming under fire and the tenacious persistence of prewar organizational culture conspired to constrain efforts to copy the Germans.[4] Moreover, each army's approach contained unique doctrinal and organizational elements.

[2]Eliot A. Cohen, "A Revolution in Warfare," *Foreign Affairs* 75, no. 2 (March–April 1996): 46; James R. FitzSimonds and Jan M. van Tol, "Revolutions in Military Affairs," *Joint Force Quarterly*, no. 4 (spring 1994): 24; Andrew F. Krepinevich, "Cavalry to Computer: The Pattern of Military Revolutions," *National Interest*, no. 37 (fall 1994): 36.

[3]Barry R. Posen, "Nationalism, the Mass Army, and Military Power," *International Security* 18, no. 2 (fall 1993): 82; João Resende-Santos, "Anarchy and the Emulation of Military Systems: Military Organization and Technology in South America, 1870–1930," *Security Studies* 5, no. 3 (spring 1996): 196; Kenneth N. Waltz, *Theory of International Politics* (Reading, MA: Addison-Wesley, 1979), pp. 124–28.

[4]Organizational culture is "the pattern of basic assumptions that a given group has invented, discovered, or developed in learning to cope with its problems of external adaptation and internal integration, and that have worked well enough to be considered valid, and, therefore, to be taught to new members as the correct way to perceive, think, and feel in relation to those problems." See Edgar H. Schein, "Coming to a New Awareness of Organizational Culture," *Sloan Management Review* 25, no. 2 (winter 1984): 3. See also Gareth Morgan, *Images of Organizations* (Newbury Park, CA: Sage, 1986), ch. 5; William G. Ouchi and Alan L. Wilkins, "Organizational Culture," *Annual Review of Sociology* 11 (1985), pp. 457–83; Andrew M. Pettigrew, "On Studying Organizational Cultures," *Administrative Science Quarterly* 24, no. 4 (Dec. 1979); Edgar H. Schein, *Organizational Culture and Leadership*, 2d ed. (San Francisco: Jossey-Bass, 1992); Linda Smircich, "Concepts of Culture and Organizational Analysis," *Administrative Science Quarterly* 28, no. 3 (Sept. 1983), pp. 339–58.

The Development of Armored Warfare

World War I's battles demonstrated that tanks would have a role to play in future conflicts. They did not reveal how armies should organize, train, and equip themselves to best exploit the potential of the tank. Nor did they indicate how they should employ armored forces. Should tanks serve as an auxiliary to the infantry, as they had in World War I, or should independent mechanized formations become the centerpiece of combat operations? Should armies employ armored forces independently, or in combination with other combat arms? It was these questions that armies attempted to answer in the 1920s and 1930s.

Great Britain

In 1918, Great Britain led the world in tank technology and doctrine. On Armistice Day the British tank corps could muster twenty-five battalions, eighteen of which were deployed in France.[5] In the years that followed, the British army, like its counterparts around the world, sought to determine the most effective use of the tank on the battlefield. While most officers believed that the tank was a valuable weapon, the organization and training of armored forces spawned considerable debate. Armor enthusiasts such as Basil H. Liddell Hart, J. F. C. Fuller, and Giffard LeQ. Martel argued that the tank had brought about a radical change in the conduct of warfare. In the words of Robert H. Larson, British armor enthusiasts sought nothing short of "a revolution in military thought, a comprehensive philosophy of war explicitly designed to destroy the theoretical and practical foundations of the strategy of attrition that grew out of the experience of the Napoleonic Wars and dominated the strategic doctrines of the postwar period."[6] Fuller saw the development of the tank as "the greatest revolution that has ever taken place in the history of land warfare" and believed that independent, fast-moving tank units would restore strategic mobility to the battlefield through a combination of firepower, shock, and protection unattainable by the traditional combat arms.[7]

While armor enthusiasts were vocal, at times excessively so, they did not represent the views of the British army leadership. Most British officers, especially those from infantry and cavalry regiments, viewed armor as a means to enhance the effectiveness of traditional combat arms. Recalling the success of the tank in

---

[5]A. J. Smithers, *Rude Mechanicals: An Account of Tank Maturity during the Second World War* (London: Leo Cooper, 1987), p. 1.

[6]Robert H. Larson, *The British Army and the Theory of Armored Warfare, 1918–1940* (Newark, NJ: University of Delaware Press, 1984), p. 82.

[7]J. F. C. Fuller, *Armored Warfare: An Annotated Edition of Lectures on F.S.R. III* (Harrisburg, PA: Military Service, 1943), p. 2.

World War I, many believed that armor should support the advance of the infantry. Others advocated using tanks to carry out traditional cavalry missions such as reconnaissance and screening.[8] Not surprisingly, an officer's branch affiliation strongly affected his views of armored warfare: an overwhelming number of officers from the infantry and cavalry opposed mechanized forces, while many from the technical branches favored them.[9] Infantry and cavalry officers looked down upon tankers, while the "rude mechanicals" regarded officers from the traditional combat arms as obstructionist and unprogressive.

On 1 September 1923 the army established the Royal Tank Corps (RTC) composed of four tank battalions.[10] In the years that followed, tank corps officers fostered the belief that large armored formations operating independently of the traditional combat arms held the key to victory on the battlefield. British tankers tended to emphasize the mobility of tanks over their striking power while ignoring their substantial shortcomings. While such an approach greatly simplified the task of developing armored doctrine, the neglect of combined-arms tactics limited the flexibility of tank formations.[11]

In 1926 the incoming Chief of the Imperial General Staff (CIGS), Field Marshal Sir George F. Milne, authorized an Experimental Mechanized Force (EMF) to assess the performance of large armored formations. The unit included a light tank battalion for reconnaissance, a medium tank battalion for assault, a machine-gun battalion for security, five motorized or mechanized artillery batteries, and a motorized engineer company.[12] The EMF's 1927 and 1928 maneuvers demonstrated the effectiveness of armored formations. Particularly in its 1927 exercises, the mechanized force acquitted itself well against infantry and cavalry units. The maneuvers were hardly an unblemished success, however: they also revealed the difficulty of coordinating infantry, cavalry, tank, and air forces.[13] Indeed, they strengthened the conviction among many tank corps officers that armor had to be employed separately from traditional combat arms.[14]

---

[8]Brian Bond, *British Military Policy between the Two World Wars* (Oxford: Clarendon, 1980), pp. 130–32.

[9]Barton C. Hacker, "The Military and the Machine: An Analysis of the Controversy over Mechanization in the British Army, 1919–1939," Ph.D. dissertation, University of Chicago, 1969, pp. 86–87.

[10]Bond, *British Military Policy*, pp. 132–33.

[11]Captain Jonathan M. House, U.S. Army, *Toward Combined Arms Warfare: A Survey of 20th-Century Tactics, Doctrine, and Organization*, Combat Studies Institute Research Survey No. 2 (Fort Leavenworth, KS: U.S. Army Command and General Staff College, Aug. 1984), pp. 48–49; Richard M. Ogorkiewicz, *Armor: A History of Mechanized Forces* (New York: Frederick A. Praeger, 1960), p. 58.

[12]House, *Toward Combined Arms Warfare*, p. 48.

[13]Bond, *British Military Policy*, p. 144; Harold R. Winton, *To Change an Army: General Sir John Burnett-Stuart and British Armored Doctrine, 1927–1938* (Lawrence: University Press of Kansas, 1988), pp. 80–81.

[14]House, *Toward Combined Arms Warfare*, p. 48.

In 1937 the army took its first step toward mechanizing its cavalry by forming a single "Mobile Division" composed of two mechanized cavalry brigades, one tank brigade, and divisional troops. Like the horse cavalry, the mobile division was designed for reconnaissance and screening rather than assault. The use of armor to perform traditional cavalry functions made tanks acceptable to the army's traditionalists. As Richard Ogorkiewicz has written: "The 'cavalry' concept represented the price exacted by traditional thinking for the existence of a tank branch untied to the infantry. It reflected the influence of this thinking and it confined the development of mechanized forces to the limited roles of the cavalry of the late 19th and 20th centuries."[15] The army also consolidated tank units for infantry support into army tank brigades composed of three tank battalions each.[16] The army began to procure fast, lightly armored "cruiser" tanks for cavalry roles and slow, heavily armored "infantry" tanks for infantry support.

In early 1939 the army reorganized the Mobile Division and renamed it the 1st Armoured Division. Its composition reflected the British army's emphasis upon tank-pure formations: it included two armored brigades totaling 321 tanks and a Support Group composed of a motorized rifle battalion, a small motorized artillery regiment, and an engineer company.[17] The army also raised the 7th Armoured Division in Egypt. In all, the army planned to field three armored divisions, five army tank brigades for infantry support, and a single armored car regiment for cavalry missions.[18] This program reflected the continuing split within the army over the proper organization and employment of tanks on the eve of World War II.

Soviet Union

The Soviet Union pursued a markedly different approach to armored warfare during the interwar period. The Red Army was a new institution largely unencumbered by tradition.[19] Moreover, the Soviet Union's position as a revolutionary state surrounded by enemies necessitated an innovative approach to combat. In the 1920s, Soviet military theorists, like their counterparts throughout Europe and the United States, attempted to discern the shape of future wars through the fog of peace. Some, such as Aleksandr A. Svechin, argued that a modern conflict would be a replay of the attritional struggle that marked World War I. Others,

[15]Ogorkiewicz, *Armor*, p. 17.
[16]Ibid., p. 58.
[17]Ibid., p. 59.
[18]Smithers, *Rude Mechanicals*, p. 34.
[19]Many Red Army officers, the so-called military specialists, had previously served in the Russian army. From the mid-1920s to the mid-1930s, however, forty-seven thousand officers, most of them military specialists, were forced from service. David M. Glantz, *Stumbling Colossus: The Red Army on the Eve of World War* (Lawrence: University Press of Kansas, 1998), p. 27.

such as Mikhail N. Tukhachevskiy and Vladimir K. Triandafillov, argued that a future war would be characterized by offensive maneuver. While exponents of the former view pointed to the deadlock on the Western Front during World War I, the latter school drew lessons from the experience of the Russian Civil War, a conflict fought by small, mobile armies operating over vast distances.

In the late 1920s, Tukhachevskiy, Triandafillov, and their adherents at the War Academy began to articulate the concept of "deep operations," a vision of warfare that differed considerably from the experience of World War I. By seizing the offensive during the initial period of a future war, the Red Army would break through an adversary's defensive front, pushing the war onto his territory.[20] To them, operational maneuver required the coordinated employment of all combat arms. In *The Character of Operations of Modern Armies*, Triandafillov argued that tanks were capable of conducting mobile, offensive operations in conjunction with the traditional combat arms.[21]

In July 1929 the army created its first experimental mechanized unit, which grew rapidly into a "corps" (actually a small division) composed of two tank and one rifle brigades containing five hundred tanks and two hundred other motorized vehicles.[22] Throughout the 1930s the mechanized corps became progressively larger and more tank-heavy, so that by 1936 the Red Army possessed four mechanized corps with up to a thousand tanks each.[23] Each was a self-contained combined-arms force capable of conducting deep operations. The army also equipped its rifle divisions with tank companies or battalions, and its horse cavalry divisions with tank regiments.[24]

While British armor advocates favored independent tank operations, their Soviet counterparts viewed tanks within the context of combined-arms operations. By 1936 the army had codified the concept of deep operations in its doctrine. A future battle would unfold in two phases. The first would consist of a massed, echeloned attack along a narrow front by mechanized divisions operating in conjunction with infantry, artillery, and aviation. Once through the front lines, this force would attempt to convert the tactical breakthrough into an operational success by penetrating into the enemy's rear areas, disrupting his command and control, and destroying his reserves.[25]

---

[20]Earl F. Ziemke, "The Soviet Armed Forces in the Interwar Period," in Allan R. Millett and Williamson Murray, eds., *Military Effectiveness*, vol. II: *The Interwar Period* (Boston: Unwin Hyman, 1988), p. 16.

[21]V. K. Triandafillov, *Kharakter Operatsii Sovremennykh Armii* (Moscow: Voenizdat, 1936).

[22]John Erickson, *The Road to Stalingrad: Stalin's War with Germany* (New York: Harper and Row, 1975), p. 32; Ziemke, "The Soviet Armed Forces in the Interwar Period," p. 29.

[23]C. J. Dick, "The Operational Employment of Soviet Armour in the Great Patriotic War," in J. P. Harris and F. H. Toase, *Armoured Warfare* (New York: St. Martin's, 1990), p. 91.

[24]House, *Toward Combined Arms Warfare*, p. 66.

[25]George F. Hofmann, "Doctrine, Tank Technology, and Execution: I. A. Khalepskii and the Red

While the Red Army excelled in the theoretical aspects of armored warfare, it experienced difficulty turning theory into practice. The army's 1935 Kiev and 1936 Minsk maneuvers, designed to examine deep operations, revealed a host of practical difficulties with conducting combined-arms warfare, including command and control, coordination of different combat arms, and logistics. In the opinion of many commanders the Red Army was unable to implement the concept of deep operations.[26]

The Red Army's experience during the Spanish Civil War further undermined the position of Soviet armor advocates. The war convinced the commander of the Soviet tank force in Spain, Dmitri G. Pavlov, that mechanized corps had no useful role to play on the modern battlefield. Rather, tanks were best used in support of the infantry.[27] Indeed, the war convinced even Tukhachevskiy that technology had strengthened the defense and that all arms—including armor—should be tied to the advance of the infantry.[28]

Stalin's purges further weakened the armor advocates. On 12 June 1937, Tukhachevskiy and eight of his closest associates were executed. In the ensuing four years, Stalin and his cronies imprisoned or executed at least 20 percent of the officer corps, including 3 of 5 marshals of the Soviet Union, 14 of 16 army commanders, 60 of 67 corps commanders, 139 of 199 division commanders, and 221 of 397 brigade commanders.[29] In their place, a group of Stalin's cronies, many of whom were cavalry veterans, rose to power within the army's leadership.[30] Indeed, the cavalry was Stalin's favorite combat arm. The people's commissar of defense, Kilment Ye. Voroshilov, was a strong proponent of horse cavalry and ensured that it remained equal with other combat arms.[31] Voroshilov also opposed the formation of tank units larger than brigades.[32] In November 1939 the Red Army's Main Military Council ordered the army's five mechanized corps disbanded and their tanks dispersed.[33] Thus it was that as World War II began, the Red Army, once a leader in combined-arms armored doctrine, possessed the largest cavalry force in the world—thirty-two divisions—and was dismantling its last tank divisions.[34]

Army's Fulfillment of Deep Offensive Operations," *Journal of Slavic Military Studies* 9, no. 2 (June 1996): 284, 287, 304.

[26]Ibid., pp. 311–13.
[27]Erickson, *The Road to Stalingrad*, p. 32.
[28]Ziemke, "The Soviet Armed Forces in the Interwar Period," p. 16.
[29]Glantz, *Stumbling Colossus*, p. 30.
[30]A number of Red Army leaders were veterans of Marshal S. M. Budenny's 1st Cavalry Army during the Russian Civil War.
[31]Hofmann, "Doctrine, Tank Technology, and Execution," p. 317.
[32]Erickson, *The Road to Stalingrad*, p. 15.
[33]Dick, "The Operational Employment of Soviet Armour," p. 91.
[34]Ziemke, "The Soviet Armed Forces in the Interwar Period," pp. 28–29.

United States

The U.S. Army's approach to armored warfare during the interwar period reflected its limited experience in World War I, as well as foreign debates over the proper employment of tanks on the battlefield. The National Defense Act of 1920 gave the infantry control of tank units. Infantry doctrine, reflecting the experience of the American Expeditionary Force, saw armor as a means to protect soldiers as they advanced across the battlefield.[35] Tanks were parceled out, with a company assigned to each infantry division.

Not all army officers agreed with the infantry's approach to armored warfare. A number believed that army doctrine should emphasize the mass employment of tanks. Several, including Dwight D. Eisenhower and George S. Patton, Jr., wrote articles on the subject.[36] Patton argued for a separate armored corps, writing that grafting tanks onto the traditional combat arms would be "like the third leg to a duck—worthless for control, for combat impotent."[37] The army leadership did not embrace such sentiment, however. As Eisenhower later wrote: "I was told that my ideas were not only wrong but were dangerous, and that henceforth I was not to publish anything incompatible with solid infantry doctrine."[38]

The army's only attempt to field an armored force prior to World War II was short-lived. In 1927, Secretary of War Dwight F. Davis witnessed a demonstration of the British army's Experimental Mechanized Force. Impressed by the performance of British units, he asked the army to undertake similar experiments.[39] The army assembled a mixed brigade-sized unit at Ft. Meade, Maryland, in 1928, and a second force at Ft. Eustis, Virginia, in 1930. Despite the fact that the units possessed obsolescent equipment, they demonstrated the importance of combined-arms operations, including the striking power of tanks, the need for motorized infantry to support them, and the command and control problems created by dispersed forces.[40] Despite such success, the cost of the project, coupled with doctrinal disputes, organizational politics, and limitations of contemporary tank technology conspired to end the experiment. In May 1931 Army Chief of Staff Douglas MacArthur ordered the force disbanded and directed the cavalry and infantry to create their own mechanized units.

[35]Allan R. Millett and Peter Maslowski, *For the Common Defense: A Military History of the United States of America* (New York: Free Press, 1984), p. 381.
[36]Captain D. D. Eisenhower, "A Tank Discussion," *Infantry Journal* 17, no. 5 (Nov. 1920): 453–58; Colonel George S. Patton, Jr., "Tanks in Future Wars," *Infantry Journal* 16, no. 11 (May 1920): 958–62.
[37]Patton, "Tanks in Future Wars," p. 962.
[38]Dwight D. Eisenhower, *At Ease: Stories I Tell to Friends* (Garden City, NY: Doubleday, 1967), p. 173.
[39]Timothy K. Nenninger, "The Experimental Mechanized Forces," *Armor* 78, no. 3 (May–June 1969): 33.
[40]John T. Hendrix, "The Interwar Army and Mechanization: The American Approach," *Journal of Strategic Studies* 16, no. 1 (Mar. 1993): 78–82.

The mechanized cavalry's approach to armored warfare during the interwar period reflected the horse cavalry's tradition of maneuver. The army envisioned using mechanized units composed of fast tanks to envelop enemy positions and attack targets in his rear.[41] The cavalry's mechanized unit, the 7th Cavalry Brigade (Mechanized), grew into a prototype armored force that included two mechanized cavalry regiments, an artillery battalion, and observation aircraft. Under the command of Colonel Adna R. Chaffee, the brigade compiled an impressive record of achievements during maneuvers against traditionally organized units. The infantry, for its part, formed a Provisional Tank Brigade that contained seven of the branch's eight infantry battalions. During the climax of the army's 1940 Louisiana Maneuvers, the two brigades joined to form an improvised mechanized division.[42] The United States nonetheless possessed no armored divisions at the outbreak of World War II. Neither the infantry nor the cavalry favored the establishment of an armored force. Moreover, the United States faced no threat urgent enough to demand a break with past practice.

Germany

The German army developed an approach to armored warfare that differed markedly from those of other major armies. To some extent, German concepts of armored warfare grew out of the Prussian military tradition of mobile operations as well as the storm-troop tactics developed during the final phases of World War I. The novelty consisted of marrying new technology—the tank, radio, and close-support aircraft—with combined-arms doctrine and an organization—the Panzer division—that allowed the German army to translate a tactical breakthrough into an operational penetration of enemy positions.

At first glance, it is surprising that it was Germany that mastered armored warfare. During World War I, the German army possessed a minuscule tank force, over half of which was composed of captured British vehicles.[43] Moreover, the Versailles Treaty explicitly denied Germany the right to possess armored vehicles. Despite these restrictions, tanks had considerable appeal among German army officers. Armored forces offered the German army leverage against its more numerous and powerful adversaries, such as France and Poland. An armored thrust also seemed to offer Germany its best chance of regaining the territory it had lost at Versailles.

During the late 1920s, Heinz Guderian, Germany's most prominent armor ad-

---

[41]By law, only the infantry was allowed to possess tanks. As a result, the mechanized cavalry acquired "combat cars" that were often indistinguishable from the infantry's tanks.

[42]Christopher R. Gabel, *The U.S. Army GHQ Maneuvers of 1941* (Washington, DC: U.S. Army Center of Military History, 1991), p. 23.

[43]The World War I German tank force included fifteen Type A7V tanks, commonly called the "Elfriede," as well as twenty-five captured British Mark IV tanks.

vocate, studied World War I tank and cavalry operations, postwar British armor experiments, as well as the writings of Fuller and Liddell-Hart.[44] By 1929 he concluded that the key to harnessing the effectiveness of the tank lay in organizing armored divisions as combined-arms formations. He envisioned each unit as a self-contained armored and motorized team tailored to allow tanks to fight with full effect. The armored division would have the means not only to breach enemy lines but also to exploit the breakthrough. Tanks needed reconnaissance units to lead the way and screen flanks; mechanized infantry and artillery to reduce bypassed centers of resistance, support tanks, and hold ground; and combat engineers to sustain mobility.

During the 1920s, Germany began secretly cooperating with the Soviet Union.[45] In the spring of 1927, Germany and Russia agreed to open an armor training and test center at Kazan. The Heavy Vehicle Experimental and Test Station, as it was known, trained German armored officers, tested tank prototypes, and evaluated foreign vehicles. The Reichswehr conducted company- and battalion-level exercises at Kazan and observed the Red Army's 1930 armored maneuvers. The following year, the two armies conducted joint exercises. By the time the center closed in the summer of 1933, it had trained a cadre of more than fifty German and sixty-five Soviet armored specialists.[46]

The German army created its first tank unit, equipped with fifty-five Panzerkampfwagen I training tanks, on 1 November 1933. In July 1935 the army's ordnance office organized a demonstration involving motorcycle, tank, antitank, and armored car platoons for an enthusiastic Hitler at the training area of Kummersdorf. A month later it held four weeks of exercises with an improvised armored division at the training ground at Munster. The maneuvers were designed to demonstrate the command and control of large, rapidly moving armored units as well as cooperation between armored, infantry, and artillery forces.[47]

On 15 October 1935 the army established three Panzer divisions. Each included a tank brigade, a motorized infantry brigade, a motorized artillery regiment, an antitank battalion, a reconnaissance battalion, an engineer company (soon ex-

[44]Azar Gat, "British Influence and the Evolution of the Panzer Arm: Myth or Reality? Part I," *War in History* 4, no. 2 (Apr. 1997), pp. 150–73.

[45]Aleksandr M. Nekrich, *Pariahs, Partners, Predators: German Soviet Relations, 1922–1941*, ed. and trans. Gregory L. Freeze (New York: Columbia University Press, 1997); Manfred Zeidler, *Reichswehr und Rote Armee, 1920–1933* (Munich: R. Oldenbourg Verlag, 1994).

[46]Corum, *The Roots of Blitzkrieg: Hans von Seeckt and German Military Reform* (Lawrence: University Press of Kansas, 1992), pp. 191–95; Nekrich, *Pariahs, Partners, Predators*, p. 61.

[47]Wilhelm Deist, Manfred Messerschmidt, Hans-Erich Volkmann, and Wolfram Wette, *Germany and the Second World War*, vol. I: *The Build-up of German Aggression*, ed. Militärgeschichtliches Forschungsamt, trans. P. S. Falla, Dean S. McMurry, and Ewald Osers (Oxford: Clarendon, 1990), p. 433.

panded to a battalion), and signal and service units.[48] The Panzer division was a versatile organization that combined striking power and mobility and was capable of either smashing or outflanking the opposition. Moreover, it could divide into Kampfgruppen, combined-arms battle groups generally of regiment strength.

The German army continued to refine the organization of the Panzer division in the years leading up to World War II. In 1938 the army raised an additional two divisions, bringing the total to five. Maneuvers indicated that the initial organization lacked sufficient infantry. As a result, in 1938–39 the army increased the strength of the infantry element of the Panzer division to four battalions. Moreover, it reorganized the division's tank battalions into one medium and two light companies, totaling 320 tanks.[49]

The Spread of Armored Warfare

On 10 May 1940, 135 German divisions supported by 2,750 aircraft and airborne forces launched an attack upon France, Belgium, and the Netherlands. Three days later, Guderian's XIXth Panzer Corps opened up a sixty-mile gap in the Allied center and breached French positions guarding the Meuse. A week later, the first German unit reached the Channel coast, splitting Allied forces. In the process, the German Panzers swept aside six French armored and mechanized divisions, a British armored division, and many infantry tank battalions. The German army occupied Paris on 14 June, and a cease-fire took effect on 25 June. The entire campaign lasted forty-six days but was effectively decided in ten. The Allies lost half their forces on the Continent, as well as three-quarters of their first-line equipment.[50] The speed and decisiveness of France's defeat stunned participants and onlookers alike. The success of Germany's Panzer division showed clearly how tank forces should be employed and forced other armies to adopt similar methods.

Britain

The fall of France stunned the British. As Winston Churchill confessed in his memoirs: "I did not comprehend the violence of the revolution effected since the last war by the incursion of a mass of fast-moving heavy armour. I knew about it, but it had not altered my inward convictions as it should have done."[51] The campaign also demonstrated the inadequacy of British armored organization. In

---

[48]Larry H. Addington, *The Blitzkrieg Era and the German General Staff, 1865–1941* (New Brunswick, NJ: Rutgers University Press, 1971), p. 35.

[49]Ogorkiewicz, *Armor*, p. 74.

[50]Addington, *The Blitzkrieg Era*, pp. 97–109; Matthew Cooper, *The German Army, 1933–1945: Its Political and Military Failure* (New York: Stein and Day, 1978), p. 217.

[51]Winston S. Churchill, *Their Finest Hour* (New York: Houghton Mifflin, 1949), p. 43.

March 1941, Major General Giffard LeQ. Martel, director of the Royal Armoured
Corps, compiled a report documenting lessons drawn from the army's experience in France. In it, Martel noted that the British armored division lacked the
flexibility of combined-arms Panzer divisions, which "always had the right tool
available to deal with whatever opposition appeared before them."[52] He felt that it
was too late for Britain to adopt German methods; its best hope was to adapt to
the situation.[53]

The British army lacked time for sober reflection and reform. While some political and military leaders recognized the need for organizational and doctrinal
change, the immediacy of the threat facing Britain limited the army's ability to
adapt. After the fall of France, the British army was divided between units in the
British Isles and North Africa. The former had to contend with the possibility of a
German invasion of Britain, while the latter faced the reality of combat against
the Italians. While the demonstrated effectiveness of German armor provided a
motivation for Britain to adopt combined-arms armored warfare, Britain's high
threat environment actually impeded the army's ability to reform its doctrine and
organization.

The failure of British armor in France was partially offset by its success in
North Africa. Between December 1940 and February 1941, British forces under
the command of General Richard N. O'Connor defeated Marshal Rodolpho Gra-
ziani's numerically superior Tenth Army.[54] The British army believed that the
campaign vindicated its prewar concepts, including the use of tank-heavy armored divisions for independent offensive operations. In November 1941, for example, the 7th Armored Division contained 480 tanks in three armored brigades,
each of which possessed more tanks than an entire Panzer division. To compensate for the tank-heavy armored division, the British assigned heavy but slow infantry tank regiments and brigades to support infantry formations. Unfortunately, the British had considerable difficulty coordinating the action of infantry,
artillery, and tank forces. The Germans exploited this weakness to separate British
forces and destroy them piecemeal.

In March 1941, General Erwin Rommel's Afrika Korps attacked. For the first
time, British forces in North Africa faced flexible combined-arms units led by
able commanders. The British, by contrast, had considerable difficulty coordinating combat arms. The British predilection for tank-on-tank combat played
into the hands of the Germans. In particular, tank units found it difficult to resist

[52]MID 2017–822/109, "Lessons in Employment of Armoured Units," 12 March 1941, *Military Intelligence Division Correspondence, 1917–1941*, Box 641, Record Group (RG) 165, National Archives (hereafter
referred to as NA), p. 1.

[53]Ibid., pp. 2–3.

[54]Correlli Barnett, *The Desert Generals* (London: William Kimber, 1960), chs. 1–3.

the temptation to charge German positions before their antitank guns had been neutralized.[55]

In late 1941, driven in part by the demonstrated effectiveness of Rommel's Panzer divisions, the British army began to move from tank-pure units to combined-arms formations. The commander of the 8th Army, Field Marshal Sir Claude J. E. Auchinleck, and others began formulating measures to create a more balanced division and give brigades enough supporting arms to make them tactically self-sufficient. The British were keen to mimic German doctrine as well, especially the integration of tank and antitank assets.[56] In early 1942, Martel visited British troops in North Africa. After observing their performance in the desert, he ordered the reorganization of British forces and the adoption of combined-arms doctrine.[57]

While the British grasped the fact that combined-arms organization gave the Germans considerable tactical flexibility, it was not until October 1942 that the 7th Armoured Division was reorganized into a force of one armored brigade, one light armored brigade, one infantry brigade, and divisional assets. Even after the 8th Army had reconfigured its armored forces, it still had difficulty developing combined-arms doctrine. The British lacked an organization to analyze combat performance and then develop and disseminate appropriate doctrine.[58] The lack of appropriate organizational mechanisms impeded the army's implementation of combined-arms armored doctrine.

While the 8th Army fought the Italians and Germans, the rest of the army began to reform its doctrine and organizations in response to the success of German combined-arms armored operations. Following the evacuation from Dunkirk, the army decided to retain the bifurcation of its mechanized forces into armored divisions for mobile operations and army tank brigades for infantry support.[59] It also began fielding a growing number of armored divisions. While prewar plans called for the formation of three armored divisions, after the fall of France the government revised them dramatically upward, creating nine and eventually eleven.[60] In addition, the War Office requested that Canada and Australia each raise an armored division.[61] South Africa converted its tank corps into two armored divisions.[62]

[55]House, *Toward Combined Arms Warfare*, p. 94.

[56]P. G. Griffith, "British Armoured Warfare in the Western Desert, 1940–43," in Harris and Toase, *Armoured Warfare*, pp. 80–81.

[57]House, *Toward Combined Arms Warfare*, p. 94; Smithers, *Rude Mechanicals*, p. 118.

[58]Griffith, "British Armoured Warfare in the Western Desert," pp. 81–83, 86.

[59]Smithers, *Rude Mechanicals*, p. 55.

[60]Ogorkiewicz, *Armor*, pp. 60, 76, 79, 98–99.

[61]John A. English, *Failure in High Command: The Canadian Army and the Normandy Campaign* (Ottawa: Golden Dog, 1995), p. 68.

[62]Neil Orphen, *War in the Desert* (Cape Town: Purnell, 1972), p. 461; H. J. Martin and Neil D. Orphen, *South Africa at War* (Cape Town: Purnell, 1979), pp. 226–32.

The army also reorganized its armored divisions to increase their flexibility.[63] The General Staff, studying the campaign in North Africa, identified the need to implement combined-arms organization below the divisional level.[64] To do so, the army would need to increase the proportion of infantry in its armored divisions. In 1942, the War Office reduced the number of armored brigades assigned to each armored division from two to one, cutting the number of tanks per division from 368 to 188. In addition it replaced the divisional support group with a motorized infantry brigade and a divisional artillery element. For the first time, the armored division's infantry battalions outnumbered its tank battalions.[65] The result was a more balanced armored organization, one that more closely mirrored the Panzer division.

Soviet Union

While the fall of France had robbed Britain of the opportunity to reorganize its armored forces, cooperation between Hitler's Germany and Stalin's Soviet Union gave the Red Army greater time to adapt. The Soviets followed the early campaigns of World War II closely. Those armor enthusiasts who had managed to survive Stalin's purges understood the importance of combined-arms armored warfare, but their assessments did not spread beyond the General Staff and War Academy.[66]

Despite the demonstrated success of armored warfare in France, the Red Army's leadership remained skeptical. On 25 September 1940, at the close of the army's annual maneuvers, Marshall Semion K. Timoshenko, who had replaced Voroshilov as the people's commissar of defense, told participating commanders that there was "no such thing as blitzkrieg."[67] In December he told military district commanders that Germany's victories raised no new considerations for Soviet strategy.[68]

The success of German armor nonetheless prodded the Red Army into action. On 6 July 1940 the People's Commissariat of Defense ordered the creation of nine mechanized corps; in February–March 1941, the commissariat authorized twenty more.[69] Each corps was significantly larger than its predecessors and included two tank and one motorized rifle divisions, with a total of 36,000 men, 1,031 tanks, 268 armored cars, and 358 artillery pieces.[70] On paper, each corps was stronger than its German counterpart. While German armored forces struck a balance between

[63]Ogorkiewicz, Armor, p. 60.
[64]House, Toward Combined Arms Warfare, p. 87.
[65]Ibid., p. 89; Ogorkiewicz, Armor, p. 60.
[66]Ziemke, "The Soviet Armed Forces in the Interwar Period," p. 18.
[67]Ibid., p. 27.
[68]Ibid., p. 18.
[69]Glantz, Stumbling Colossus, pp. 116–17.
[70]Dick, "The Operational Employment of Soviet Armour," p. 91.

striking power and mobility, the Soviet mechanized corps was designed more for weight than maneuver.

By January 1941, the Red Army possessed—on paper—twenty-nine mechanized corps. However, Moscow had neither the men nor the materiel to equip them. The first nine mechanized corps, deployed along the Soviet Union's western border, were the most complete. Even those units, however, were not combat-ready. They were manned at 75 percent and had only 53 percent of required equipment. Because the Red Army formed mechanized corps out of existing cavalry corps and divisions, officers and men lacked even basic familiarity with armored operations.[71] Only five had begun to receive the new T-34 and KV tanks, and all suffered from poor training.[72]

Germany's attack on the Soviet Union in June 1941 caught the Red Army unprepared. Surprised and outnumbered, the Soviets were forced onto the strategic defensive. Battles along the Soviet frontier led to the annihilation of 28 of the Red Army's 303 divisions, with an additional 70 divisions sustaining in excess of 50 percent losses in personnel and equipment.[73] Counterattacks by armored forces were poorly planned and executed. The lack of fuel and spare parts created mounting losses.

On 15 July 1941, the Red Army once again disbanded its mechanized corps, forming them instead into seven tank divisions as well as separate tank brigades for infantry support. In part, this was an effort to reduce the commander's span of control. It also represented a recognition that the Red Army's tank assets were limited.[74] By winter, however, the army leadership realized that it lacked armored formations powerful enough to conduct an armored breakthrough and exploitation. As a result, in April 1942 the army began to field a new tank corps composed of three tank and one motorized rifle brigades plus supporting arms.[75] In May, two of the new corps saw action during the Khar'kov offensive, with poor results. As a result, by November, the Soviets had fielded fifteen tank corps with increased combat power, reconnaissance capability, and sustainability.[76]

The Soviets organized increasingly large tank units. In the summer of 1942 the Commissariat of Defense consolidated existing tank corps into three tank armies. Each included two to three tank corps, one to three rifle divisions, a separate tank

[71]Glantz, *Stumbling Colossus*, p. 118.

[72]Steven J. Zaloga, "Technological Surprise and the Initial Period of War: The Case of the T-34 Tank in 1941," *Journal of Slavic Military Studies* 6, no. 4 (Dec. 1993), p. 637.

[73]Dick, "The Operational Employment of Soviet Armour," p. 92.

[74]Ibid., 93–94.

[75]House, *Toward Combined Arms Warfare*, p. 100.

[76]Each corps included three tank brigades, mortar, multiple rocket launcher, motorcycle, and armored car battalions, transportation, engineer, and tank and artillery repair companies. Dick, "The Operational Employment of Soviet Armour," p. 96.

brigade, a multiple rocket launcher regiment, and air defense and light artillery.[77] The tank army was designed to break through enemy defenses and exploit. However the units lacked common mobility. While the infantry was motorized, the lack of armored personnel carriers made it difficult to keep up with tanks. Air defense and logistics were also wheeled, but some artillery was horse-drawn. As a result, tank corps routinely outran their infantry and artillery support and suffered accordingly.

In September 1942, the Red Army formed a new, more manageable mechanized corps composed of three mechanized and one tank brigade, one or two separate tank regiments, and combat service support elements. The corps possessed three times the infantry and more tanks than a tank corps, but lacked armored infantry carriers. In January 1943, it finally fielded a successful tank army, a corps-sized unit composed largely of tanks. By the end of 1943, the Soviet Union possessed twenty-four tank armies and thirteen mechanized corps.[78] Thus by the middle of World War II, through a process of trial and error, the Red Army had finally adopted a modified combined-arms approach to armored warfare.

United States

The opening battles of World War II forced the U.S. Army, like its counterparts in Britain and the Soviet Union, to revise its assumptions about the character and conduct of warfare. France's capitulation served as a dramatic demonstration of the effectiveness of the combined-arms approach to armored warfare. It also marked a pivotal point in the development of U.S. armored forces. Over the next two years, the army focused its efforts upon achieving or containing such spectacular armored breakthroughs.

The United States was in a unique position to adopt combined-arms warfare. U.S. military attachés stationed in Berlin and other European capitals gathered an impressive amount of information on the development of German armored warfare in the years between the two world wars.[79] Army intelligence officers also filed detailed and insightful reports on the performance of German forces in both Poland and France.[80] U.S. neutrality allowed the army to view the campaigns

[77]Ibid., p. 96.

[78]Ibid., pp. 96, 100; House, *Toward Combined Arms Warfare*, p. 102.

[79]One particularly insightful report, filed by Lieutenant Colonel Truman Smith, the U.S. military attaché in Berlin, is MID 2016–1206/35, "The Panzer Division," 24 Nov. 1937, *Military Intelligence Division Correspondence, 1917–1941*, Box 626, RG 165, NA.

[80]On Poland, see, for example, "The German Campaign in Poland, September 1–October 5, 1939," 1 February 1940, Special (Bulletins), Operations in European War, A.W.C. File 236-F, Copy No. 1, Part 1, U.S. Army Military History Institute (hereafter referred to as USAMHI). On France, see, for example, "German Principles of Employment of Armored Forces," 30 August 1940, Tentative Lessons from the Recent Active Campaign in Europe, vol. 1, AWC File 236-F, USAMHI.

from both sides. It also gave Washington time to study and implement the lessons of the early phases of the war. Moreover, the United States had both the
materiel and resources necessary to implement changes. Thus, at least in theory,
the United States should have been able to adjust quickly to the emergence of
armored warfare. In fact, while the U.S. Army recognized that German combined-arms armored warfare represented a revolution in military affairs, its organizational culture prevented it from adopting the concept rapidly.

On 10 June 1940, Major General Frank M. Andrews, the army's assistant chief
of staff for operations, advanced a proposal to create two combined-arms
mechanized divisions by restructuring the infantry's light tank units as mechanized cavalry and leavening them with cavalry veterans. While the infantry and
cavalry would nominally share responsibility for the development of the division's components, mechanized cavalry officers and tactics were expected to
dominate. Exactly a month later, the adjutant general approved the plan over the
objections of both the cavalry and infantry, creating an Armored Force composed
of two armored divisions and a headquarters reserve battalion.[81] Five days later,
the 7th Cavalry Brigade (Mechanized) became the 1st Armored Brigade and the
Provisional Tank Brigade the 2nd Armored Brigade. The army chose the term
"armored" to avoid existing infantry and cavalry terms for their armored units
("tank" and "mechanized," respectively).[82]

To some U.S. officers, the advent of tank combat signaled a transformation of
warfare. Lieutenant Colonel Thomas R. Phillips, writing in the infantry's semi-
official journal, argued that the tank had replaced the cavalry as the decisive arm
of maneuver and that armored forces should be organized into combined-arms
formations. As he wrote: "[In] the continuing evolution of tactics the tank and
the airplane can be expected to become the decisive elements of war. . . . Unar-
mored infantry and artillery both complete what the tank has effected and prepare the way for the tank."[83] For the U.S. Army to enjoy success in war, it would
have to break from tradition and embrace armored warfare. As he put it: "[T]he
greatest conquerors in history have ridden on the wings of war's transformations;
the losers of battles and wars have been the administrators and traditionalists."[84]

The success of armored operations in Europe and North Africa raised larger
questions regarding the army's composition. The war indicated to some observ-

[81]Robert S. Cameron, "Americanizing the Tank: U.S. Army Administration and Mechanized Development within the Army, 1917–1943," Ph.D. dissertation, Temple University, 1994, p. 493.

[82]Gabel, *The U.S. Army GHQ Maneuvers of 1941*, p. 24.

[83]Lieutenant Colonel Thomas R. Phillips, "Traditionalism and Military Defeat," *Infantry Journal* 49
(Mar. 1941): 23.

[84]Ibid., p. 27.

ers that armored forces backed by tactical aviation had displaced the infantry and artillery as the centerpiece of ground combat. On 1 March 1941, Brigadier General Sherman Miles, the head of the army's Military Intelligence Division, wrote Army Chief of Staff George C. Marshall that the fall of France constituted "a revolutionary turning point in the history of warfare, for in it the tank and airplane deprived the infantry and artillery of their former supremacy." He argued that army doctrine, organization, tactics, and equipment should be altered to reflect the new reality: "Either the infantry division must increase its anti-tank defense to a point which will permit it to check a tank division . . . or the tank-air team must supplant the infantry-artillery team as the fighting nucleus of the army of the future."[85] To U.S. armor advocates, the latter option was clearly preferable.

Not all army leaders favored the establishment of a separate armored force. Major General Lesley McNair, the GHQ chief of staff, argued that the infantry-artillery team remained sound, and he opposed altering the structure of the army.[86] Instead, he believed that the GHQ should establish a pool of independent tank battalions that it could employ as it saw fit.[87] In the end, the army chose to create large armored formations without disrupting the existing administrative organization. It was, in effect, a compromise between creating a new, independent combat arm and relying upon a temporary collection of units from existing arms.

Intelligence officers were the Armored Force's main conduit of information regarding the performance of foreign armies. The Armored Force sent U.S. military attachés extensive lists of questions regarding the organization and employment of German armored forces in both Poland and France.[88] Attachés returning from Germany gave lectures on the war to the force's leadership as well.[89] The army also assigned a succession of former attachés in Berlin as the unit's senior intelligence officer.

The organization and doctrine of German tank forces influenced the development of the Armored Force in several significant ways. First, the U.S. armored

[85]MID 2016–1297, "Evaluation of Modern Battle Forces," *Assistant Chief of Staff, Intelligence, Intelligence Branch Estimates, 1940–1942*, Box 1, RG 319, NA.

[86]Cameron, "Americanizing the Tank," p. 577.

[87]Badsey, "The American Experience of Armour," in J. P. Harris and F. H. Toase, *Armoured Warfare* (New York: St. Martin's, 1990), p. 135.

[88]"Statement by Major Percy G. Black, Field Artillery, Who Recently Returned from Germany Where He Has Been on Duty as Assistant Military Attache," 5 Dec. 1939, in The Willis D. Crittenberger Papers, USAMHI, pp. 1–5; Scott Alan Koch, "Watching the Rhine: U.S. Army Military Attache Reports and the Resurgence of the German Army, 1933–1941," Ph.D. dissertation, Duke University, 1990, pp. 231–32.

[89]Larry I. Bland, ed., *The Papers of George Catlett Marshall*, vol. 2: *"We Cannot Delay," July 1, 1939–December 6, 1941* (Baltimore: Johns Hopkins University Press, 1986), p. 120.

division was an explicit attempt to replicate the German Panzer division. U.S. armored divisions were geared toward fighting the German army: during World War II, all but one was employed in the European Theater of Operations.[90] The structure of the 1940 U.S. armored division was nearly identical to that of the prewar German Panzer division. It included a reconnaissance battalion; an armored brigade with two light and one medium tank regiments, a field artillery regiment, and an engineer battalion; an infantry regiment; field artillery battalion; and service and maintenance units.[91] As the army received information regarding changes in the composition of the Panzer division, it altered that of the armored division accordingly.

Second, U.S. armored doctrine bore many similarities to that of the German tanks corps. In part, this represented a continuation of a cavalry tradition that emphasized speed, maneuver, and initiative. It was also the result of a conscious effort to emulate German doctrine because of its success on the battlefield. The U.S. Army, like the Wehrmacht, viewed armored divisions as general-purpose units. Army doctrine envisioned using armored divisions against objectives deep in the enemy rear. They would seize critical areas, envelop and encircle enemy units, and pursue retreating forces. Tactical aircraft would support their forward movement while motorized infantry units would consolidate their gains.[92]

In some cases, the Armored Force attempted to copy German tactics explicitly. Lieutenant Colonel James C. Crockett, the assistant military attaché in Berlin, sent the force a translation of a German textbook on the training and tactics of a tank platoon, including lessons derived from the campaign in France. The Armored Force duplicated the document, replaced the illustrations of German tanks and vehicles with those of their U.S. counterparts, and used it as a basis of a field manual governing U.S. tank platoons.[93]

U.S. tank forces also bore the enduring mark of the mechanized cavalry. Indeed, more than fifty thousand cavalrymen served in armored units during World War II.[94] Consistent with their cavalry background, armored commanders such as Chaffee valued the speed and mobility of armored forces over their firepower and protection.[95] The Armored Force, like the mechanized cavalry before it, did not expect to get into a slugging match with enemy armored vehicles. As a result, the army acquired tanks that emphasized speed at the expense of armor and firepower. U.S. tanks were thus generally inferior to their German counter-

[90] The 1st Armored Division served in North Africa and Italy.
[91] Gabel, *The U.S. Army GHQ Maneuvers of 1941*, p. 24.
[92] Cameron, "Americanizing the Tank," pp. 545–46.
[93] Ibid., p. 565.
[94] Badsey, "The American Experience of Armour," p. 135.
[95] Gabel, *The U.S. Army GHQ Maneuvers of 1941*, p. 25.

parts, especially the Mark V Panther and Mark VI Tiger, which appeared in the later stages of the war.[96]

The rise of armor led to a redistribution of resources within the army. The arm that lost the most was the cavalry, which saw armor usurp many of its traditional missions. This is ironic, considering that it had been the cavalry that had first embraced the style of mobile armored warfare that came to dominate World War II. The Office of the Chief of Cavalry was abolished on 9 March 1942, although armor did not officially replace the cavalry until the Army Organization Act of 1950.[97]

The mechanization of the U.S. Army also spawned a new combat arm: the antitank forces. On 29 June the army activated divisional antitank battalions. Six months later, the War Department removed the antitank battalions from divisional control, renamed them "tank destroyer" battalions, and designated them as GHQ assets. In the process, the tank destroyer arm became virtually autonomous.[98]

The army planned to use independent antitank units composed of fast, lightly armored tank destroyers to counter massed armor attacks. Indeed, the infantry used the promise that tank destroyers could free the battlefield from the threat of tanks to defend its autonomy and argue against expanding armored forces and reorganizing around the tank.[99] The experience of the war, however, soon demonstrated that the weapon best suited to destroying a tank was another tank. U.S. forces seldom met massed German armored formations, so tank destroyers were rarely used in the role envisioned.[100]

In March 1942 the army reorganized its armored divisions. Like the British and Germans, the United States realized the need for its armored formations to possess more infantry. Each division lost one of its three armored regiments and gained a battalion each of self-propelled artillery and infantry.[101] The result was a leaner, more balanced division. The army also established two tactical headquarters (known as Combat Commands) directly subordinate to the division headquarters. These commands allowed a general to split his division into two combined-arms teams.[102] The combat commands institutionalized the concept of

[96]Martin Blumenson, *Breakout and Pursuit* (Washington, DC: U.S. Army Center of Military History, 1961), pp. 44–45.

[97]James E. Hewes, Jr., *From Root to McNamara: Army Organization and Administration, 1900–1963* (Washington, DC: U.S. Army Center of Military History, 1975), pp. 66, 392.

[98]In virtually every other army, antitank forces were organic to tactical formations. The United States, however, pooled its tank destroyers as a reserve at the divisional, corps, and army level.

[99]Cameron, "Americanizing the Tank," pp. 610, 887.

[100]Charles M. Baily, *Faint Praise: American Tanks and Tank Destroyers during World War II* (Hamden, CT: Archon, 1983), p. 6.

[101]Badsey, "The American Experience of Armour," p. 136.

[102]Ogorkiewicz, *Armor*, pp. 89–90.

battle groups (Kampfgruppen) that the Germans had employed successfully since the beginning of the war.

By 1943 the army had created sixteen armored divisions and sixty-five independent tank battalions.[103] The army also planned to form armored corps of two armored and one motorized infantry divisions, much like the Panzer corps.[104] In 1943, however, it dropped the idea. In September the War Department announced a new, smaller armored division composed of three battalions each of tanks, armored infantry, and armored field artillery, plus supporting troops. The division's battalions could be allocated between three Combat Commands to facilitate combined-arms operations.[105]

Acquiring an armored force was one thing; developing the doctrine, tactics, and procedures to conduct successful combined-arms operations was quite another. As late as the fall of 1943, the United States had little experience with massed armored operations. The 1st and 2nd Armored Divisions fought in Tunisia and Sicily but did not encounter large armored formations. The other fourteen U.S. armored divisions remained in the United States.[106] The rapid expansion of the U.S. Army and lack of combat experience limited the effectiveness of its armored divisions. The performance of U.S. forces in their first action with German armored forces at Kasserine Pass in February 1943 demonstrated the army's difficulties. During the battle the Germans divided and then routed the U.S. II Corps, using combined-arms teams to destroy it piecemeal. U.S. armor, infantry, and artillery, by contrast, were unable to act in concert.[107] Despite possessing combined-arms formations, the army had yet to master armored warfare.

Conclusions

By the end of World War II, all major European armies had become mechanized. Germany entered the war with six Panzer divisions; at the peak of its power, it fielded thirty-three. Britain and France together had a handful of armored units at the outbreak of the war, while the United States had none. By the end of the war, the Western Allies had twenty-eight armored divisions: sixteen U.S., five British, three French, two Canadian, one South African, and one Polish. The Red Army, which formed its first tank division in 1940, possessed six tank

[103]Badsey, "The American Experience of Armour," p. 131.
[104]House, *Toward Combined Arms Warfare*, p. 108.
[105]Badsey, "The American Experience of Armour," p. 140.
[106]Ibid., p. 139.
[107]The corps suffered 20 percent casualties—six thousand men killed, wounded, or missing compared with less than a thousand Germans. The corps lost 183 tanks, 194 half-tracks, 208 artillery pieces, and 512 trucks and jeeps. See Martin Blumenson, *Kasserine Pass* (Boston: Houghton Mifflin, 1967), pp. 303–4.

armies by the end of the war.[108] As Allied armies reorganized and retrained their armored forces, Germany's battlefield dominance began to wane. By late 1942, German armored forces were unable to match their earlier successes. Indeed, some operations became immobile for weeks or months. The Normandy campaign featured fighting on a fairly static front for almost two months, and Monte Cassino six.

Cooperation between arms increased as well. Armies initially reorganized their armored divisions into combined-arms formations resembling the Panzer division. As the war progressed, tank units began integrating separate arms at lower levels. By 1945 several armies had adopted combined-arms brigades; the U.S., British, and German armies possessed combined-arms battalions.[109]

The war also witnessed the employment of progressively larger armored formations. While its opening battles featured the use of tank units against nonarmored forces, the latter phases were marked by large-scale actions between tank forces, such as those between the Wehrmacht and the Red Army on the Eastern Front. On the Western Front, Patton's Third Army's pursuit of German forces through France exemplified the new style of warfare.

The British, Soviet, and U.S. armies adopted combined-arms armored warfare out of necessity rather than choice. The fall of France, more than any other event, demonstrated the effectiveness of combined-arms methods and convinced them of the need to imitate German practices. The fact that Britain was at war with Germany and that the Soviet Union and United States faced the prospect of war further heightened the urgency of action. After the fall of France, the British, Soviet, and U.S. armies expanded their armored forces considerably and began moving toward combined-arms doctrine and organization.

Despite the fact that the British, Soviet, and U.S. armies possessed tanks, contained prominent armor advocates, had experimented with various approaches to armored warfare, and witnessed the effectiveness of combined-arms operations, each had difficulty developing the doctrine and organization necessary to compete with Germany. The record of all three armies reveals a considerable amount of trial and error, adaptation and resistance. In the end, developing the capability to wage combined-arms armored warfare proved to be neither rapid nor smooth.

In part, the Allies' difficulty in adopting combined-arms armored doctrine was a result of the circumstances they faced. World War II not only compelled the Allies to emulate German doctrine and organization but also interfered with their

---

[108]House, *Toward Combined Arms Warfare*, pp. 111–13.
[109]Ibid., pp. 107–8.

efforts. During the early years of World War II, Britain, and then the Soviet Union, faced the difficult task of reorganizing their armored forces and reformulating their doctrine while engaged with the enemy. The United States, by contrast, had the opportunity to observe the war's opening campaigns, raise combined-arms armored divisions, and develop appropriate doctrine. What the British and Soviet armies had but the U.S. Army lacked was the opportunity to train under realistic conditions.

The organizational culture of the three armies was another constraint. Prewar practices hindered the British and Soviet armies from adopting combined-arms doctrine and organization; they arguably fostered the acceptance of such methods within the U.S. Army. Of the three, only the British possessed a well-established tank corps at the beginning of World War II. The Royal Armored Corps' enthusiasm toward tank-pure units and doctrine continued to shape the British approach well into World War II. Indeed, the 8th Army's experience against the Italians in North Africa seemed to vindicate the tank-pure method. It was only after the army faced Rommel's Afrika Korps that it was forced to reform its approach to tank operations.

In the Soviet Union, the horse cavalry's domination of the Red Army left its mark upon the employment of tanks. The preference of Stalin and his cronies for cavalry forces led to the dissolution of the Soviet Union's mechanized corps on the eve of World War II. When Moscow reconstituted its armored formations after the fall of France, they were led by former cavalry officers with little experience in tank warfare. The Red Army spent much of the early phases of World War II relearning how to conduct combined-arms warfare.

In the United States, the mechanized cavalry had spearheaded the development of armored warfare. When the Armored Force was formed in 1940, it inherited an approach to tank warfare that emphasized speed, maneuver, and tactical initiative. The mechanized cavalry's prewar approach to armored warfare nonetheless proved dysfunctional in several important ways. The preference for speed over firepower and protection, for example, manifested itself in tanks that were inferior to their German counterparts. Nor was the infantry willing to concede that the intrusion of armor onto the battlefield had permanently changed the conduct of war. Instead, it looked to tank destroyers to sweep enemy armor from the battlefield, returning war to a more familiar and comfortable form.

The hypothesis that a competitive security environment will promote the spread of military innovations thus requires several important modifications. First, it is exceedingly difficult to determine the effectiveness of new approaches to warfare in peacetime. Absent a convincing demonstration, such as that provided by the fall of France, the utility of new ways of war is purely theoretical.

Second, a highly competitive environment can actually hinder attempts to adopt new approaches to warfare. Finally, the character of military organizations may prevent, delay, or shape the adoption of innovations. Indeed, it is unlikely that a military organization will adopt a foreign innovation completely. Rather, its approach will contain unique doctrinal and organizational elements.

There are many similarities between the situation that armed forces confronted during the interwar period and today's strategic environment. Then, as now, the great powers faced considerable uncertainty over both the source of potential threats and the shape of future wars. The position of the United States as the world's sole superpower considerably complicates the situation. The U.S. Army had the luxury of observing the early stages of World War II from afar and adapting accordingly. The challenge we face today is one considerably more difficult. Today, U.S. armed forces are deployed across the globe. When the next major war occurs, it is likely that the United States will be involved from the outset. This posture puts a premium on our ability to identify and adapt to innovations before they appear on the battlefield, something that has traditionally proven difficult.

The position of the United States as the world's sole superpower bears an intellectual burden as well. In the past, the U.S. armed forces were willing to learn from others, as the U.S. Army did in emulating the Panzer division. Today, armed with the world's most competent armed forces, we face the danger of hubris. While the United States currently leads the world in many emerging warfare areas, such a situation is not destined to continue indefinitely. A number of states are exploring new approaches to warfare. We would do well to observe and learn from their experience, just as others are attempting to learn from ours.

The U.S. armed forces in the Gulf War achieved one of the most lopsided battlefield victories in military history, establishing in the process a new paradigm of warfare. We should not, however, expect other states to field carbon copies of the U.S. armed forces. Most states have needs that differ considerably from those of the United States. Few, for example, require the ability to project power rapidly across the globe. Similarly, few have the technological, economic, and human resources necessary to wage the type of high-technology, high-tempo operations at which the United States excels. While many states—adversary and ally alike—will adopt the trappings of modern warfare, few will embrace its substance. The natural conservatism of military organizations will constrain many from adopting emerging concepts of warfare, especially those that have yet to prove their value in combat. The Gulf War demonstrated the importance of information technology, stealth, precision-guided munitions, and space systems. It did not, however, reveal how best to employ those technologies. If history is a guide, it will take a decisive battle, such as the fall of France in 1940, to demonstrate the effectiveness of new ways of war.

CHAPTER TEN

Receptivity to Revolution

Carrier Air Power in Peace and War

EMILY O. GOLDMAN

Over a twenty-five-year period from the introduction of aircraft at sea to the Battle of Midway, air power revolutionized naval warfare. The British attack on the Italian harbor at Taranto hinted at the potential of the carrier as an offensive strike element. The Japanese attack on Pearl Harbor demonstrated the importance of carrier-based strikes and presaged an era when carriers would be the real targets. The Battle of Midway demonstrated that the aircraft carrier had become the new capital ship of modern navies.

From the end of World War I to the end of World War II, navies integrated air power into their naval operations, organizations, and doctrines at different rates and to differing degrees.[1] Great Britain relegated carrier air power to a defensive role. The United States and Japan adopted the carrier as an offensive strike weapon. In Germany and Italy, the development of naval air power was stunted early on and never recovered. Integration of air power in naval operations did not occur smoothly or easily, even in the most competitive of environments. Each naval power operated under a set of constraints that restricted the scope, pace, and efficiency with which new ideas and practices were adopted.

This essay evaluates the motivations and capacity to adopt revolutionary maritime airpower practices between 1920 and 1945 in light of predictions about the scope, pace, and patterns of diffusion made by three schools of thought: neo-

I would like to thank Brian Sullivan, Thomas Mahnken, Leslie Eliason, Andrew Ross, and Gregory Wilmoth for their comments on this paper. Leo Blanken provided valuable research assistance.
[1] My chief interest is in the evolution of the fast fleet carrier as distinct from the smaller, slower escort and trade protection carriers.

realism, bureaucratic functionalism, and organizational sociology. The dynamics of diffusion during this period of rapid technological change are assessed, and suggestions are offered for practitioners and scholars studying and tracking military diffusion today.

Motivations to Adopt

Several motivations stimulate the spread of innovative military practices. Neorealism emphasizes the pervasive competitive pressure that governs the international system as the most powerful factor in the spread of military innovations. National survival depends on battlefield success, so states are driven to become as militarily effective as possible. For neorealists, states are like firms that "emulate successful innovations of others out of fear of the disadvantages that arise from being less competitively organized and equipped. These disadvantages are particularly dangerous where military capabilities are concerned, and so improvements in military organizations and technology are quickly imitated."[2] States therefore adopt the most successful practices (e.g., the best model) given mission requirements.

Competition and strategic necessity were powerful drivers behind carrier development. Most efforts to exploit the offensive potential of carrier air occurred in the late 1930s and during World War II, a high threat environment. These are the conditions under which we would expect neorealist predictions to be most robust. The higher the threat environment, the more rapid and efficient is the diffusion.

Bureaucratic functionalists emphasize the inherently conservative nature of large organizations.[3] Organizations are risk averse. They adopt technologies and strategies that support existing goals, defined in terms of protecting resources, autonomy (jurisdiction and independence), and organizational essence (the views on missions and capabilities held by the dominant group in the organization). Changes that do occur tend to be incremental adjustments that are consistent with existing tasks. The theory also posits that military organizations have an institutional interest in adopting offensive technologies and strategies because these enhance the organization's resources, autonomy, and essence.[4] Yet the ten-

[2]Joao Resende-Santos, "Anarchy and Emulation of Military Systems: Military Organizations and Technology in South America, 1870–1930," *Security Studies* 5, no. 3 (spring 1996): 196; Kenneth N. Waltz, *Theory of International Politics* (Reading, MA: Addison-Wesley, 1979), p. 127.

[3]Graham T. Allison, *Essence of Decision: Explaining the Cuban Missile Crisis* (Boston: Little Brown, 1971), p. 78–94; Richard Cyert and James March, *A Behavioral Theory of the Firm* (Englewood Cliffs, NJ: Prentice-Hall, 1963); Herbert Kaufman, *The Limits of Organizational Change* (Tuscaloosa: University of Alabama Press, 1971); James Q. Wilson, *Bureaucracy: What Government Agencies Do and Why They Do It* (Basic Books, 1989).

[4]Jack Snyder, "Civil-Military Relations and the Cult of the Offensive, 1914 and 1984," *International Security* 9, no. 1 (summer 1984): 108–46; Stephen Van Evera, "The Cult of the Offensive and the Origins of the First World War," *International Security* 9, no. 1 (summer 1984): 58–107.

dency to preserve existing roles and routines will often override this incentive when change involves great risk and when the technological environment is highly uncertain.

Organizational sociologists see competition as a driver of military diffusion, but they argue that institutional pressures also stimulate the spread of forms and practices across organizations in the same sector. Organizations adopt particular practices to attain legitimacy as well as to increase efficiency, while socialization pressures stimulate the spread of practices within a profession.[5] Since professional networks are key avenues for transfer of information, the theory predicts that organizations will pattern their practices after those with whom they interact. Accordingly, familiar and accessible models, as well as legitimate ones, should spread. In either case, the most efficient model given strategic circumstances will not necessarily be adopted.

In the carrier case, socialization pressures and networks were important. Operating together in the Pacific late in the war assisted the British in adopting America's offensive carrier practices. Legitimacy pressures, which tend to have a greater impact on later adopters, were also at play. Corum argues that the quest for prestige and status accounts for Germany's interest in carriers.[6] Herrick attributes the "big Navy" enthusiasm in the Soviet Union in the late 1930s to deterrence and prestige.[7] More recently, efforts by the Chinese to acquire an aircraft carrier have been attributed as much to the belief that a "nation cannot be a real power without seapower" as to strategic requirements in the South China Sea.[8]

Capacity to Adopt

While competition may drive diffusion, assimilating new practices requires considerable skill and adaptation. Neorealist theory is silent on the question of whether military organizations can easily adopt practices developed and honed outside their borders. Militaries may be "rational shoppers," choosing from the

[5]Paul J. DiMaggio and Walter W. Powell, "The Iron Cage Revisited: Institutional Isomorphism and Collective Rationality in Organizational Fields," *American Sociological Review* 48 (April 1983): 147–60; Barbara Levitt and James G. March, "Organizational Learning," *Annual Review of Sociology* 14 (1988): 329–30.

[6]James S. Corum, *The Luftwaffe: Creating the Operational Air War 1918–1940* (Lawrence: University Press of Kansas, 1997), pp. 178–79.

[7]Given that Stalin planned a wartime strategy of active defense, an oceanic navy was superfluous. But "[i]nasmuch as the Spanish Civil War had demonstrated that even the world's largest submarine force was not enough to command respect as a naval power for the USSR, big ships would be built as necessary in the effort to acquire international prestige considered to be a prerequisite to success for Soviet diplomacy." Robert Waring Herrick, *Soviet Naval Strategy: Fifty Years of Theory and Practice* (Annapolis: Naval Institute Press, 1968), p. 35.

[8]Jun Zhan, "China Goes to the Blue Waters: The Navy, Seapower Mentality and the South China Sea," *Journal of Strategic Studies* 17, no. 3 (Sept. 1994): 180–208.

best practices available, but is assimilation rapid and faithful? The Imperial Japanese and U.S. navies varied in speed and faithfulness of adoption, and in developing the doctrine and fine tuning the techniques for employing self-contained naval air power. Both navies planned for war in the Pacific. The U.S. Navy required offensive power projection across vast ocean expanses, a mission for which the carrier was ideally suited. Japan's overall naval strategy was defensive, but in the 1930s the Japanese adopted a preemptive strategy to punch out a defensive ring in the Pacific. Yet neither assimilated offensive carrier air power practices quickly. Even after the Japanese attack on Pearl Harbor had demonstrated the power of carriers operating independently of the battle line, adoption lagged. Carrier-based air also had strategic utility for offensive operations in the Mediterranean, but aside from the attack on Taranto the British did not adopt offensive carrier strikes until late in the war, when the Royal Navy launched strikes against German warships in Norwegian waters. Nor did the Italians succeed in surmounting the myriad internal obstacles to adopting the vessel that was crucial to challenging British naval supremacy in the Mediterranean.

Bureaucratic functionalism predicts that military diffusion will be difficult, particularly when a new mission for a combat arm challenges existing roles and routines. Routines and standard operating procedures determine action. Bureaucratic inertia ensures that change will be incremental. Revolutionary change should be rare, because it by definition challenges existing tasks. The carrier challenged the central position of the battleship in naval operations, and the dominant battleship subculture in all interwar navies.

Organizational sociologists focus on several factors that affect the ease and faithfulness of adoption. The new practice may not be "compatible" with valued local patterns;[9] the organizational set—necessary supporting organizations and institutions such as schools and industries—may be inadequately developed to facilitate adoption;[10] reformers may lack authority to institute change in the face of pressures to conform to existing social institutions.

Compatibility captures the degree to which a new practice meshes with deeply held attitudinal patterns and implicit understandings of the organization's roles and structures. It influences the level of legitimacy accorded a new idea or practice, and the degree of opposition that reformers must surmount to establish the legitimacy of a new way of waging war. When compatibility between local values and imported practices is low, reformers face opposition, and they must garner

[9]Everett M. Rogers, *Diffusion of Innovations*, 3d ed. (New York: Free Press, 1983), pp. 14–16; James M. Blaut, "Two Views of Diffusion," *Annals of the Association of American Geographers* 67, no. 3 (Sept. 1977): 346.

[10]D. Eleanor Westney, *Imitation and Innovation: The Transfer of Western Organizational Patterns to Meiji Japan* (Cambridge: Harvard University Press, 1987), pp. 28–31.

TABLE 10.1

Assumptions and Predictions about Diffusion

	Neorealism	Bureaucratic Functionalism	Organizational Sociology
Diffusion driver	Competition	(n/a—system level forces exogenous)	(i) Competition; (ii) Socialization
Incentives to adopt/emulate new technology/practice	Strategic necessity (e.g., external threat, ambition, defeat in war)	Increase organization's resources, autonomy, essence	(i) Efficiency; (ii) Legitimacy and identity
Barriers to adoption	(i) Resources (pure neorealism) (ii) State power (state-centered realism)	Organizational inertia (due to protection of essence, roles, and/or standard operating procedures)	Compatibility, administrative authority, organizational set
Capacity to adopt	High	Low	Moderate
Likelihood of adoption	High	Low	Partial emulation or adaptation
Speed of emulation	Rapid	Slow	Moderate
Model selected	Most successful	Offensive	Most accessible and legitimate
Faithfulness to model	High	High	Low

legitimacy for the new methods while mobilizing resources to effect organizational change. Surmounting these obstacles depends not only on the resources (e.g., money) at their disposal but also on leadership commitment to the new practice and the authority of reformers to implement change. Alternatively, an adopter may far surpass the initial innovator because of higher compatibility, more effective supporting institutions, and greater administrative authority. Any challenge to the dominant battleship orthodoxy in all interwar naval establishments automatically threatened deeply held ideas and beliefs. Compatibility problems existed in all navies, though in some states reformers succeeded in surmounting political and intellectual barriers.

Compatibility also affects how an innovation is incorporated into existing organizational practices and routines, thereby influencing the extent, faithfulness, and speed of adoption. While the bureaucratic approach predicts that adoption will be difficult but faithful, organizational sociologists expect that even if obstacles are surmounted, there will be divergence from the original model, particularly when values, expectations, and experience in the importing social system diverge from those of the exporting social system. For example, the British were the initial innovators and the chief source of Japan's information about carrier operations. The Japanese initially adopted British practices but altered them on account of different needs, values, and experiences. They pushed the offensive envelope of the aircraft carrier, which was more consistent with the Imperial Navy's offensive biases and wartime experiences in China in 1937–38.

Key assumptions and predictions of the theoretical approaches appear in Table 10.1.

Modeling Successful Practices

Military practices have multiple facets—technological, conceptual, doctrinal, and organizational. These facets need not diffuse at the same rate. The key issue on the eve of World War II was putting the pieces of the carrier paradigm together to maximize offensive potential and defensive safety.

From the end of World War I until the end of World War II, the role of the aircraft carrier in naval warfare was dramatically reformulated from a vision of carrier as reconnaissance and support element to the main battle line, to carriers as independent strike platforms, massed together to maximize offensive firepower and defensive protection. On the eve of World War II, the right mix of offensive potential and defensive safety was hotly contested. The proper mix would change midway through the Pacific war with the resurgence of the tactical defense in 1942 because of developments in radar and shipboard antiaircraft.[11]

Three episodes of carrier air power provided key demonstrations of the new weapon system's potentially revolutionary impact: Britain's attack on the Italian fleet at Taranto on 11 November 1940; Japan's attack on the U.S. fleet at Pearl Harbor on 7 December 1941; and Japan's defeat at the Battle of Midway on 4 June 1942. The British attack on the Italian naval base at Taranto was the first time an assault was made on a large, heavily defended harbor facility, and the first time that such an attack was carried out with carrier-based planes. Although the Swordfish biplanes used in the attack were virtually obsolete, the British succeeded in inflicting a major blow against the Italian fleet and overturning conventional wisdom that ships could not be torpedoed inside their own harbor at night.[12] The attack, some have speculated, may have made an impression on Isoroku Yamamoto, commander in chief of the Combined Fleet of the Imperial Japanese Navy, who had long been thinking about how to neutralize the U.S. fleet at Pearl Harbor. Taranto indicated that a fleet in a shallow, heavily defended har-

[11]CPT Wayne P. Hughes, Jr., *Fleet Tactics: Theory and Practice* (Annapolis: Naval Institute Press, 1986), pp. 85–110.
[12]Harry Stegmaier, Jr., "The British Naval Effort in World War II: 1939–1945," in James J. Sadkovich, ed., *Reevaluating Major Naval Combatants of World War II* (New York: Greenwood, 1990), p. 29. One modernized pre–World War I battleship was sunk and never returned to service. Two other battleships suffered moderate damage. One was back in service by March 1941; the other in May 1941. Jack Greene and Alessandro Massignani, *The Naval War in the Mediterranean 1940–1943* (London: Chatham, 1998), pp. 101–9.

bor could be sunk by carrier attack. Whether or not Taranto directly inspired Yamamoto, the potential revealed at Taranto was realized one year later and twelve thousand miles away at Pearl Harbor.[13]

Midway represented an important turning point in the tactics for operating carriers. Attempting to lure the U.S. fleet into a decisive battle, Yamamoto dispersed his fleet over the North Pacific, ultimately leaving it vulnerable to a weaker U.S. force, which decisively defeated it. By 1942 the balance had shifted in favor of ship defense, making concentration of the striking force essential to sink a carrier while maximizing defensive firepower. Midway also signified a conceptual shift, as the aircraft carrier came to be seen as the primary capital ship in Britain, the United States, and Japan. The British continued work on *Vanguard* but it was their last battleship; the Japanese converted the incomplete *Shinano* to a carrier; in the United States, the Montana-class battleships were canceled in favor of carrier construction.[14]

While all navies of the day realized that air power had an important defensive role to play in spotting, scouting, and reconnaissance, the real issue was whether carrier air power could operate effectively as an offensive strike element, and whether the target of such an attack should be the battle line or the enemy's carriers. Collectively, Taranto, Pearl Harbor, and Midway compelled naval leaders to revise how they viewed the relationship between the battleship, battle line, and aircraft carrier. The carrier had a vital role to play operating independently of the battle line. Moreover, the battleship should protect the carriers rather than vice versa. This required overturning the big ship–big gun doctrine that had been deeply entrenched in naval thinking since the days of Mahan. This conceptual shift proved difficult for naval leaders in all countries to accept so long as battleships could survive air attack.[15]

At the tactical level, the question was the proper balance between concentration and dispersal in order to provide sufficient defenses without compromising the weight of attack. Concentrated forces permitted the massing of air attack and maximized offensive potential. Yet privileging offense at the expense of defense could prove disastrous. Carriers were the Achilles' heel of a system for waging offensive war. Only after the protection problem was solved would it be possible

[13]Thomas P. Lowry and John W. G. Wellham, *The Attack on Taranto: Blueprint for Pearl Harbor* (Mechanicsburg, PA: Stackpole, 1995).

[14]Robert Gardiner, ed., *The Eclipse of the Big Gun: The Warship 1906–45* (London: Conway Maritime, 1992), p. 32.

[15]Once it was shown that airborne weapons could sink battleships, navies that had separated carrier operations from the battle line—the United States and Japan—had an easier time abandoning the battle line concept than did navies—Great Britain—that continued to operate carriers as part of the battle line. Norman Friedman, *Carrier Air Power* (New York: Routledge, 1981), p. 27. This suggests how transformation can cascade.

to realize the carrier's offensive potential. The solution was threefold: concentration, whereby several carriers operating in one concentrated tactical formation could mutually defend each other against air attack with a greater volume of anti-aircraft fire; protection of carriers by battleships; and increasing the number of antiaircraft batteries on carriers, escorting cruisers, and destroyers.

Exploitation of a new method of warfare also depended upon organizational reforms. Carriers could assume a new and central role in the system for waging naval war to the extent that naval aviators had representation in the highest administrative and civilian echelons, wielded significant influence over the prosecution of war, and had a career path that allowed them to advance professionally. Air-centered naval warfare also benefited to the extent that naval aviation was free from oversight and control by land-based air organizations.

Certain design elements facilitated realization of the carrier's potential. Chief among them were those relating to carrier capacity. Certain tactics were viable only if a carrier had a large capacity, which U.S. carriers did by virtue of Washington Naval Treaty provisions on carrier conversion. They allowed the U.S. Navy to convert two huge, uncompleted battle cruisers, *Lexington* and *Saratoga*, displacing thirty-three thousand tons each, into carriers. By contrast, the British were strapped with smaller carriers of World War I vintage. The British preference for greater protection on their carriers, given the need to operate in close proximity to enemy land-based air, also limited carrier capacity,[16] as did their practice, shared by the Japanese, of stowing aircraft in hangars. The U.S. practice of deck storage, on the other hand, maximized carrier capacity. "Usable flight deck area was inevitably much greater than hangar area, so the aircraft complement of a carrier which permanently parked her aircraft on deck could greatly exceed that of a carrier limited to hangar stowage."[17]

Aircraft performance was also critical. The Japanese Zero was far superior to any U.S. fighter until the introduction of the Hellcat in 1942.[18] A comparison between the attack craft used by the British at Taranto—the Fairey Swordfish—and those used by the Japanese at Pearl Harbor—the Val and Kate—show dramatic differences in speed, maximum range, and maximum ceiling.[19] British air

[16]Barry Watts and Williamson Murray, "Military Innovation in Peacetime," in Willliamson Murray and Allan R. Millett, eds., *Military Innovation in the Interwar Period* (Cambridge: Cambridge University Press, 1996), pp. 399–400, explain why the fundamental measure of offensive carrier effectiveness was the number of aircraft it could launch for a given mission.

[17]Norman Friedman, *British Carrier Aviation* (London: Conway Maritime, 1988), p. 19. Of course, this was not a real option for countries that had to operate carriers under the harsh weather conditions of the North Sea.

[18]Clark G. Reynolds, *The Fast Carriers: The Forging of an Air Navy* (New York: McGraw Hill, 1968), p. 57.

[19]Lowry and Wellham, *The Attack on Taranto*, pp. 103–4.

TABLE 10.2

Carrier/Naval Aviation Design and Practice

Air Power–Battle Fleet Paradigm	Carrier Doctrine	Organizational Structure	Carrier Design and Aircraft Complement	Model Emulated
BRITAIN				
Aircraft for scouting and reconnaissance	Carriers operate singly with escorts 1945: Pacific Fleet adopts concentration and U.S. fast-carrier doctrine	Navy loses control of Fleet Air Arm 1918–39	Armored decks; airplane capacity 44 per ship; hangar storage; multi-purpose aircraft design	mid-1940s, adopts U.S. offensive doctrine and design with open hangar and lightly armored deck
UNITED STATES				
Air battle precedes battle fleet action	1927: carriers operate independently June 1943: multi-carrier task force and tactics of concentration adopted	1921: Bureau of Aeronautics 1926: Morrow Board gives navy aviation high-level civilian representation; commanding officer of carrier must be aviator 1943: Deputy—CNO (Air) established	Unarmored flight decks; airplane capacity 91 per ship; open hangars; deck park with crash barriers; large capacity; single purpose types	*Midway* class emulated Britain's armored decks
JAPAN				
Carriers for preemptive strike against enemy carriers	1928: First Carrier Division 1938: single strike force 1941: box formation July 1942: Combined Fleet reorganized around carriers post-Midway: compromise between concentration and dispersal little attention to fleet air defense	1927: independent naval air headquarters; must qualify as aviator to achieve flag rank; 1936: high-level aviation representation 1938: combined naval air wing April 1941: First Air Fleet	High-speed, light-weight, light armor; airplane capacity 61 per ship; hangar park; attack aircraft (dive bombers and torpedo planes) favored over fighters and scouts; aircraft emphasize offensive (speed, maneuverability, range)	1920s: British carrier design 1936: U.S. carrier design 1941: British armored flight deck 1930s: U.S. dive-bombing aircraft design and manufacture assistance from Britain and Germany
GERMANY				
Carriers for offensive attacks against merchant shipping; not effective against battle fleet	1942: offensive doctrine preferred but no carriers operational	Independent air force; no separate naval air arm	n/a	U.S. and Japanese offensive doctrine; *Graf Zeppelin* design follows Britain's *Courageous*
ITALY				
Convoy protection, ASW, reconnaissance	n/a	Independent air force; no separate naval air arm	n/a	n/a

craft were inferior because they were designed to be multipurpose. The United States, though it entered the war with inadequate aircraft, eventually produced planes that were far superior to their British counterparts and to most shore-based aircraft because they were single-duty.

Table 10.2 summarizes carrier design, doctrine, and practices adopted in five navies and captures differences in the scope and pace of adoption. Three different institutional responses to naval air power are evident.[20] The Americans and Japanese adopted the offensive carrier air power paradigm. They made air power the centerpiece of their navies, transitioned to air-centered naval organizations and operations, and concentrated and operated carriers independently in carrier battle groups. The British grafted air power onto existing doctrine, keeping the carrier in a defensive role, subordinate to and part of the battle line. They used carriers to hunt down enemy raiders and supply ships, escort convoys, attack special land targets, conduct ocean sweeps and patrols, and ferry land-based aircraft to fighting zones.[21] They never developed the offensive potential of naval air power hinted at in their raid on Taranto. The Italians failed to see any role for naval air power until very late in the war, convinced that Italy was an unsinkable carrier and that land-based air could adequately support fleet operations. German views about the role for the carrier evolved from that of fleet reconnaissance, escort, and coastal patrol to a view of the carrier as an offensive strike element. But like the Italians, German realization of the offensive potential of the aircraft carrier came too late for any carriers to enter service in World War II.

Great Britain: From Leader to Follower

The British were the most advanced in carrier design and development by the end of World War I, the first to assign aircraft to ships on a regular basis, and the inventors of the aircraft carrier. But the British rapidly lost their lead in the interwar years. Late in World War II, the British Pacific Fleet adopted U.S. practices for managing carrier task forces when the two navies operated alongside one another in the Okinawa campaign of 1945.[22] Only then did the Royal Navy come to

[20]Based on strategic necessity, Italy has more in common with Great Britain than with Germany. Based on bureaucratic constraints, however, Britain has much in common with Germany and Italy, all three being characterized by strong independent air forces and no purely naval air arm (in the case of Britain for most of the interwar period). I group states based on carrier practices, or by "dependent" variable, rather than based on some particular "independent" variable such as strategic necessity or bureaucratic context. My interest is in accounting for variation in outcomes.

[21]James H. Belote and William M. Belote, *Titans of the Seas: The Development and Operations of Japanese and American Carrier Task Forces during World War II* (New York: Harper and Row, 1975), pp. 30–31.

[22]Friedman, *British Carrier Aviation*, 10; Marc Milner, "Anglo-American Naval Cooperation in the Second World War, 1939–45," in John B. Hattendorf and Robert S. Jordan, eds., *Maritime Strategy and*

fully appreciate the offensive potential of carrier air power. Why did these pioneers in carrier development and aviation fail to develop the potential of this new weapons system?

Any explanation of the British approach to integrating air power into their naval operations must start from the fact that Britain was a maritime empire. British power depended on the defense of trade and communications, and the primary naval mission through mid-1944 was keeping the sea-lanes open. To defend the empire's sea lines of communication, Royal Navy carriers played a defensive role in trade protection. Escort carriers provided convoy protection, trade-protection carriers searched and made limited strikes against surface raiders, and trade-protection fighter carriers protected shipping from air attack outside the range of British land-based fighters. Main fleet carriers operated singly with escorts to keep the sea-lanes open, spotted and scouted for the battle fleet, and controlled the air over a battle area so that the battleships could achieve maximum effectiveness.[23]

The British confronted three enemies with different implications for mission requirements and force structure. In their primary theater of operations—Europe—the British faced the Germans and Italians, both lacking aircraft carriers. The more immediate threat was Germany, so the British adopted a "Germany-first" strategy and made every effort to avoid war with Japan. Greater resources were allocated to the Royal Air Force (RAF) and its land-based strategic bomber force to meet the more proximate threat to the homeland of enemy land-based air attacks by the Luftwaffe. Offensive carrier operations would not protect London from bombing or protect shipping, except to the extent that carriers enhanced the battle fleet's ability to secure sea control.[24] Naval air conditions in the European theater, specifically the proximity of enemy land-based air, compelled the British to armor their carriers.[25] Given treaty limits on displacement, the effect of armor was to shrink the overall size of the ship and limit the volume of the hangar, thereby limiting the carrier's airplane capacity, since it was British practice to store aircraft in the hangar. There was a direct trade-off between protec-

---

the Balance of Power (New York: St. Martin's, 1989), p. 262. By contrast, the United States and Britain jointly developed escort carriers.

[23]Friedman, *British Carrier Aviation*, p. 9.

[24]Offensive carrier operations had greater applicability to the Mediterranean than the British imagined, as Taranto foreshadowed. Till points out that carrier and land-based maritime aircraft "offered the most effective means of protecting shipping—merchant as well as military—against air and submarine attack, and proved to be a potent means of attack on enemy warships and maritime commerce." Geoffrey Till, "Adopting the Aircraft Carrier: The British, American, and Japanese Case Studies," in Murray and Millett, *Military Innovation in the Interwar Period*, p. 191.

[25]The British "armored flight deck" carriers had armored hangars, which meant armoring part of the flight deck above the hangar and the sides of the hangar. There was also a belt for side protection to resist cruiser shellfire.

tion and offensive carrier capacity. Had the British not faced a Continental threat, they might have made greater strides in offensive carrier warfare.

Still, the British had strategic interests where offensive carrier air power could have been used effectively. British strategy depended on access to Middle East oil. Keeping open the sea lines to Suez required neutralizing the Italian fleet at a time when land-based air support was not an option because of German operations on the Continent and Italy's entry into the war. The British were the only navy with firsthand experience with offensive naval air power when their small carrier force reversed the local balance of sea power at Taranto in 1940. The effects of Taranto may have been short-lived, for by mid-1941 Italy's Regia Marina was fighting hard to regain control of the Central Mediterranean. Nonetheless, naval forces, it was demonstrated, could exert disproportionate influence because their mobility permitted greater concentration than land-based aircraft, which were tied to fixed bases and more difficult to concentrate.[26] Yet paradoxically, "the British remained more relaxed than the American and Japanese despite the threatening situation."[27]

Strategic necessity in itself was inadequate to bring about change in British operations. Resource shortages caused by severe financial difficulties contributed to deficiencies in the carrier program, but the U.S. and Japanese navies also faced reduced resources, competing demands, and uncertainty over which naval aviation option to choose. As Hone and Mandeles point out, World War I experience as seen or observed by all naval officers "had shown the value of airships and blimps, seaplanes, land-based multiengined bombers and scouts, and aircraft carriers."[28] Why did each navy decide to organize naval aviation as it did, and what factors shaped the Royal Navy's choices? Three factors explain British choices: the status of supporting institutions and existing force structure, compatibility, and administrative authority.

The inferior quality of British aircraft can be traced to severe structural problems in the aircraft industry. The British economy was shipping-based, so shipbuilding remained strong. While air communication played a vital role in economic development of the United States, Russia, and Germany, the British had limited need for civilian aircraft orders. Thus the aircraft industry relied on the military. During the interwar period, the British reduced military aircraft orders because of budget restrictions and disarmament pressures. By the time rearmament became necessary, the aircraft industry was crippled. It had little experience

[26]Friedman, *British Carrier Aviation*, p. 10.
[27]Till, "Adopting the Aircraft Carrier," p. 203.
[28]Thomas C. Hone and Mark D. Mandeles, "Interwar Innovation in Three Navies: U.S. Navy, Royal Navy, Imperial Japanese Navy," *Naval War College Review* 40, no. 2 (spring 1987): 64.

in mass production.[29] It was incapable of meeting all orders once war began, and first priority went to the RAF. Entering World War II, the United States and Japan had more than 600 front-line aircraft at sea. The British had 230.[30] Inability to properly manufacture aircraft, coupled with the RAF's unwillingness to design and fund adequate naval aircraft, dramatically hampered the capacity of naval aviation to perform existing missions, let alone to make the leap from ancillary weapon system to fleet backbone.

Existing force structure also hampered subsequent innovation. The British were the only power to leave World War I with carriers in their possession. Those early carriers were designed for smaller, lighter airplanes. With limited deck and hangar capacity, they quickly became obsolete as aircraft design improved. Since the British possessed a larger number of carriers than either the U.S. or the Japanese, it was difficult for the Admiralty to make its case to the Treasury for new carrier construction.[31] Force structure decisions made early on, based on available technology, hampered subsequent adaptation to the rapidly changing capabilities of aircraft between the wars.

The inferior quality and low numbers of carrier aircraft clearly hampered carrier potential. Some have attributed this directly to the loss of navy control over the Fleet Air Arm (FAA) between 1918 and 1939. Till blames dual control for Admiralty decisions to build multipurpose aircraft, which performed neither the fighter nor bomber role well and which made it very difficult for naval aviation to make the "quantum jump from being a supportive, ancillary weapon of the Fleet to being a dominant and even decisive one."[32] Dual control also is blamed for the low number of aircraft, which made it impossible to launch massed air attacks.[33]

Others, like Hezlet, argue that even once the Admiralty regained control of the naval air arm, they did not demand any change in the method of supplying aircraft for the Fleet Air Arm, which remained under Air Ministry control. He notes that the Fleet Air Arm fell behind the U.S. Navy in numbers and the performance of aircraft, but disagrees with the view that poor performance of British carrier aircraft was the legacy of divided control. "The Admiralty had laid down the requirements for Fleet Air Arm aircraft for the whole period between the wars, and

[29]Malcolm Smith, *British Air Strategy between the Wars* (Oxford: Clarendon, 1984), pp. 229, 312.

[30]Geoffrey Till, *Airpower and the Royal Navy 1914–1945* (London: Jane's, 1979), pp. 95, 187.

[31]Till, "Adopting the Aircraft Carrier," p. 199.

[32]Till, *Airpower and the Royal Navy*, pp. 86–87, 187.

[33]Hone and Mandeles, "Interwar Innovation in Three Navies," p. 66, also point out that the FAA's failure to obtain high-performance aircraft led the Royal Navy to settle for ineffective arresting gear on their carriers. "The absence of an arresting gear system meant that British carriers could not demonstrate their true military potential as platforms for mounting larger strike missions with high performance aircraft."

in their case for control of the Fleet Air Arm made before 1937 did not ask for any change in the system and also appeared satisfied with the results." He concludes: "There is little doubt that the British inferiority was more due to the Admiralty air policy than the period of dual control between the wars."[34]

Admiralty air policy toward the adoption of innovative carrier practices was also hampered by compatibility problems. Although all navies of the period were imbued with the Mahanian big ship–big gun naval orthodoxy, the Royal Navy had the strongest "gunnery" bias.[35] The British realized the danger of air attacks on battleships but believed that with better protection, better antiaircraft gunnery, and fleet fighter aircraft, the battleship could compete. The British opted to rely nearly exclusively on the gunnery solution. Offensive carrier practices did not mesh easily with deeply held attitudes in the Royal Navy.

Compounding this bias was the fact that, Taranto aside, offensive carrier practices clashed with the preponderance of Britain's wartime experience. In World War I, the Royal Naval Air Service had little success bombing German U-boat bases in Belgium. Nor had aircraft destroyed any ship of importance during the war.[36] Jutland reinforced belief in the value of the main fleet. Reconnaissance planes would locate the enemy fleet, torpedo bombers would slow the enemy, and the battle line would close for the decisive engagement.[37] The value of naval air power lay in forcing a foe out of the security of its ports and defeating an enemy in harbor.[38]

Finally, aviation reformers in Britain never attained the administrative authority to establish the requisite legitimacy of new models in the face of the low compatibility between new practices, and existing traditions and past experiences. All navies of the period had pockets of reformers who foresaw the potential of air power at sea. Yet in Britain, the big gun–big ship orthodoxy was overwhelmingly dominant. "Royal Navy officers overestimated the power of defenses against carrier aircraft and underestimated the strike value of their own planes."[39] The consensus in Britain was that the relatively small numbers of aircraft the Royal Navy could launch at any one time was the best that could be achieved. The British continued to doubt American claims about the number of aircraft that could operate off U.S. carriers.[40] Several factors converged to undermine the administrative authority of an internal lobby that could argue the aviators' case.[41]

---

[34]Sir Arthur Hezlet, *Aircraft and Sea Power* (New York: Stein and Day, 1970), pp. 127–28, 135.

[35]Ibid., p. 111.

[36]Ibid., p. 112.

[37]Friedman, *British Carrier Aviation*, p. 13.

[38]See Till, "Adopting the Aircraft Carrier," p. 215.

[39]Hone and Mandeles, "Interwar Innovation in Three Navies," p. 66.

[40]Watts and Murray, "Military Innovation in Peacetime," p. 403.

[41]See Till, "Adopting the Aircraft Carrier," pp. 205–9.

The long period of RAF control over the Naval Air Arm depleted naval aviation of its top leaders and trained personnel, and prevented the establishment of a solid career path for new personnel.[42] General service officers, not aviators, commanded British carriers. The RAF retained control over research, experimentation, development, and supply of aircraft, undermining the chances for carrier proponents to prove their case.

The British entered World War II fielding obsolete carriers and an inadequate number of nearly obsolete aircraft. The lull in British innovation and investment during the interwar years allowed the United States and Japan to close the carrier gap. The British were instrumental in the early diffusion of carrier technology and practices. They helped the Japanese and Americans in carrier design and played a significant role in the design of the naval air wing of the Imperial Japanese Navy. The Japanese sent technical missions to Britain to observe carrier construction and operations in 1920. The British sent a technical mission to Japan to train flight cadets, design carrier aircraft, and help design the first carrier, the *Hosho*.[43] The United States was also a beneficiary. British ideas heavily influenced development of the first true U.S. carriers, the *Lexingtons*.[44] By the end of World War II, the British had ceded their lead, were copying U.S. carrier designs, and were relying on the United States for escort carriers. In 1945 the United States was assisting the British in the tactics and doctrine of fast carrier operations as the two navies sought to achieve interoperability in the Pacific.[45]

The United States and Japan: Competitive Emulators

Carrier practices developed in the Japanese and U.S. navies in parallel. Eventually, U.S. practices diffused back to Great Britain. The Japanese preceded the United States in implementing the multicarrier task force concept, but the United States honed the practice, in large part because by late 1942, the United States could produce more carriers and planes than Japan.

It has been widely documented how U.S. fleet exercises and war-gaming helped the U.S. Navy put together the pieces of the carrier air power paradigm. Fleet Problem VII of 1927 demonstrated the value of carrier mobility. Planes from

[42]In 1918 a unified Royal Air Force secured control over all air assets. The Fleet Air Arm was transferred back to the Navy in 1937, but it took two years for the transfer to be completed; centralization of aircraft procurement persisted. By 1938 the RAF was receiving the lion's share of service allocations, and it dramatically reduced funding for naval aircraft, championing seaplanes and long-range land-based aircraft as cheaper alternatives to sea-based aircraft. Friedman, *British Carrier Aviation*, p. 17.

[43]Till, *Airpower and the Royal Navy*, p. 64; see also John Ferris, "A British 'Unofficial' Aviation Mission and Japanese Naval Developments, 1919–1929," *Journal of Strategic Studies* 5 (Sept. 1982): 416–39.

[44]Gardiner, *Eclipse of the Big Gun*, p. 39.

[45]Milner, "Anglo-American Naval Cooperation," p. 262; Reynolds, *Fast Carriers*, p. 305.

the *Langley,* a converted collier-made carrier that was assigned to protect the battle line, were of no help during a surprise attack by twenty-five land-based aircraft. The results led to the recommendation that carrier admirals be given complete freedom of action in employing carrier aircraft. Even before the United States employed fast carriers, they came to appreciate the value of freedom of movement from the battle line.[46]

Fleet Problem IX of 1929 showed the value of the large, fast fleet carrier as a separate offensive striking force. Admiral Pratt's carrier task force achieved a devastating surprise attack on the Panama Canal with the *Lexington* and *Saratoga.* But it was still not clear how to make the most effective use of the carrier, or how to achieve an optimal number within the limits imposed by international treaties. Nor did experience point in one direction. Fleet exercises between 1929 and 1933 demonstrated that carriers could be successful at launching devastating air strikes but were vulnerable when operating independently of the battle line. These exercises cautioned against developing carrier task forces that operated independently of the battle line in the 1930s, even though Fleet Problem XVIII of 1937 revealed how restricting carrier mobility also led to vulnerabilities and large carrier losses.[47] The United States entered World War II with the carrier still subordinate to the battle line.

In Japan, carrier doctrine developed incrementally, based on strategic requirements, experimentation, and the capabilities of aircraft. In the early 1930s, the Japanese still had no clear sense of how aircraft would be used in fleet action. Until 1937 most naval officers envisioned the carrier in supporting roles of reconnaissance, gunnery observation, aerial attack, and cover.[48] By the mid-1930s, as the range and power of aircraft increased, the concept of preemptive carrier strikes against the enemy's carriers through mass aerial attack was accepted by naval aviators. The question became whether carriers should be dispersed or concentrated.

Japanese naval tradition called for ships to disperse to trap the enemy, and for planes to converge over the target.[49] War games favored dispersal. In 1937 carrier-based aircraft secured local air superiority and bombed Shanghai, revealing the importance of carriers for projecting power over great distances and the value of massing attack aircraft for bombing and defense against fighters.[50] Based on this combat experience, the Japanese concluded that carrier forces should be con-

[46]Reynolds, *Fast Carriers,* p. 17.
[47]Till, "Adopting the Aircraft Carrier," pp. 197, 221.
[48]Toshiyuki Yokoi, "Thoughts on Japan's Defeat," *U.S. Naval Institute Proceedings* 86, no. 10 (Oct. 1960): 72.
[49]Reynolds, *Fast Carriers,* p. 7.
[50]Till, "Adopting the Aircraft Carrier," p. 221.

centrated.[51] They experimented with concentration of carriers in a "box" formation to enhance tactical effectiveness and minimize strategic risk. Trials led to unification of all carrier and land-based air units under a single command, the Eleventh Air Fleet, in January 1941. Three months later, carriers were massed in the First Air Fleet. Fleet maneuvers in spring 1940 had already shown that torpedo planes could successfully attack surface ships. With the organizational apparatus now in place, the feasibility of a Pearl Harbor was no longer in question.[52]

Immediately after the Pearl Harbor attack, the fast carrier task force of the First Air Fleet disbanded, and carriers resumed their traditional role of supporting army operations, in line with Japan's continental strategy. Between December 1941 and March 1942, the First Air Fleet did not deploy for fleet action.[53] When Japan's fast carriers reassembled at Midway, some carriers were enlisted to provide air cover for landing operations, which violated the tactical requirement of mobility and deprived the carrier task force of defensive cover through concentration. After losing four carriers at Midway, the Japanese made the conceptual leap to seeing the carrier as the backbone of the fleet, with all available battleships, cruisers, and destroyers used to protect the carriers. In July 1942 the navy reorganized the Combined Fleet around its carriers, forming multicarrier task groups, each consisting of one light carrier to provide fighter protection for two big carriers that would launch attack missions.[54] The task groups were widely separated from each other to achieve a better balance between concentration and dispersal.[55] Despite severe carrier losses, the Japanese were committed to a fleet with the carrier as its nucleus.

The U.S. Navy reacted to the loss of its carrier *Yorktown* at Midway quite differently. CNO King determined that the tactic of concentration was unsafe and forbade any two carriers to operate together in the same screen in tactical concentration. Each carrier would have its own escorting screen of cruisers and destroyers.[56] King's decision to operate carriers singly after Midway made sense so long as the Japanese retained air superiority. After Midway, the Japanese had superior aircraft and pilots while the Americans had a paucity of carriers and inferior aircraft. Reynolds views the single task formation as a measure of expediency

[51]David C. Evans and Mark R. Peattie, *Kaigun: Strategy, Tactics, and Technology in the Imperial Japanese Navy, 1887–1941* (Annapolis: Naval Institute Press, 1997), p. 347; Reynolds, *Fast Carriers*, p. 6, claims that Japan's leading naval tactician, Commander Minoru Genda, saw a newsreel in late 1940 of four U.S. carriers operating together and realized the superiority of this tactical arrangement.

[52]Clark G. Reynolds, "The Continental State upon the Sea: Imperial Japan," in Clark G. Reynolds, ed., *History and the Sea: Essays on Maritime Strategies* (Columbia: University of South Carolina Press, 1989), p. 148.

[53]Reynolds, *Fast Carriers*, pp. 9–10.

[54]Richard B. Frank, *Guadalcanal* (New York: Random House, 1990), p. 160.

[55]Evans and Peattie, *Kaigun*, p. 349; Reynolds, *Fast Carriers*, pp. 11–13, 28–30.

[56]Reynolds, *Fast Carriers*, pp. 29–30.

given the navy's many defensive missions (defending Allied bases and lines of communication; convoy escort) relative to the number of available carriers.

Yet strategic necessity and available resources do not fully explain the pace at which the U.S. and Japanese naval establishments came to accept the transformation in naval warfare represented by the aircraft carrier. The Japanese made the conceptual leap in July 1942 and formed multicarrier task groups, despite a shortage of carriers. U.S. carriers, on the other hand, were denied mobility and forced to operate singly and in confined waters during the Guadalcanal campaign of August–November 1942. Navy leadership continued to endorse the single carrier task force concept, and as late as May 1943 the United States had no permanent multicarrier task force trained to operate together.[57]

As early as the mid-1920s, aviation advocates in the U.S. Navy had lobbied for a doctrine of mobility to realize the full offensive potential of the aircraft carrier, but inferior aircraft and limited numbers of carriers weakened their cause. The arrival of new carriers in the summer of 1943, the introduction of advanced communications equipment (radar and radio) to coordinate antiaircraft fire, improved antiaircraft guns with proximity-fuzed shells, and the Hellcat fighter that was superior to the Japanese Zero—all made the acceptance of the multicarrier task forces more feasible.[58] The Americans did adopt a fleet doctrine emphasizing mobility of carriers in June 1943.[59] But the first Fast Carrier Task Force was not formed until after the Gilberts campaign in December 1943. During that campaign, carriers were often tied to the beaches as shore artillery, assigned to defensive sectors denying them the necessary defensive mobility when within range of land-based air.

How can we account for the different ways that the Japanese and U.S. navies responded to the lessons of Midway, and the different rates at which they adopted the carrier as the backbone of the fleet? Each faced a compelling strategic necessity for which the carrier offered a solution. The Imperial Navy planned to punch out a defensive ring in the Pacific before returning to its secondary roles of troop and supply transport for army operations. Once the Japanese came to view the United States as the chief threat to their expansionist aims in the Far East, they adopted a preemptive strategy to eliminate the U.S. carrier force. Offensive carrier air power had a clear role to play, and by 1932 the concept of mass aerial attack began to shift carrier air away from defense of the main battle force to attacks over the horizon. In the United States, the whole basis of peacetime planning rested on the assumption that the navy must take the offensive into the Pa-

---

[57]Ibid., p. 36.

[58]See ibid., pp. 53–59, for a discussion of these new weapons.

[59]The Americans invoked the British success at Taranto. Friedman, *British Carrier Aviation*, p. 10; Till, "Adopting the Aircraft Carrier," p. 221.

cific. By the mid-1930s, the navy adopted the strategy of "progressive advance" across the Pacific through the mandates.[60] Implicit in the concept of progressive advance was the notion of offensive war, and in the course of solving the problems of transoceanic passage, the U.S. Navy focused on offensive technologies and doctrines, including a new offensive role for the carrier.

Two factors help to explain variation in rates of adoption of offensive carrier doctrine and practices between Japan and the United States: administrative authority and compatibility. Both the Japanese and U.S. navies retained control over their air assets. Both made administrative changes early on that paved the way for an air-centered vision of naval operations. Yet organizationally and administratively, the Japanese possessed a more receptive organizational environment for developing offensive carrier doctrine. The army had failed to assert bureaucratic control over navy air assets,[61] and in 1927 an independent naval air headquarters was created that assumed control over training; airframe, engine, and ordnance development; and technical research. Under this centralized direction, the Japanese Naval Air Service matured rapidly. All junior officers were required to take a short course in aviation instruction, and officers had to qualify as aviators to achieve flag rank. Crucially, by 1936, naval aviation achieved representation at the highest operational and political echelons. Maritime and shore-based air power were concentrated in a combined naval air wing in 1938.[62] Three years later, the Imperial Japanese Navy created the First Air Fleet, the first permanent carrier force whose pilots trained together, the most powerful concentration of naval air power in the world.

It would be inaccurate to say that the Imperial Navy suffered no administrative constraints. The army dominated policy-making and the navy often suffered in budget battles because of its inferior status. When army deficit spending increased fivefold after 1932, the navy was forced to choose between a fighting fleet and an escort fleet.[63] Interservice rivalry reached such heights that the army actually stockpiled large quantities of materiel that the navy needed throughout the war.[64] The army also was given higher priority for industrial workers, so the navy suffered severe manpower shortages, especially in the fields of shipbuilding and aircraft production, which affected the aircraft industry and the training of new pilots. Yet even though the Imperial Navy lacked the resources to sustain its carrier capacity, operationally it had accepted the transformation in war and contin-

[60]Maurice Matloff, "The American Approach to War," in Michael Howard, ed., *The Theory and Practice of War* (London: Cassell, 1965), p. 221.

[61]Richard J. Samuels, *Rich Nation, Strong Army* (Ithaca: Cornell University Press, 1994), p. 114.

[62]Till, "Adopting the Aircraft Carrier," pp. 211–12.

[63]Francis W. Hirst, *Armaments: The Race and the Crisis* (London: Cobden-Sanderson, 1937), p. 148; Andrieu D'Albas, *Death of a Navy* (New York: Devin-Adair, 1957), p. 89.

[64]D'Albas, *Death of a Navy*, p. 270.

ued to operate its forces accordingly. Resources affected implementation but not ideational adoption. In the United States, on the other hand, even with more resources, adoption was constrained because reformers lacked operational and policy control.

Like the Imperial Navy, the U.S. Navy retained control over its flyers, but it lagged behind the Japanese in making the organizational and administrative shift manifested in the First Air Fleet. The U.S. Navy possessed a very effective air lobby that had paved the way for expansion of naval aviation in the 1920s. This lobby was instrumental in creating a strong bureaucratic machine to promote the interests of aviators.[65] The Bureau of Aeronautics was established in 1921, and, in 1926, Morrow Board legislation created the position of assistant secretary of the navy aeronautics, giving navy aviation top representation in civilian echelons. The board required all commanding officers of carriers, seaplane tenders, and naval air stations to be aviators. Thus aviators remained in the navy and became eligible for promotion to senior ranks, increasing the likelihood that their ideas would eventually triumph.

Yet U.S. carrier forces were organized in an ad hoc manner for exercises until well into the Pacific war, in large part because of the absence of reformers in top-level administrative, policy-making, and command positions.[66] Aviation advocates enjoyed civilian support at the very highest levels, from President Franklin Roosevelt and Under Secretary of the Navy James Forrestal. But the airmen did not make their first inroads into the top command until March 1943, when two key staff positions crucial for prosecuting the offensive in the Pacific—the posts of operations officer and air officer to Cominch (the office that controlled fleet command, administration, policy, and strategy)—were filled by airmen.[67] In August 1943, the post of deputy CNO (air) was created, a positive step in principle but a position that was confined to issues of readiness and logistics, not policy and operations. Reformers continued to press their case throughout 1943 with limited success.[68] The new fleet doctrine PAC 10 was adopted in June 1943, but applying it in battle was a problem since the central Pacific commander was not an aviator and had no aviation advisors. At the highest levels of military administration, Admiral Nimitz disputed the aviators' argument that commanding officers of large naval forces should either be aviators or have aviators as senior staff officers. Despite the success of fast carrier doctrine in the Marcus and Wake Island strikes, Nimitz and his staff, all nonairmen, wanted carriers tied to the beaches as offshore artillery in the Gilberts campaign, vulnerable to land-based

[65]Till, "Adopting the Aircraft Carrier," pp. 210–11.
[66]Reynolds, Fast Carriers, p. 110.
[67]Ibid., pp. 40–41.
[68]See ibid., pp. 42–77, for a discussion of the debates in Washington and the Pacific.

air attack. Eventually, successes in November 1943 in the Gilberts campaign demonstrated the importance of operational flexibility and convinced even the most conservative naval officers that aviation and the carrier had become the dominant elements in the fleet. The Fast Carrier Task Force finally emerged in December 1943, two and a half years after the Japanese had formed the First Air Fleet and one and a half years after they had reorganized their Combined Fleet around carriers.

Compatibility considerations also affected rates of adoption. Scholars have identified an offensive bias in the Imperial Navy that enhanced the compatibility between offensive carrier practices and the organization's tradition. The offensive bias manifested itself in several ways:[69] composition of air groups weighed heavily in favor of attack aircraft—dive-bombers and torpedo planes—over fighters; little provision for carrier scouting; sacrifice of armor protection to maximize carrier offensive potential; neglect of fleet air defense. The navy's fixation on offensive operations seems to have led the Japanese to press the envelope of the carrier's offensive potential, even to their detriment at Midway, where the concentration of carriers proved disastrous. Nonetheless, this may account for why the Japanese first introduced the revolutionary operational concept of massing carriers for attack and increased the administrative authority of aviation reformers.

More generally, the aircraft carrier challenged traditional naval doctrine. Despite the insights gained from war games and fleet exercises in peacetime and from clashes in wartime, every navy believed, and with good reason, that only battleship fire could sink a modern battleship. It was by no means clear that carrier aircraft could perform this vital task. All the maritime powers, the nations one would imagine most open to offensive carrier operations, remained wedded to the big ship–big gun naval orthodoxy well into World War II. But the Japanese, pioneers in use of the multicarrier task force, began to make the conceptual shift as early as 1941 and definitely by mid-1942, while it took the Americans until the end of 1943 and the British until 1945. It has even been argued that the development of fast carrier task forces in the U.S. Navy would have been slower because of the battleship orthodoxy and dominant positions of battleship admirals had not Pearl Harbor so devastated the U.S. battleship force.[70]

By the end of 1943, largely because of greater industrial prowess, the United States had displaced Britain as the carrier power to be emulated. The United States began to supply the British with aircraft and carriers. U.S. Lend-Lease aircraft became such an integral part of British carrier operations that after 1942, the British incorporated the U.S. hangar deck requirements into planned carriers in

[69]See Evans and Peattie, *Kaigun*, ch. 9.
[70]Thomas C. Hone, "Battleships vs Aircraft Carriers: The Pattern of U.S. Navy Operating Expenditures, 1932–1941," *Military Affairs* 41, no. 3 (Oct. 1977): 133–41.

order to properly accommodate U.S. aircraft.[71] The British were also copying U.S. carrier designs by the end of the war. The unfinished Malta-class fleet carriers were designed with U.S.-style open hangars on the deck because of the advantages of increased launching intervals.[72] The United States, however, also copied British designs. The Midway-class fleet carriers, introduced at the end of the war, were designed with armored flight decks in recognition of the value of added protection against Japanese kamikaze pilots.[73]

The British Pacific Fleet also learned by operating with the United States after 1945. The British emulated America's mix of aircraft type, weapon loads, and procedures for coordinating between decks during an air strike.[74] British carrier squadrons used U.S. planes and trained at the Quonset Point Naval Air Station, Rhode Island; British observers were placed with key U.S. staffs in the Pacific; and U.S. liaison, communications, and air combat intelligence officers operated aboard British ships in the Pacific to standardize tactical procedures.[75] Although the British *Argus* had been the model for all future carriers in 1918,[76] by 1943 the United States was setting the standard.

Late Learners: Italy and Germany

Lack of adequate air cover was one of the major failures of Italian naval strategy. Greene and Massignani conclude that "Italy's failure to adopt the aircraft carrier before the war, to rapidly pursue the conversion of merchant ships once the war had started and to develop a powerful naval air arm were important contributing factors to her defeat in the Mediterranean naval war."[77] The Italian experience with aircraft carriers reflected an overly ambitious strategy that the country could not support financially or materially, an intellectual conservatism that paralyzed innovation, lack of a supportive leadership, and rampant bureaucratic disputes among the services. Mussolini's goal was to dominate the Mediterranean and Red Seas, and establish Italy as a world power. These ambitious goals "required an offensive aeronaval strategy and the ability to project and maintain the army's forces overseas."[78] Mussolini may have conceived of the navy's func-

[71]Gardiner, *The Eclipse of the Big Gun*, p. 53.
[72]Ibid., p. 40.
[73]Ibid., p. 46.
[74]Milner, "Anglo-American Naval Cooperation," p. 262.
[75]Reynolds, *Fast Carriers*, pp. 305–9.
[76]Gardiner, *Eclipse of the Big Gun*, p. 38.
[77]Greene and Massignani, *The Naval War in the Mediterranean*, p. 109.
[78]Brian R. Sullivan, "The Italian Armed Forces, 1918–40," in Allan R. Millett and Williamson Murray, *Military Effectiveness*, vol. II: *The Interwar Period* (Boston: Allen and Unwin, 1988), p. 169. I am deeply indebted to Brian Sullivan for his guidance on the Italian case.

tion as offensive, performing the role of a "fleet-in-being" against Great Britain,[79] but he never equipped the navy to perform that role in the age of Midway.

Prior to 1935, the Regia Marina planned for war against the French fleet in alliance with the small Yugoslav navy. Carriers seemed unnecessary because France's one carrier was judged to be inferior to Italian land-based aircraft, and any war would be fought in the Western Mediterranean where Italy possessed many islands with airfields from which the Regia Aeronautica could operate. When the British Home Fleet redeployed to the Mediterranean in mid-September 1935, the Italians were forced to confront their naval deficiencies, not only for dominating the Mediterranean but also for supporting the army's land offensive in North Africa by controlling the Central Mediterranean and keeping supply lines open to Libya.

In 1935 the Navy Ministry's proposed plan to augment Italy's naval capability by 300,000 tons included three aircraft carriers.[80] In January 1936 the war plans office concluded that the Italians could expel the British from the Mediterranean only by launching a vast building program of ten capital ships, four aircraft carriers, thirty-six cruisers, and forty-six to seventy-five submarines. With Italy's limited industrial base, the "escape fleet" would take twenty-four years to complete. The alternative *programma minimo* would continue the existing policy of replacing old vessels with new ones.[81] Although unable to control the exits of the Mediterranean, an alliance with Germany would allow Italy to focus on the central regions of the sea while the Germans engaged the British in the Atlantic. With the British forced to disperse their forces widely, hegemony of the Mediterranean would lie within Italy's reach. The *programma minimo* proposed no carriers, on the assumption that land-based air units would support fleet operations within the confines of the Mediterranean.

With the de-escalation of tensions with Britain in August 1936 and Mussolini's rejection of the excessively expensive escape fleet, the Navy Ministry abandoned the aircraft carrier, even though navy planners remained convinced of the utility of carriers for operations within the Mediterranean.[82] Navy planners urged Mussolini to build at least one 15,000-ton carrier in 1938–40, and a second carrier in 1941–43.

As events would show, assumptions about the utility of aircraft carriers held by those who shaped Italian naval policy were seriously flawed. Without adequate

[79]MacGregor Knox, *Mussolini Unleashed 1939–1941: Politics and Strategy in Fascist Italy's Last War* (Cambridge: Cambridge University Press, 1982), p. 19.

[80]Ibid., pp. 19–20.

[81]Robert Mallett, *The Italian Navy and Fascist Expansionism 1935–1940* (London: Frank Cass, 1998), pp. 54–55.

[82]Ibid., pp. 102–3.

air cover, six of Italy's seven battleships were damaged or destroyed, many by air attacks in 1940. The Royal Navy's ability to shoot down enemy reconnaissance aircraft and launch air strikes was vital to its success at Matapan. "The high speeds of Italian warships offered little protection against strikes by British aircraft. . . . Italy was to find itself blockaded within the Mediterranean Sea, and at the mercy of enemy fleets, which could attack decisively all lines of communication between the mainland and colonies."[83] The range of shore-based planes proved insufficient for reconnaissance and fleet operations beyond the central Mediterranean, so air reconnaissance of British bases was entirely inadequate.[84]

Throughout the rearmament period, the naval plans office continued to press for the construction of aircraft carriers. There were Italian admirals and lower-ranking naval officers who believed that the navy needed carriers. Mussolini even thought that carriers might be a good idea and so did, very briefly, the very conservative navy chief of staff from 1933 to 1940, Admiral Domenico Cavagnari. He proposed one in 1935 at the height of Italo-British tension, probably to expand his service as much as anything else.[85] But deep skepticism about aircraft carriers pervaded Italian naval thinking, just as it did in all interwar navies. In 1925 the navy's admirals' committee rejected building any carriers allotted to Italy under the Washington Naval Treaty limits on the grounds that the main theaters for fleet operations could be covered by land-based air.[86] Cavagnari emphatically opposed carriers from 1936 onward. Mussolini dismissed Cavagnari after Taranto and ordered conversion of the transatlantic liner *Roma,* but the navy staff resisted, citing cost and technical problems, a lack of planes with folding wings, and German problems with converting the *Graf Zeppelin.*[87] Only in July 1941, four months after Italy's defeat at Matapan, did the Navy Ministry undertake conversions of the *Roma* and *Augustus*. Neither was completed before the Italian Armistice on 8 September 1943.[88] Even if completed, the carriers could not recover fighter aircraft. Nor did Italy possess torpedo or bomber aircraft to operate off its carriers.[89]

The British and Italian cases share many similarities. In each a compelling case for carriers can be made based on strategic necessity, yet a confluence of factors—resource constraints, weak administrative authority, low compatibility, and bureaucratic rivalry—combined to ensure that carriers did not take on the

[83]Ibid., p. 78.

[84]Ibid., p. 62.

[85]Ibid., p. 54.

[86]Ibid., pp. 60–61.

[87]Greene and Massignani, *The Naval War in the Mediterranean*, p. 113; Clark G. Reynolds, "Hitler's Flattop—The End of the Beginning," *U.S. Naval Institute Proceedings* 93, no. 1 (1967): 48.

[88]MacGregor Knox, *Hitler's Italian Allies: Royal Armed Forces, Fascist Regime, and the War of 1940–1943* (Cambridge: Cambridge University Press, 2000), p. 59.

[89]Greene and Massignani, *The Naval War in the Mediterranean*, pp. 113–14.

prominent role that strategic necessity would dictate. This is particularly notable for Italy given two facts: the Italians were the only naval power to have neither a carrier nor plans to build one at the start of the war; even the navy with the most extreme fixation on the decisive battle fleet engagement, Japan, with its 69,000-ton super-battleships, created the world's most effective carrier strike force.[90]

Clearly the Italians were materially and financially constrained. They had serious industrial-technological deficiencies, possessed minuscule raw materials, and imported nearly all their oil.[91] Mussolini's rejection of the "escape fleet" in 1936 indicated his appreciation that it was entirely beyond Italy's material, industrial, and economic capabilities. But that does not explain why not even a single aircraft carrier was commissioned.

Mussolini is partly to blame. He was determined to subordinate all the armed services to himself, and his belief in Douhet's doctrine of strategic bombing as the weapon for poorly armed nations to oppose the greatest industrial powers was virtually law. Convinced that the air force could operate efficiently against land and sea targets with the same aircraft, weapons, and courses of action, Mussolini established a unified air force in 1923 and placed all aircraft under its chain of command.[92] He appointed himself minister of each service in 1925 with the intent of commanding the armed forces in wartime.[93] Yet Mussolini had little insight into naval warfare, while his senior military advisor, Marshal Pietro Badoglio, was a conservative army officer with no understanding of modern naval warfare or air operations.[94]

Sullivan notes that despite Mussolini's intentions, the armed services retained operational independence. So it was with the Regia Marina and Regia Aeronautica, the hostile relations between them, and their conservative leadership where the fate of carriers in the Italian navy truly lay. Navy planners were fully aware of the possibilities of an air wing for reconnaissance and appreciated the capabilities of torpedo aircraft. In 1922 the Italian Institute of Maritime Warfare conducted studies similar to those carried out at the U.S. Naval War College and concluded that carriers were a necessity.[95] But that opinion was not shared in the highest echelons of the navy. Carriers were seen as vulnerable to modern gunnery and torpedo attack and of little threat to modernized battleships, while high-speed, long-range, shore-based aircraft, it was believed, would provide more adequate air support than carrier-borne aircraft in the geographically restricted theater of the Mediterranean. Navy chief of staff Cavagnari in particular opposed construc-

[90]Ibid., p. 160.
[91]Sullivan, "The Italian Armed Forces," pp. 172–73.
[92]Admiral Romeo Bernotti, "Italian Naval Policy under Fascism," *U.S. Naval Institute Proceedings* 82, no. 7 (July 1956): 723.
[93]Sullivan, "The Italian Armed Forces," p. 178.
[94]Ibid., pp. 178, 193.
[95]Bernotti, "Italian Naval Policy," p. 724.

tion of carriers. A staunchly conservative battleship admiral, he believed that any naval war would be a Jutland-style battle fleet engagement fought within proximity of Italy's bases with adequate support from land-based air.[96]

Sullivan contends the navy actually fared the best of the services between the wars in terms of budget share, in large part because the army and air force were forced to devote substantial resources to the wars in East Africa and Spain.[97] But rivalry between the Regia Marina and Regia Aeronautica had a crippling effect on aeronaval operations and on the ability of the navy to realize its full potential. The Regia Aeronautica exercised virtual monopoly over military aircraft after 1928 and was enamored with the doctrine of strategic bombing. Air Commander Italo Balbo declared in 1929 that he would never allow the navy to build carriers.[98] Balbo made it clear that fighter and bomber units would remain under control of the air force, with air units released to the army and navy in wartime. In 1934 procedures stipulated that the naval high command would inform the air force of its proposed operations and the air general staff would decide "whether or when to intervene in naval actions on a completely independent basis."[99] The result was a navy that lacked aircraft of its own for reconnaissance, torpedo attack, air defense, or offensive strikes. The impact might not have been as devastating had not the air force refused to train directly with the fleet, or to provide its bomber forces with the specialized crew training and equipment needed for maritime reconnaissance and strike missions.[100] The Italians went to war in 1940 with no effective air cover. Greene and Massignani conclude that the failure of Italy's air force to give proper support to the navy was its main failure in World War II.[101]

Air-sea cooperation did improve after Matapan, and the Italians along with their German allies did control the Central Mediterranean from mid-1941 to late 1942. From the spring of 1941 on, the Regia Marina began to stress the use of submarines, torpedo boats, mines, and aeronaval cooperation, in partnership with the Germans, for offensive operations. The navy realized that its primary responsibility was not to go looking for a Mediterranean Jutland but to ensure the arrival of cargoes to North Africa and Albania-Greece, and to neutralize Malta. Still, absence of a joint air force–navy operations center and planning staff led to navy complaints of insufficient air cover, while a shortage of aircraft after 1941 further hindered joint action.[102]

[96]Mallett, *Italian Navy and Fascist Expansionism*, pp. 62, 78, 111–12, 147.
[97]Sullivan, "The Italian Armed Forces," pp. 171–72.
[98]Bernotti, "Italian Naval Policy," p. 728.
[99]Sullivan, "The Italian Armed Forces," p. 197.
[100]Knox, *Hitler's Italian Allies*, p. 130.
[101]Greene and Massignani, *The Naval War in the Mediterranean*, p. 112.
[102]Knox, *Hitler's Italian Allies*, pp. 133, 151.

The Italians did benefit from German technology transfers and assistance. They received Ju-87 Stuka dive bombers, Daimler Benz engines, 88 mm anti-aircraft/antitank guns, type VII U-boats, radar sets, sonar, and other technology transfers. The Germans supplied the Italians with catapults from the *Graf Zeppelin*, arrester gear, and plans for other components to assist in carrier conversions.[103] An Italian naval mission worked at the Luftwaffe's War Experience Training Center from 6 to 15 March 1942, and the Italians learned a great deal from fighting alongside the Germans in the Mediterranean, North Africa, and Russia.

Nonetheless, organizational set issues posed obstacles to the Italians in taking greater advantage of these opportunities. Italian military industrial firms successfully fought to limit the licensing of foreign designs at their expense. As early as 1936, shipbuilding firms with vested interests urged Mussolini to fund capital ship construction over carriers because the former were more profitable.[104] A limited industrial base and inefficient wartime mobilization could never have produced new designs in quantities required to fight a modern war. Nor could the Italians assimilate this new equipment and these new ideas because their industry was so backward, and the Germans supplied such a small amount of war materiel.

Knox argues that even had the Italians commissioned a carrier, "no solutions were as yet in sight for critical issues that had taken the British, US and Japanese carrier forces the best part of a decade to resolve—from specialized pilot training to the provision of launching catapults, 'navalized' fighters, and the dive-bombers and light torpedo-bombers that the aircraft industry had so far been unable to produce."[105] Crucially, the Italians never had an aircraft carrier to train with to solve these problems.

Greene and Massignani conclude that while an Italian aircraft carrier probably would not have helped escort convoys to North Africa, if a carrier had arrived by late 1941 or early 1942, it might have had a significant impact. The British had lost five carriers in 1941 and early 1942. Their only carriers available for the Mediterranean also had to serve in the Atlantic. An Italian carrier at this time would have offered immediate, direct fighter protection to the Italian fleet, driven off enemy air reconnaissance, widened Italy's area of operations, and importantly, boosted the fleet's morale.[106]

As with Italy, one of the greatest deficiencies in Germany's naval preparedness was the absence of a carrier component. Of the major navies, writes Knox, "only the Germans were as retrograde in carrier development and air cooperation" as

---

[103]I thank Gregory Wilmoth for bringing this to my attention; Greene and Massignani, *The Naval War in the Mediterranean*, p. 114.

[104]Mallett, *Italian Navy and Fascist Expansionism*, p. 61.

[105]Knox, *Hitler's Italian Allies*, pp. 59–60.

[106]Greene and Massignani, *The Naval War in the Mediterranean*, p. 114.

the Italians.[107] Germany's naval air doctrine scarcely advanced beyond the most elementary notions of seaplanes for fleet reconnaissance and coastal defense. Germany faced many of the same obstacles that Italy did in exploiting the potential of naval air power: a geographic situation that made the case for carriers seem less compelling; absence of a strong airpower constituency in the navy; strong battleship bias among naval leaders; lack of political support at the highest civilian levels; and poor cooperation between the navy and air force.

For most of the interwar years, Germany perceived its main enemies to be its Continental opponents: France, Poland, and Czechoslovakia. Germany's first naval rearmament plan in 1934 was designed primarily with France in mind. It called for 5 pocket battleships, 7 cruisers, 36 seaplanes for fleet reconnaissance, and 2 medium-size aircraft carriers of between 18,000 and 20,000 tons, carrying 50 to 60 aircraft. Admiral Raeder, commander in chief of the navy, envisioned a battleship-dominated fleet for use against Great Britain, even though it was Hitler's policy until May 1938 to reach an accommodation with Britain. Other naval leaders favored a cruiser and submarine strategy against Britain. Admiral Rolf Carls, commander-in-chief of the fleet, melded the two views together, proposing four strong naval groups operating permanently in the world's oceans, in addition to the German home fleet, in order to defeat Britain in an oceanic war and achieve world supremacy.[108] The navy's planning committee responded by developing a program of 10 battleships; 15 pocket battleships; 5 heavy, 24 light, and 36 small cruisers; 8 carriers; and 249 submarines. Realizing that this was an enterprise for the long run, Raeder chose to focus on the cruiser and submarine programs in the short run. In 1939, Hitler intervened to accelerate naval rearmament and battleship construction. The Kriegsmarine adopted the Z-Plan, which called for 25 battleships and pocket-battleships, 4 aircraft carriers, and 222 U-boats by the end of 1947. War broke out before the "fleet-in-being" could be completed or any of the carriers commissioned.[109] In any event, the Z-Plan was too ambitious for Germany's resources. Germany went to war woefully unprepared for conflict with Britain, with only 2 battleships, 3 pocket battleships, 1 heavy and 6 light cruisers, 21 destroyers, 12 torpedo boats, and 57 U-boats.[110] Raeder lamented that the only thing the fleet could do is sink honorably.[111]

[107]Knox, *Mussolini Unleashed*, p. 20.

[108]Manfred Messerschmidt, "German Military Effectiveness between 1919 and 1939," in Millett and Murray, eds., *Military Effectiveness*, vol. II, p. 234.

[109]Keith W. Brid, *German Naval History—A Guide to the Literature* (New York: Garland, 1985), pp. 529, 556.

[110]Jürgen E. Förster, "The Dynamics of *Volksgemeinschaft*: The Effectiveness of the German Military Establishment in the Second World War," in Millett and Murray, eds., *Military Effectiveness*, vol. III, p. 192.

[111]Messerschmidt, "German Military Effectiveness," p. 234.

While Italy had no plans to build an aircraft carrier between the wars, Germany had plans to build two by 1941 and two more by 1947. Funds for the *Graf Zeppelin* were authorized in 1936; the carrier was launched in December 1938, and it was 85 percent complete when Germany went to war in September 1939. Originally designed to displace 19,250 tons and carry fifty aircraft, its displacement rose to 34,000 tons and its final aircraft complement was to be twelve fighters and twenty-eight bombers, suggesting a shift in mission from reconnaissance and escort to attack.[112] The Germans, however, never developed the doctrine for using the carrier against the Royal Navy or Britain's merchant fleet. In Corum and Muller's estimation, "the lack of a naval air doctrine was the weakest aspect of German airpower thinking."[113]

This was not for lack of appreciation for naval air power developments abroad. Suchenwirth describes how as early as 1923, the Navy Command set up agencies to keep the aviation idea alive, supervise air training, and examine the tactical employment of naval aircraft. A monthly magazine, *Marine-Flotten-Rundschau* [*Navy Fleet Review*] "printed the best foreign articles on naval air forces, accompanied by very fine photographs, sketches and news articles, . . . to keep Navy personnel up to date with respect to technological advances in the field of naval aviation and warfare."[114] When designing the *Graf Zeppelin*, the Germans consulted open-source literature including Weyer's *Taschen der Kriegsflotten* and Jane's *Fighting Ships*, along with other foreign aircraft journals.[115] In 1935, as part of their exchange agreements with Japan, they were given access to the 26,900-ton carrier *Akagi* and received nearly one hundred detailed blueprints of the ship's flight deck apparatus,[116] although they adopted the British design of the 22,500-ton carrier *Courageous*. The Germans were also aware that the U.S. Navy was successfully developing dive-bombing and torpedo aircraft that could operate from carriers, and that carrier construction was underway in the United States, Britain, Japan, and France.[117] Nonetheless, naval airpower was viewed primarily as an auxiliary weapon for coastal patrol and strategic reconnaissance.

Germany's geographic location and continental orientation partly explain the defensive approach to naval air power. The rough North Sea weather cast doubt on the feasibility of flight deck operations, and the proximity to Great Britain

[112]Reynolds, "Hitler's Flattop," p. 48.

[113]James S. Corum and Richard R. Muller, *The Luftwaffe's Way of War: German Air Force Doctrine 1911–1945* (Baltimore: Nautical and Aviation Publishing Company of America, 1998), pp. 8–9.

[114]Richard Suchenwirth, *The Development of the German Air Force, 1919–1939*, USAF Historical Studies, no. 160 (Montgomery, AL: Air University, 1968), pp. 41–42.

[115]Reynolds, "Hitler's Flattop," p. 42.

[116]Ibid.

[117]Corum, *The Luftwaffe*, p. 178.

made survivability a serious problem.[118] The German Naval Staff worried about the lack of foreign bases and the difficulty of entering and exiting the North Sea passages. Collectively, these concerns contributed to the start-and-stop development of the *Graf Zeppelin*, preventing a full-scale effort to complete and integrate the carrier into German naval operations. In 1939 the German Naval Staff prioritized U-boat construction over construction of the *Graf Zeppelin*, already in an advanced stage.[119] Yet given that Naval Command had operated under the assumption that Britain should be treated as an enemy as early as 1934, Messerschmidt judges the navy's greatest operational gap to have been its carrier component, which made operations by larger ships in the Atlantic against Britain very risky.[120]

Absence of a strong airpower constituency within the Kriegsmarine, and the attendant bureaucratic rivalry between the navy and air force, had the same crippling effect on carrier development in Germany as it had in Great Britain and Italy. After the Nazis seized power, naval aviation was placed under control of the Luftwaffe. Admiral Raeder put up stiff resistance to losing control of naval aviation, but Field Marshal Hermann Göring, air minister and Hitler's close confidant, was determined that anything that flies should belong to the Luftwaffe. In 1935, Göring made one concession. He created a special branch for naval operations within the Luftwaffe and transferred naval officers there to form the Naval Aviation Branch. The navy was to have operational control of this naval aviation corps in wartime.[121] The Luftwaffe's air training commander was also to cooperate with naval commanders to develop appropriate training for naval aviators, a common doctrine, and common procedures for the naval aviation branch and the rest of the Luftwaffe.[122] But in 1937, Göring reclaimed full responsibility for assigning aircraft to the navy and control of all aviation assets in wartime. A compromise in 1939 created the post of "General of the Air Force with the Commander in Chief, Navy." A Luftwaffe general serving with the commander-in-chief of the navy was to control the staff of the naval air forces and its coastal and naval aircraft, but the general was also directly responsible to Göring.[123]

When Germany went to war, the Luftwaffe was responsible for all air power procurement, training, and equipping; the navy was responsible for coastal defense and air operations at sea; and the Luftwaffe was to provide the navy with the air assets to accomplish those tasks. The level of cooperation between the two

[118]Edward P. Von der Porten, *The German Navy in World War II* (New York: T. Y. Crowell, 1969), p. 25.
[119]Edward Barker, "German Naval Aviation," *U.S. Naval Institute Proceedings* 76, no. 2 (Feb. 1950): 733.
[120]Messerschmidt, "German Military Effectiveness," p. 243.
[121]Barker, "German Naval Aviation," p. 731.
[122]Corum, *The Luftwaffe*, p. 177.
[123]Messerschmidt, "German Military Effectiveness," p. 249.

services was always inadequate. Manuals addressing interservice cooperation were published in 1936 and 1938, too late according to Messerschmidt.[124] Naval command controlled no naval air units, and it had to rely on the Luftwaffe, which had its own priorities and too few planes to meet both its own needs and those of the navy. When war broke out and the Luftwaffe had to replace all its obsolescent aircraft, Göring cancelled all naval aircraft construction.[125] Barker documents how the navy's loss of administrative control of all Naval Air Units to the Luftwaffe before the war was followed by its loss of operational and tactical control during the war.[126] The air force eventually took control of all the navy's air reconnaissance operations, and by 1940 it was diverting Naval Air Units, desperately needed to cover the Norwegian coastline, to the "air blitz" over Britain.[127]

At the highest levels of authority, Hitler was a lukewarm advocate for the carrier. He supported its development only after seeing the successful attacks at Taranto and Pearl Harbor, British carrier attacks on the *Bismarck* in May 1941 and the *Tirpitz* in March 1942, and the successful handling of U.S. carriers at Midway in June 1942.[128] But Hitler was unrealistically optimistic about building carriers. Within the navy, Admiral Raeder was a battleship admiral who saw naval warfare as a Jutland-style affair. He came to see the value of carriers for escorting Germany's commerce raiders in 1942 and even succeeded in convincing Hitler to complete the *Graf Zeppelin* over the protestations of Göring. But with Raeder's replacement by Admiral Karl Doenitz in 1943, all navy resources were shifted to submarines.

Although German naval air doctrine was woefully underdeveloped, the Germans were constructing an aircraft carrier when war broke out and they had plans to build more. The consensus among scholars is that the *Graf Zeppelin* was not built to be "an integral and effective part of a combined fleet" but rather to be a "prestige ship and symbol of naval status."[129] If Germany were to be a world-class military power, it had to be a world-class naval power. The major naval powers of the period all had aircraft carriers, so Germany had to have carriers as well. According to Messerschmidt, the German Navy believed that "[t]he military power—and above all the naval power—of a people symbolized its will to maintain world prestige (*Weltgeltung*)."[130] Germany's motivation for building carriers, like its ambitious naval building plans more generally, was to garner world respect.

[124]Ibid.
[125]Barker, "German Naval Aviation," p. 732.
[126]Ibid., p. 733.
[127]Ibid., p. 736.
[128]Reynolds, "Hitler's Flattop," p. 47.
[129]Corum, *The Luftwaffe*, p. 177; Reynolds, "Hitler's Flattop," p. 49; Herbert Molloy Mason, Jr., *The Rise of the Luftwaffe: Forging the Secret German Air Weapon 1918–1940* (New York: Dial, 1973), p. 250.
[130]Messerschmidt, "German Military Effectiveness," p. 233.

Diffusion: Scope, Pace, and Faithfulness

Between 1918 and 1943, ideas about how to employ air power at sea circulated among a few military powers. The number that seriously considered the use of carriers was even smaller. The aircraft carrier was an expensive weapon system. It came of age during a period of global fiscal stringency. Accordingly, the scope of diffusion was narrow.

It may be more accurate to speak of the "circulation" rather than the diffusion of new ideas. The British were the initial source of information on carrier design and doctrine and actively disseminated information to the Japanese, Americans, French, and Germans. By the early 1920s, the Royal Navy had stopped exchanging technical data with the U.S. Navy, so secure were they about their naval aviation lead.[131] From the mid-1920s onward, experimentation with carriers accelerated in the United States and Japan, and the British rapidly lost their lead. It is not possible to identify an "emulatee" during this period. Japanese and American learning proceeded in tandem, and both navies struggled to draw appropriate lessons from battlefield experience.

Decisions about adopting carriers were driven primarily by strategic necessity and competition. Development of carrier operations took place in wartime, under the greatest pressures of strategic necessity, in the most competitive of environments when resources were scarce. Bureaucratic pressures operated as hindrances rather than facilitators of diffusion. Legitimacy and identity were also at work—to preserve the status quo in most navies except for Germany's, where the aircraft carrier was a symbol of world-class power.

The evolution of carrier practices was not as simple as neorealism suggests. Nor was the best or most "rational" model given mission requirements adopted, but frequently the most familiar and accessible. For example, Japan's doctrine of operating carriers in concentrated formations that were widely dispersed diverged from the British practice of operating carriers singly with escorts. But Japanese carrier designs were modeled on Britain's. Like the British, they stowed aircraft below in hangars, storing each landing plane below before landing the next.[132] This reduced the platform's aircraft-carrying capacity and speed of operations, thereby limiting the carrier's offensive strike potential. While the design and practice were consistent with British carrier doctrine and tactics, which did not contemplate mass air attacks,[133] they were not consistent with Japan's offensive

---

[131]Hone and Mandeles, "Interwar Innovation in Three Navies," p. 66.

[132]Permanent deck stowage made sense for the Royal Navy, since aircraft on deck tended to rapidly deteriorate in the extreme climates of the Arctic and tropics, though the British recognized the advantage of the deck park system and adopted it during World War II. Friedman, *British Carrier Aviation*, p. 19.

[133]Friedman, *Carrier Air Power*, p. 15. To increase operating capacity, the British adopted a second

TABLE 10.3
Factors Affecting Capacity to Adopt

Resources	Bureaucratic Rivalry	Administrative Authority	Compatibility	Organizational Set
		BRITAIN		
Washington Treaty limits; fiscal constraints	Competition with RAF; divided control of Fleet Air Arm	Reformers fail to secure high-level political or military support	Big ship–big gun orthodoxy; defensive bias; gunnery bias	Structural problems in aircraft industry
		UNITED STATES		
Washington Treaty limits; 1938 Naval Expansion Act	Navy retains control over aviation	FDR supports carrier development	Big ship–big gun orthodoxy; offensive war bias in strategy of progressive advance	Late 1930s expands pilot training and production of training aircraft
		JAPAN		
Washington Treaty limits	Separate Naval Air Service established 1912; army predominates over navy	Yamamoto achieves ability to influence naval strategy August 1939; Naval General Staff vs. Combined Fleet staff over strategy 1942	Big ship–big gun orthodoxy; Naval General Staff plan for traditional battleship engagement held sway 1942; defensive naval strategy in support of land operations; offensive war bias	No reserve of qualified aviators due to rigid screening and expulsion practices; aircraft industry falters; lacks access to strategic alloys and sufficient R&D base
		GERMANY		
Versailles Treaty limits	Luftwaffe secures control over all air assets	Hitler indifferent; Raeder replaced by Donitz	Offensive bias	
		ITALY		
Financial and resource constraints by 1941	Unification of military aviation in Regia Aeronautica, 1923	Mussolini opposes carriers, though reconsiders after losses at Battle of Cape Matapan	Promonopolistic, land-based air service built around high-altitude bombing; belief that land bases sufficient; defensive "fleet-in-being" strategy	Lack of skilled labor due to deficiencies in education system

carrier doctrine. The U.S. practice of deck parking, by contrast, maximized aircraft carrying capacity, speed of operations, and the platform's offensive power. Deck parking was much more in line with Japan's offensive carrier doctrine and tactics, but the Japanese had learned carrier design from the British: the British model was the most familiar, if not the most "rational" given mission requirements.

---

hangar under the original one, but delays in elevator operations still curtailed operations. The double hangar, first adopted in the British carriers *Courageous* and *Glorious*, was copied by the Japanese in the *Akagi* and *Kaga* and by the German's in the *Graf Zeppelin*.

Finally, rarely was a practice emulated in its entirety. Rather, technologies and practices were adapted to existing force structure and organizations, as organizational sociology posits. To the extent that diffusion was not a smooth process, a variety of factors collectively influenced the capacity to adopt, as well as the scope and pace of diffusion (See Table 10.3).

The three states with the greatest carrier air power experience all faced significant obstacles to adopting offensive carrier practices. The big gun–big fleet doctrine was firmly entrenched. It withstood the Japanese attack at Pearl Harbor. The U.S. Navy even backed away from the concept of a multicarrier task force after Midway. Yet the Japanese and Americans did succeed in overturning the prevailing orthodoxy, while the British, who had a strategic incentive to adopt offensive carrier practices, did not. Given low compatibility between the carrier paradigm and British naval traditions, values, and wartime experience, some variation in paradigm adoption and receptivity to new practices can be accounted for by the compatibility dimension. These patterns are further reinforced by differential capacity of reformers to establish the legitimacy of new practices over existing orthodoxy.

Organizational constraints also varied. In Great Britain, naval aviation fell under the purview of the RAF for a significant part of the interwar period. In Germany, all air assets came under the control of the Luftwaffe; in Italy, under the Regia Aeronautica. But the Japanese also had to struggle against the navy's inferior status vis-à-vis the army, while in the United States, operational and policy-making control over the actual operation of carriers was not achieved until the end of 1943.

Organizational set issues played a role too. Educational institutions are a vital part of the organizational set for any new military practice. Unlike their British counterparts, the Americans and Japanese developed flying schools and positions on naval staffs and in naval war colleges for their aviators. Methods for training aviators were also important. The Japanese Naval Air Service relied on a long-term selective process, which produced highly competent pilots that outflew their American counterparts in early air battles. The Americans opted for a short training period to create a pool of pilots to complement career pilots, and provided a nucleus for wartime expansion.[134] As war progressed, the Japanese were unable to replace their pilot elite.

A final factor influencing the adoption of the offensive carrier air power paradigm, emphasized by neorealists, is resources. Here the Japanese and British were overtaken by their U.S. counterparts in quantity and quality. Japan's aircraft may have been far superior to America's at the outset of the war, but by mid-1943 the Americans had seized the technological lead.

[134]Reynolds, *Fast Carriers*, p. 19.

A range of barriers can narrow the scope, slow the pace, and produce a range of adaptive responses to technological innovations. Thorough examination of barriers to diffusion is critical for policy-makers eager to understand which factors might inhibit our allies from adopting new practices that we would like to spread, as well as which factors might delay the spread of innovations that we would prefer our potential enemies not acquire. The carrier case suggests some of the ways in which barriers might be surmounted. Operating closely together in the Pacific allowed for the rapid dissemination of U.S. practices to the British, and for the British to overcome deep biases against independent carrier operations. Moreover, the dependence of Britain on U.S. airplanes provided the technological base for the British to adopt U.S. naval air practices.

Conclusion

Table 10.4 summarizes the relationship between motivation and capacity in the five cases examined here. From a social scientific perspective, the carrier case may seem "overdetermined." The British faced every obstacle to making the revolutionary shift to seeing the carrier as the centerpiece of the naval offensive. The Germans and Italians recognized the offensive potential of air power, but strategic imperatives and bureaucratic obstacles privileged land-based air. The Japanese and Americans had most of the forces pointing in the opposite direction—strategic necessity and autonomous naval air organizations. But carriers made strategic sense for the British, Italians, and Germans; the Japanese Navy faced immense bureaucratic obstacles with the army; and the U.S. Navy possessed a healthy skepticism that a naval revolution was underway such that aviators did not gain control over carrier operations until late in the war. In a highly competitive environment, strategic necessity is an appropriate place to begin when examining the motivation to adopt a new military practice. But the presence or absence of strategic incentives alone cannot explain the variation observed among these five cases.

A complete theory of military diffusion must also account for the capacity to incorporate the new practice into the organization's existing routines and structures. Capacity is best viewed as a composite variable influenced by resources, bureaucratic rivalry, administrative authority, compatibility, and organizational set. Each subcomponent can take on different values across cases. For example, the multicarrier task force concept meshed easily with the Imperial Navy's offensive strategic bias, but not with the navy's chief role of supporting the army. The offensive carrier air power paradigm fit the U.S. Navy's strategy, but challenged deeply held conceptual premises within the organization. A theory of capacity must also reflect a nuanced understanding of resource constraints, which will

TABLE 10.4

Motivation and Capacity to Adopt Carrier Air Power

Strategic Necessity	Facilitators and Inhibitors	Capacity	Naval Air Paradigm
	BRITAIN		
Moderate/High	Resources low Bureaucratic rivalry high Administrative authority low Compatibility low Organizational set weak	Very Low	Defensive carrier operations
	UNITED STATES		
High	Resources high Bureaucratic rivalry low Administrative authority moderate Compatibility moderate Organizational set high	Moderate/High	Offensive carrier operations
	JAPAN		
High	Resources low Bureaucratic rivalry moderate/high Administrative authority high Compatibility moderate Organizational set moderate/low	Moderate/Low	Offensive carrier operations
	GERMANY		
Moderate	Resources low Bureaucratic rivalry high Administrative authority low Compatibility low Organizational set low	Very Low	Defensive land-based air
	ITALY		
Low/Moderate	Resources low Bureaucratic rivalry high Administrative authority low Compatibility high	Low	Hypothetical offensive carrier operations

vary over the course of the adoption process. Resource constraints hindered Britain's carrier development in the prewar period, while augmentation with resources from the United States late in the war facilitated the evolution of British carrier practices. On the other hand, resource constraints became crucial for Japan by mid-1942, but the offensive carrier paradigm already was institutionalized. Only by understanding motivation and capacity can we account for the adoption of transformative practices and for the timing and extent of their assimilation.

As with other chapters in this volume, the carrier case demonstrates that new practices are rarely adopted in their entirety and that military organizations rarely remake themselves anew. Neorealism assumes that states face few obstacles to adopting ascending practices, that states can be rational shoppers in the market-place of military ideas. The "rational shopper" image of military diffusion is flawed. States must graft new practices and technologies onto existing practices, technologies, and organizations—so military change is a path dependent process. For the British, in particular, an initial lead in carrier development paradoxically harmed subsequent innovation. Finally, understanding patterns of assimilation and inefficient outcomes requires knowledge of transmission paths.

Air power revolutionized naval warfare and stimulated a wholesale transformation in naval operations, organizations, and attitudinal beliefs. It remains to be seen whether information technologies will have a similarly revolutionary impact on military operations, organizations, and ideas today. To the extent that contemporary military leaders believe this to be the case and are actively seeking to promote such a transformation, the carrier case is particularly instructive for what it suggests about how revolutions in military affairs unfold. Although many believe that the United States today has and will retain its lead in shaping the information-RMA, there was no single innovator whose ideas spread in the carrier case. Different players hit upon pieces of the solution at different times and learned from each other over the course of the transformation process. There was no efficient trajectory to a "best" practice but rather several different paths that eventually converged on a set of solutions, shaped by available resources, bureaucratic rivalries, variations in administrative authority and organizational sets, and compatibility with existing practices. Periods of military transformation are characterized by high uncertainty and risk, and low consensual knowledge. In the past, strategic necessity has often stimulated the search for new solutions, but solving the strategic problem rarely has been a rational or efficient process, as many other chapters in this volume also confirm. The carrier case urges U.S. leaders to follow closely the transformation processes elsewhere, not simply to ensure a U.S. lead or to develop offsets where appropriate but also to learn about potential ways to exploit emerging technologies from the trajectories being pursued by others.

PART IV

Diffusion in the Information Age

CHAPTER ELEVEN

Creating the Enemy

Global Diffusion of the Information Technology– Based Military Model

CHRIS C. DEMCHAK

The decade since the end of the Cold War has witnessed a decline in the level of military competition between states.[1] Despite the relative reduction in imminent threats over this period, military leaders around the globe have declared their intention to modernize their organizations using information technologies (IT) and precision weapons in smaller, more professional force structures.[2] The most prominent model for this approach emerges from the U.S. military's "Revolution in Military Affairs" (RMA) debate. The "revolution" reflects an emerging consensus on what constitutes a modern military: small, highly skilled, and rapidly deployable forces using advanced IT, that are more flexible and putatively much more lethal. This force is defined by five characteristics: "doctrinal *flexibility*, strategic *mobility*, *tailorability* and modularity, joint and international *connectivity*, and the versatility to function in war and OOTW (Operations Other Than War)" [emphasis mine].[3] A new norm is emerging, promoted in part by the widely reported success of the "modern force" during the 1991 Persian Gulf War. A wide variety of states, seeking to garner the respect of their own citizens and the community of nations, are pursuing this modernization strategy.

[1]John Mueller, *Retreat from Doomsday: The Obsolescence of Major War* (New York: Basic, 1989); John Lewis Gaddis, "International Relations Theory and the End of the Cold War," *International Security* 17, no. 3 (1992): 5–58.

[2]*Force XXI Operations: A Concept for the Evolution of Full-Dimensional Operations for the Strategic Army of the Early Twenty-First Century* (Washington, DC: U.S. Army Training and Doctrine Command, 1994).

[3]Ibid.; *United States Army Training and Doctrine Command: The Airland Battle and Corps.* Operational Concepts Series (Washington, DC: U.S. Army Training and Doctrine Command, 1995).

The global spread of the information warfare (IW) concept coupled to a fully modernized force differs from established historical patterns of iterative changes in militaries orchestrated by leaders who perceived threats to their regional, technological, or economic security.[4] Standard theories of "adaptive rationalizations" cannot explain this current modernization trend. Neither security threats nor internal economic pressures are forcing states to change their militaries. Nor do coercive diplomacy and dependence on a superpower explain the rapid spread of the "modern force" paradigm that is sweeping through countries with vastly different resources and threat environments. Rather, the trend belies a process of "institutional isomorphism"[5] in which countries copy (more or less) the example set by a leading state (in this case, the United States) because this new form is perceived as legitimate and modern.

While militaries have often copied each other, the emerging breadth and commonality of vision about what constitutes "modern" is unique. Most nations need only constabulary forces. Achieving ambitious modernization plans, especially electronic warfare capabilities and integrated platforms, comes at great cost to their existing military organizations and societies. The enormous expense of investing in new technologies and organizational adaptation to these new technologies takes its toll on other priorities including education, health care, and even economic growth. More significantly for international peace, the organizational changes, costs, and imported social constructions of modernization will affect defense organizations' perceptions and readiness to act.[6] Developing a global community of putatively rapidly deployable IW militaries will dramatically change future threat assessments, in turn influencing the willingness to use military force in a crisis.

Unprecedented Diffusion of Modernization Model

A 1996 GAO report asserted that more than one hundred nations were planning to modernize their militaries.[7] In September 1996, using more stringent coding than GAO, fifty modernizing states were identified. In April 1997, that number had increased by 18 percent to sixty-eight nations. It is extraordinary for

[4]James Adams, *Engines of War: Merchants of Death and the New Arms Race* (New York: Atlantic Monthly, 1990); Lynn White, Jr., *Medieval Technology and Social Change* (Oxford: Oxford University Press, 1978).

[5]Paul J. DiMaggio and Walter Powell, "The Iron Cage Revisited: Institutional Isomorphism and Collective Rationality in Organizational Fields," *American Sociological Review* 48 (Apr. 1983): 147–60.

[6]Franklin C. Spinney, *Defense Facts of Life: The Plans/Reality Mismatch* (Boulder, CO: Westview, 1985); Chris C. Demchak, *Military Organizations, Complex Machines: Modernization in the U.S. Armed Services* (Ithaca, NY: Cornell University Press, 1991); International Institute for Strategic Studies, *Strategic Survey* (London: Brassey's, 1992).

[7]Rick Maze, "The War that Will Never Be Won," *Army Times* 56 (5 Aug. 1996): 26.

so many distinctly different defense leaders to espouse so similar a vision of what a future "modern" military would entail, a vision based on the highly electronic, precision-guided military under construction by the United States.

It is unprecedented to have countries as disparate as those listed in the Appendix on a similar and relatively simultaneous evolutionary path.[8] A process has accelerated among the world's militaries especially over the past five years, redefining the taken-for-granted minimum in military capabilities to mean intensive use of IT and precision munitions. Military leaders are not seeking these capabilities to meet objective or potential security threats. They view an IT-intensive, small, professional military to be the norm for the twenty-first century. Map 11.1 indicates the breadth of this diffusion.

Countries such as Malaysia with few possible enemies—and none sophisticated—have been as interested in having modern computer-enabled delivery systems as any country in Central Europe. Governments of the newly democratizing nations in Central Europe as well as those in economically developing Asia have been persuaded that the "real" military of a sovereign nation is one that is electronically enabled. Although surrounded by democracies and having a relatively homogeneous societal ethnic mix, the Czech Air Force in the early to mid 1990s had ambitious modernization goals for 2005, while the ground forces were to obtain the most modern equipment affordable for their new rapid reaction brigade.[9] Poland, Slovakia, and Romania have had similar plans.[10]

Other nations have also been planning on modernizing, though they have few sophisticated enemies, even in worst-case scenarios, and suffer from pressing economic problems. Myanmar's military leaders have proven eager to modernize with advanced aircraft, even though the intended foes continue to be rebel armies within national boundaries.[11] Having already purchased antisubmarine corvettes, Thailand has been seeking submarines, displaying a modernization program more akin to power projection than the coastal defense typical of Australasian nations.[12]

The following statement by the chief of the People's Army of Vietnam best captures the taken-for-granted aspect of this process over this decade. Despite

[8]Robert L. O'Connell, *Of Arms and Men: A History of War, Weapons and Aggression* (New York: Oxford University Press, 1989).

[9]Paul Beaver, "The Czech and Slovak Air Force (CSAF) Plans to Completely Modernize Its Aircraft and Air Defence Systems," *Jane's Defence Weekly* (16 May 1992): 864; "Slovak General on Czech, Slovak Air Forces," *Daily Report,* Radio Free Europe, Munich (25 Aug. 1994); "Slovak Army Commander on Plans for Modernization," *Daily Report,* Radio Free Europe, Munich (25 Aug. 1994); "Slovak Defense Developments," *Daily Report,* Radio Free Europe, Munich (24 May 1994).

[10]Brigitte Sauerwein, "Special Report: Focus on East-Central Europe," *International Defense Review* 27 (12 Sept. 1994).

[11]Bertil Lintner, "Myanmar Has Been a Major Recipient of Chinese Weapons, But Is Now Trying to Diversify Its Source," *International Defense Review* 27 (1994): 11, 23.

[12]Robert Karniol, "Coming Out of the Cold," *Jane's Defence Weekly* 21 (2 Apr. 1994): 21.

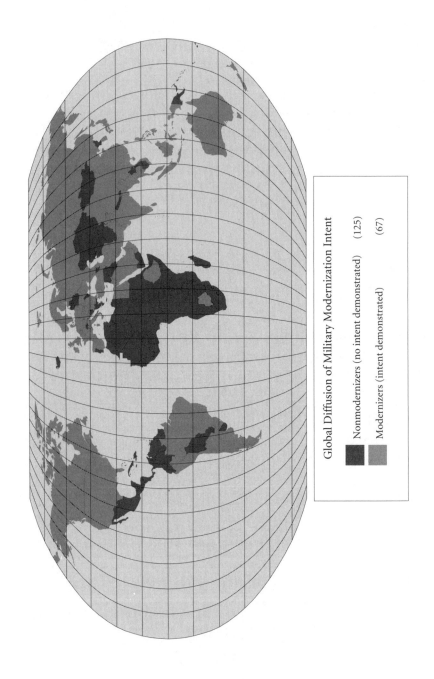

Global Diffusion of Military Modernization Intent

Nonmodernizers (no intent demonstrated) (125)

Modernizers (intent demonstrated) (67)

MAP 11.1. Massive Natural Experiment—Global Modernization

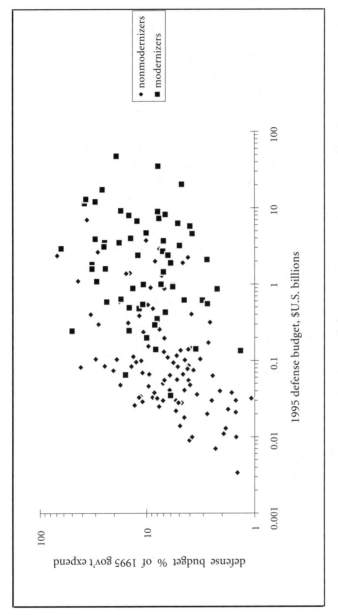

FIG. 11.1. Defense's command of national resources: Defense budget in $U.S. billions versus Defense budget as percentage of government expenditures

having armed forces of 600,000 members (halved between 1988 and 1994), Vietnam declared its intent to have "regular and modern armed forces by the end of the century."[13] A long-time intelligence analyst specializing in East Asia dismissed regional competition as a factor in the modernization programs of East Asian countries, characterizing the process as "keeping up with the neighbors" with no perception of hostile intent by neighboring countries.[14] Finally, Botswana's modernization plans are particularly curious. Its geographic location suggests no opponent aside from South Africa, which, though it could pose a sophisticated threat, has recently experienced a peaceful transition to democracy.

Not only do these countries vary greatly in their individual security conditions but, in addition, there is no discernible pattern based on resource potential. Figure 11.1 shows the lack of pattern across these nations in terms of defense demands on national finances as percentages of government expenditures. Poorer countries such as Myanmar have been planning on modernizing, as are wealthy nations such as Denmark. Indeed, relative poverty is not a strong inhibitor of modernization. Western eagerness to sell new technologies is likely to produce generous coproduction or "offset" agreements, which transfer production expertise to recipient countries.[15] Local production makes acquisition of modern defense systems more affordable.[16]

What explains the unprecedented spread of this "modernization model" at a time when international conflicts necessitating advanced systems have receded? Reasons other than efficiency or organizational goals appear to be at work. Instead, the "modernized" army has become the primary option viewed as legitimate by the international community of militaries. In the sociology literature, the term used is "structuration": a process whereby organizations in the same line of business become more similar. The diffusion of the U.S. military model to the world's military organizations, through structuration, represents an evolutionary path not based on survival of the fittest (most efficient or most effective given the state's political, economic, and security environment), but rather on what is necessary for militaries to see themselves—and to be regarded by other militaries—as legitimate.

[13]Robert Karniol, "PAVN Strives to Modernize in a Climate of Austerity," *Jane's Defence Weekly* 19 (3 Apr. 1993): 18.
[14]"Interview: Hugh White, Deputy Secretary, Strategy and Intelligence, Australian DoD," *Jane's Defence Weekly* 25 (24 July 1996): 32.
[15]"Offsets are Hurting U.S. Companies, Says GAO," *Jane's Defence Weekly* 25 (23 July 1994): 10; Christopher D. Jones, "Czechoslovakia and the New International System," in Jeffrey Simon, ed., *European Security Policy after the Revolutions of 1989* (Washington, DC: National University Press, 1991), pp. 307–30.
[16]Aaron S. Klieman, *Israel's Global Reach: Arms Sales as Diplomacy* (Washington, DC: Pergamon-Brassey's, 1985); Edward A. Kolodziej, *Making and Marketing Arms: The French Experience and Its Implications for the International System* (Princeton: Princeton University Press, 1987).

Method

This chapter uses comparative case studies and quantitative analysis of available data collected from a wide range of countries in order to paint a general overview of the global modernization process and some of the possible attributes associated with the countries that are pursuing this modernization model.

Comparative case study analysis combines "most similar systems" and "most different systems" approaches with the U.S. Army as a control case.[17] Cases selected were either most like or most unlike the U.S. case. Variables related to internal and external motivations to modernize, including expeditionary policy, the extent of recent combat experience, experience with advanced military technologies, and civil-military relations, further define the sample set.[18] Initial case study research focusing on ground forces revealed a surprising trend in which disparate nations regarded the U.S. RMA as an ideal type to be emulated. Even before the three Central European states were assured entry into the North Atlantic Treaty Organization (NATO), senior political leaders announced plans to "modernize" by upgrading their militaries with computerized systems used by the United States.

These initial findings spurred a broader investigation into the rate of likely explanations for the global diffusion of the U.S.-based "modernization model." Social network analysis suggested that diffusion "networks" were developing. Investigating this phenomenon requires more than a single sample or a single time period. Furthermore, if institutional features of the system—such as alliances like NATO—were an important part of the explanation, conventional sampling might not provide adequate data to confirm or reject that these institutions played a role in the process of diffusion.[19]

[17]David M. Fetterman, *Ethnography: Step by Step* (Newbury Park, CA: Sage, 1989); Adam Przeworski and Henry Teune, *Logic of Comparative Social Inquiry* (New York: Wiley Interscience, 1970).

[18]From 1989 to 1994, the author collected ethnographic interview and observational material along with documentary and textual evidence from the ground forces of Israel, Germany, Britain, Hungary, the Czech Republic, and Poland. These earlier findings have been published in a series of articles. For a summary piece, see Chris C. Demchak, "Coping, Copying and Concentrating: Organizational Learning and Modernization in Militaries (Israel, Germany and Britain)," *Journal of Public Administration Research and Theory* 5 (fall 1995): 345–76.

[19]Joseph A. Maxwell, *Qualitative Research Design: An Interactive Approach* (Thousand Oaks, CA: Sage, 1996). Nonetheless, there are numerous shortcomings in using national level data and in using surrogate indicators on which statistical analysis is conducted. Two manipulations of the basic data set need to be noted. First, missing data were accommodated by estimating values from linear regression analyses of states without missing data. In particular, of the 192 nations identified, government expenditure data were missing for 25 nations, and defense budgets for 16. Only two nations fell into both categories. Since a total of 39 nations out of 192 were so hindered—and some of them are prominent world players—regressing $GDP against government expenditures in nations with both data permits an estimation of what government expenditures could feasibly be in countries lacking such data. Using this presumed relationship, I constructed an estimate for missing annual data. The results are noted in Table 11.1 as values in shaded cells, and the regression equations are listed in the notes therein. Forcing data

The large group of middle states provides the best test of varying explanations for the diffusion of the U.S. "modernization model." Both modernizers and nonmodernizers are in this group. This two-tailed truncation is reasonable. The cost of advanced weapon systems prevents the extremely poor nations from participating in the modernization trend, even if they were to put all of their disposable national income to the task. Extremely wealthy nations, on the other hand, can pursue the elements of modernization quite easily. The interesting set of nations are, therefore, those interspersed in between these sets. Leaders of these countries have a choice whether or not to modernize based on their assessments of their country's security, economic, and other needs. After states with no defense budget, the poorest island nations, and the five neo–Great Powers were removed from the original 192 countries, 158 states were left in the database on which this research is based.

The years under review and the level of generality of the data are driven by the realities of a global data set. First, it takes time for social phenomena to ripple through global social networks. Underpinning this research is the fact that the global military community is more interconnected than in previous eras, and thus it is not unreasonable to expect discernible social effects within a five-year period. The years 1993–97 have been chosen because they sufficiently lag a key catalyst, the 1991 Gulf War, while the period is long enough for structurating tendencies to gain discernible momentum.

Second, the data are largely attributional and ideational, since the breadth and dispersion of basic population preclude the in-depth relational data available from the widespread interviews or surveys of traditional social network analysis. This data is taken from both the national and institutional levels. Since the standard theoretical explanations of military modernization do not identify indicators at both the national and institutional levels of generality, surrogate indicators are identified in the national level data and are used in the cluster and ANOVA analyses. Basic indicators for each nation appear in Table 11.1.

Modernization intent is the dependent variable. It is possible to qualitatively

---

into equations demonstrating central tendencies tends to reduce the variability of the dataset. However, given the importance of some of the countries that would otherwise not be included on other measures, it was preferable to include constructed data, recognizing the loss of real variation implicit in not having the actual figures. Although some nations are of interest individually, they are excluded from the wider database as logical outliers. These include countries for which a decision to modernize their military is not feasible because their defense is left to the responsibility of a larger power, or minuscule island nations whose entire GDP would not buy a single advanced technology weapon system. Also excluded is the likely emergent constellation of "neo–Great Powers" according to a ranked array of government expenditures (1995 $U.S.). This set includes four nations whose level of wealth implies no real fiscal burden in modernizing: Japan, China, Germany, and Russia. Each can acquire advanced systems easily with or without serious intent to modernize their forces. In addition, the U.S. is the most often noted role model for this new military and, for control purposes, was removed from calculations.

identify in an ordinal metric only the level of similarity among national militaries that are modernizing. Discarding conscription is, for example, a singular hallmark of nations that want to modernize; but not all modernizers have conscription at the outset. However, one can use qualitative data extracted from professional statements published by the defense industry sources to demonstrate the intent to modernize and the extent of actions in that direction so far. I sought to understand the meanings assigned by leaders of military organizations in similar structural locations to the acquisition of modern technology.[20] In the database, each nation is given a weighted and cumulative code for modernization status. The code reflects whether during the period from 1993 to 1997, a nation's defense leaders (a) declared their intent to be "modern," with smaller, technologically advanced forces, (b) were negotiating to purchase advanced command and control systems, and/or (c) were acquiring stand-alone major weapon systems (called "platforms") with advanced computer-mediated and networked internal systems. Results appear in Table 11.1.

One of two general types of cluster analysis, the "divisive" or "partitioning" approach,[21] was used to analyze the data. The entire dataset was successively subdivided into clusters to determine the levels of similarity. Since each theory tested here has a small set of indicators associated with its explanation, the single attribute method was used to test whether the modernizing nations possessed this attribute more frequently than nonmodernizing nations.[22] If a simpler approach, here a scattergram, reveals no clusters congruent with the modernizing or nonmodernizing populations, then more sophisticated analytical methods are not required. Where appropriate, ANOVA tests support the graphical displays. CART analyses revealed that no indicator that also accounted for the modernization/nonmodernization division produced a statistically significant clustering of nations.

The data demonstrate that widely disparate nations are planning military modernization including the application of advanced highly computerized systems and/or explicitly copying the U.S. model. If structuration is occurring and is the dominant explanation for this trend, other explanations for this phenomenon must be shown to be less persuasive in accounting for the diffusion of the modernization model. If a particular explanation is valid, then modernizers and

---

[20]Ronald S. Burt, *Structural Holes: The Social Structure of Competition* (Cambridge: Harvard University Press, 1992). The database is built from a comprehensive review of four years of leading defense publications: *Jane's Defence Weekly* from July 1993 to Apr. 1997, and two comprehensive annual defense publications, *The Military Balance* and the *Strategic Survey*, for the years 1993–94, 1994–95, 1995–96, and 1996–97. The data for economic figures were taken in May 1997 from the CIA World Factbook website. See Central Intelligence Agency, *World Factbook* (Washington, D.C.: U.S. Government Printing Office, 1997).

[21]See John Scott, *Social Network Analysis* (Thousand Oaks, CA: Sage, 1991).

[22]David Knoke and David L. Rogers, "A Block Model Analysis of Interorganizational Networks," *Sociology and Social Research* 64 (1979): 28–52.

nonmodernizers should cluster according to the attribute associated with that explanation. If the explanation only weakly accounts for what separates modernizers from nonmodernizers, then cases from the two general groups (modernizers and nonmodernizers) will be intermixed in the initial scattergram, with no clusters solely composed of modernizers or nonmodernizers.

Four alternative hypotheses are drawn from the respective literatures on international relations and security studies, political economy, science and technology studies, and institutionalism. For each of the four main approaches, the analysis proceeds with a qualitative discussion of the logic involved, followed by an empirical test of that explanation using an analysis of variance between modernizing and nonmodernizing states on the relevant indicators. Using this method, structuration emerges as the most plausible explanation of the emerging trend to adopt the "modernization model."

Traditional Adaptation Explanations for the Rush to Modernize

Organizations, when viewed as rational actors, respond to three major categories of dominant uncertainties in their environment: security, political-economy, and technology. Security explanations from the international relations literature can be subdivided into three groups: realism and international competition, bloc politics and signaling, and alliance constraints. Political-economy explanations focus on the economic pressures from interested actors in the military-industrial complex, and the political pressures of new wealth and emergent political coalitions inside national defense communities. Sociotechnical explanations of "technology push" propose that the acquisition of some complex equipment tends to force further acquisitions as adaptive managerial responses. Each of these explanations is discussed in greater detail in order to develop testable hypotheses.

A. Security Competition

1. Competitive Pressures and the Security Dilemma

Neorealist explanations in military history and strategic studies suggest that mimicry is based on a recognition of similarities between a successful military and its imitators.[23] Military organizations have difficulty evaluating performance and risk-benefit tradeoffs apart from actual tests during hostilities. Lacking consistent indicators of success, armies have historically copied successful opponents

---

[23]John Mearsheimer, "False Promise of International Institutions," *International Security* 19, no. 3 (1994): 7–49; Kenneth N. Waltz, "Nuclear Myths and Political Realities," *International Security* 84, no. 3 (1990): 731–46; Michael E. Brown, Sean Lynn-Jones, and Steven E. Miller, eds., *The Perils of Anarchy: Contemporary Realism and International Security* (Cambridge: MIT Press, 1995).

or similarly positioned nations as a way of reducing uncertainty in their national security dilemma. This mimicry usually occurs after a demonstration in the operating arena—a major battle or war. For example, the U.S. Army vigorously investigated armor battle techniques of the Israel Defense Forces (IDF) after the 1967 and 1973 wars in which the IDF prevailed over more numerous enemies.[24] Similarly, the apparent success of the 1991 Persian Gulf War has promulgated a strong perception that information technologies (IT) are key to dominance in international security when opponents possess numerically superior forces.[25] This explanation produces the following hypothesis.

Hypothesis 1: Competitive Security Pressures
If nation A faces a major security threat, it will emulate the modernization plan of the nation having demonstrated great success with a similar threat.

The weakness of this explanation is that the U.S. faces very different security circumstances than do most modernizing nations. Drawing upon four years of announcements by the international military community, my research shows that at least sixty-seven nations are actively seeking advanced and computerized weaponry, as shown in Figure 11.2. These states have little in common with the United States in terms of a security dilemma. In fact, one may argue that the potential enemies of most of these modernizers are at a much lower level of sophistication in military capability, and therefore that a number of these nations face no real foes that an effective constabulary force could not handle.

Moreover, if defense forces were historically as attentive (to competitive threat) as corporate competitors are and have been, the structuration process would have occurred decades ago. The historical record is much more mixed. Some militaries in some eras have paid close attention to other militaries but, by and large, most militaries throughout history have been only intermittently aware of changes in other forces. As organizations, they remained relatively isolated from each other both physically and cognitively.[26]

Nations, like corporations, have a survival instinct, and faced with a security dilemma they compete for dominance or, at least, relative stability in their security expectations.[27] This competitive pressure explanation for homogenization

[24]Edward N. Luttwak and Daniel Horowitz, *The Israeli Army, 1948–1973* (Cambridge: Abt, 1983).

[25]Gene I. Rochlin and Chris C. Demchak, "The Gulf War: Technological and Organizational Implications," *Survival* 33 (1991): 33, 260–73; Hirsh Goodman and W. Seth Carus, *The Future Battlefield and the Arab-Israeli Conflict* (New Brunswick, CT: Transaction, 1990).

[26]Examples of inflexibility and inability to adapt abound. See O'Connell, *Of Arms and Men*; Martin van Creveld, *Technology in War* (New York: Free Press, 1989).

[27]Kenneth N. Waltz, *Theory of International Politics* (Reading, MA: Addison-Wesley, 1979); Robert Art and Robert Jervis, *International Politics: Anarchy, Force, Political Economy and Decision-Making* (Boston, MA: Little, Brown, 1985).

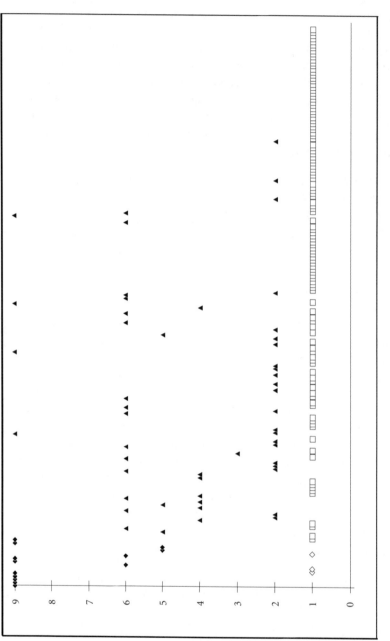

FIG. 11.2. Dispersion of states by modernization score and by region. Modernization score codes: the sum of {4 = stated declaration by senior defense leader of organizational intent to modernize using advanced technologies; 3 = signing of contract or acquisition of computer-based command-and-control (C2) systems; 2 = signed contract or acquisition of computer-enabled major weapons systems ("platforms")}; 1 = no published indication of intent to modernize. Region codes: 1 = NATO, 2 = other Europe, 3 = Middle East, 4 = Central Asia, 5 = East Asia and Australasia, 6 = Caribbean, 7 = Central America, 8 = South America, 9 = Sub-Saharan Africa, 10 = Pacific

among the world's militaries appears to be reasonable. But wars are relatively rare and unusual events for most militaries. And international wars have become even less frequent over time. The competitive drive for structuration has diminished. As shown in Figure 11.2, the fact that the bulk of modernizers have declared an intent but not yet acted to purchase equipment further suggests no pressing threat is forcing a rapid, competitive modernization. We should see less isomorphism but, in fact, just the opposite has occurred: the lag between the first occurrence of a military innovation and its adoption by other military organizations has not increased, but significantly decreased over the modern period.[28]

2. Signaling and Bloc Leadership

During the Cold War era, mirroring intentions were signals indicating loyalty to one's superpower ally.[29] Today, this explanation is obsolete. The Cold War notion of nations acting out of loyalty or coercion in a bloc is moot, overcome by the disappearance of blocs and competing superpowers able to finance the loyalties of bloc members. For example, the republics of the former Soviet Union have had enormous difficulty establishing and maintaining a mutual defense treaty; there is no latent loyalty save in a few, mostly European nations adjacent to Russia, whose assessments and participation may hinge more on minimizing what they perceive to be erratic Russian behavior than on any real desire for mutuality.[30] This theoretical explanation leads to the following hypothesis.

Hypothesis 2: Bloc Leadership

If nation A is in the perceived sphere of influence or likely to obtain associational benefits by positioning itself in this bloc, it will emulate modernization plans of the bloc leader both as a political statement of solidarity to the members of the bloc and as a deterrent statement to would-be aggressors.

Bloc leadership does not explain today's focus on emulating U.S. modernization efforts. There are no direct associational benefits to be gained by demonstrating a willingness to emulate the remaining superpower. Even among those who are part of the only remaining bloc, NATO, there is variation. As shown by Figures 11.3–11.5, NATO bloc and nonbloc nations defy ready categorization across related indicators of modernization: defense budgets, equipment choices, and overall technical development.

---

[28]Trevor N. Dupuy, *The Evolution of Weapons and Warfare* (New York: DaCapo, 1984).

[29]Stephen M. Walt, "The Renaissance of Security Studies," *International Studies Quarterly* 35, no. 2 (1991): 211–39.

[30]Sean Lynn-Jones, ed., *The Cold War and After: Prospects for Peace* (Cambridge: MIT Press, 1991).

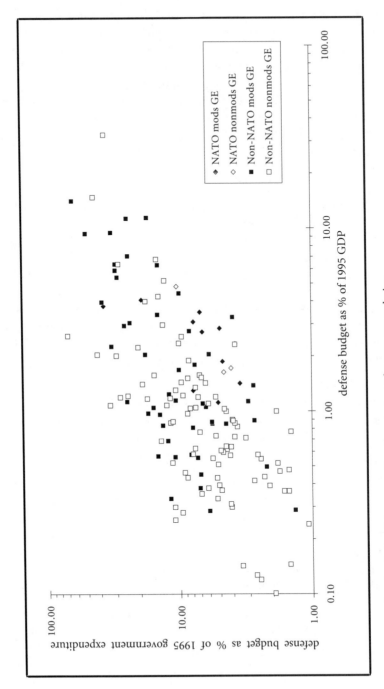

FIG. 11.3. Defense's relative claims on national resources and government choices

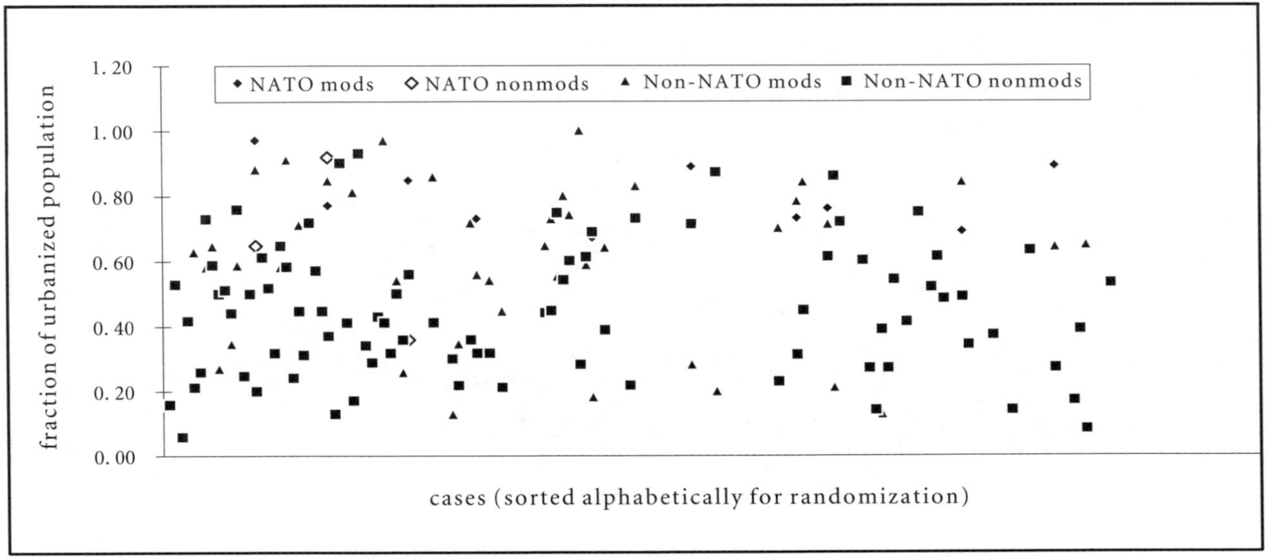

1. 20
1. 00
0. 80
0. 60
0. 40
0. 20
0. 00
fraction of urbanized population
◆ NATO mods ◇ NATO nonmods ▲ Non-NATO mods ■ Non-NATO nonmods
cases (sorted alphabetically for randomization)

FIG. 11.4. Relative levels of national modernization

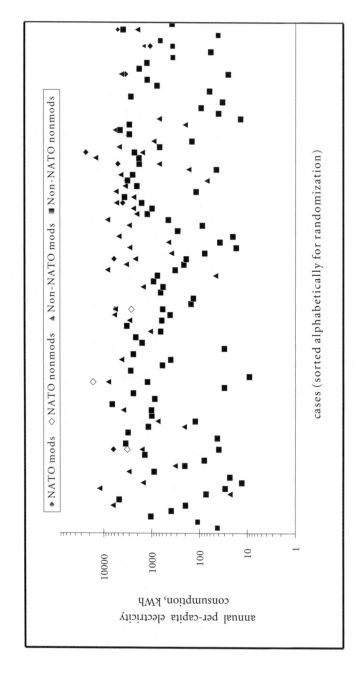

FIG. 11.5. Relative levels of national technological development

Treaty or Formal Coalition Obligations

International institutional explanations suggest that modernizers are bound by treaty requirements to be interoperable, and therefore those nations signing such treaties are bound to copy prominent military organizations in their sphere of operations.[31] The following hypothesis is suggested by this explanation.

Hypothesis 3: Alliance Obligations

If nation A is a member of a military alliance, it will pursue the modernization plans imposed by the alliance and demonstrated by the alliance leadership.

The weakness of this explanation is best exemplified by NATO, the strongest and longest enduring military alliance ever. Even in NATO, treaty requirements imperfectly, and after great delay, trickle down to national military organizations.[32] The NATO treaty explicitly excludes requirements on the military structure of the member organizations, other than to press for congruence in operations. The structure of support units is left up to the individual nation. However, interoperability in actual combat operations is spelled out, and here, despite forty years of planning to go to war together, congruence has not occurred.[33] In fact, when NATO allies fought together in the 1991 Persian Gulf War, interoperability requirements had not even extended to the terms used by native-English speaking artillery units. Until 1991, British artillery's own special term, "at priority call," simply would not be understood by NATO allies expecting the term "reinforcing."[34]

NATO is the only extant alliance with military obligations, yet Figure 11.6 shows different current levels of modernization within the organization. Not only are three NATO members not on record as modernizing—Greece, Iceland, and Portugal—but, in addition, ANOVA analyses of NATO modernizers and non-NATO modernizers show few distinguishing NATO-related features (Tables 11.1 through 11.3).

[31]John G. Ruggie, "The False Premise of Realism," *International Security* 20, no. 1 (1995): 62–70; Robert O. Keohane and Lisa L. Martin, "The Promise of Institutionalist Theory," *International Security* 20, no. 1 (1995): 39–51.

[32]Regina H. Cowen, *Defense Procurement in the Federal Republic of Germany: Politics and Organization* (Boulder, CO: Westview, 1986); Martin Edmonds, *The Defense Equation: British Military Systems Policy, Planning and Performance* (London: Brassey's, 1986).

[33]Michael Mazarr, Don M. Snider, and James A. Blackwell, *Desert Storm: The Gulf War and What I Learned* (San Francisco: WestView, 1993).

[34]Ian Durie, "The Integration of Firepower," *RUSI Journal* 137 (June 1992): 41; Rupert Smith, "The Gulf War: The Land Battle," *RUSI Journal* 137 (Feb. 1992): 1.

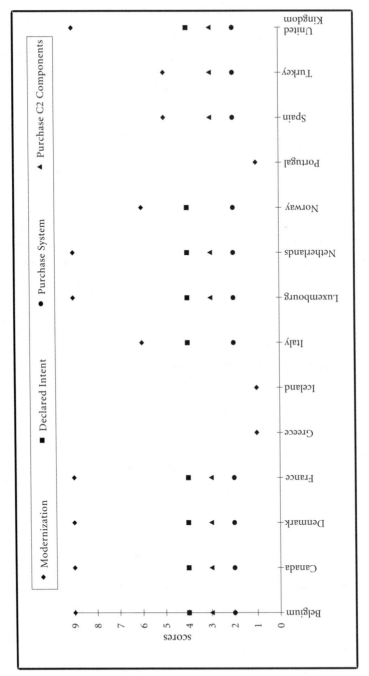

FIG. 11.6. Diversity within NATO, the remaining formal and functioning defense treaty. Scores are defined as follows: 4 = stated declaration by senior defense leader of organizational intent to modernize using advanced technologies; 3 = signing of contract or acquisition of computer-based command-and-control (C2) systems; 2 = signed contract or acquisition of computer-enabled major weapons systems ("platforms"); 1 = no published indication of intent to modernize

TABLE 11.1

Summary of Results of ANOVA Analyses on {NATO Modernizers}
vs. {Non-NATO Modernizers}, $\alpha = 0.05$

Parameter	F statistic	Fcritical	Same Means	Different Means
Defense budget	21.93	4.01		x
Government expenditure	1.25	4.01	x	
Percent defense budget of gov't expend	1.88	4.01	x	
GDP	11.68	4.01		x
Percent defense budget of GDP	0.12	4.01	x	
Percent industrial GDP	0.76	4.01	x	
Percent service sector GDP	12.28	4.01		x
Per-capita GDP	25.32	4.01		x
Percent urban population	3.11	4.01	x	
Per-capita electricity consumption	9.14	4.01		x

TABLE 11.2

Summary of Results of ANOVA Analyses on {NATO Nonmodernizers}
vs. {Non-NATO Nonmodernizers}, $\alpha = 0.05$

Parameter	F statistic	Fcritical	Same Means	Different Means
Defense budget	10.11	3.94		x
Government expenditure	24.88	3.94		x
Percent defense budget of gov't expend	0.31	3.94	x	
GDP	0.76	3.94	x	
Percent defense budget of GDP	0.26	3.94	x	
Percent industrial GDP	0.01	3.95	x	
Percent service sector GDP	5.24	3.95		x
Per-capita GDP	20.06	3.94		x
Percent urban population	2.81	3.95	x	
Per-capita electricity consumption	38.69	3.94		x

TABLE 11.3

Summary of Results of ANOVA Analyses on {Non-NATO Modernizers}
vs. {All Nonmodernizers}, $\alpha = 0.05$

Parameter	F statistic	Fcritical	Same Means	Different Means
Defense budget	33.76	3.91		x
Government expenditure	33.38	3.91		x
Percent defense budget of gov't expend	8.38	3.91		x
GDP	24.73	3.91		x
Percent defense budget of GDP	3.93	3.91		x
Percent industrial GDP	7.49	3.91		x
Percent service sector GDP	2.58	3.91	x	
Per-capita GDP	31.60	3.91		x
Percent urban population	8.14	3.91		x
Per-capita electricity consumption	20.70	3.91		x

Political-Economy Pressures

Socioeconomic explanations would argue either that, in the wake of the Cold War's demise, international competition among defense industries is providing fire sale prices of IT-enabled equipment, thereby creating strong demand among these nations to make massive purchases, or that these nations may be coming into new wealth and are therefore at liberty to buy things previously considered too expensive.[35] This theoretical explanation suggests the following hypotheses.

Hypothesis 4: Defense Industry Push

If defense industry representatives from major weapons-producing countries are desperate to make sales such that they offer "loss leaders" to entice military purchases or arrange government to government generous subsidies or offsets, then the relative reductions in effective prices will drive up the demand of otherwise uninterested countries for modern military systems.

Hypothesis 5: National Wealth Pull

If nation A is growing in national wealth, a proportionate or greater than proportionate amount will be available in the budgets of national military organizations who then can act to implement modernization plans that were previously deferred because of costs.

The shortcomings of these explanations are twofold. First, the only fire sale items available are old equipment, largely from the former Soviet Union. The new computer systems, networks, platforms, and munitions compose integrated systems of extraordinary cost even when national offsets in acquisition are considered.[36] Second, the public justification for many modernization plans is budget reductions, not increases.[37] It is often argued that the modernized military will achieve as much as it ever could, but with less force structure, more speed and accuracy, and smaller budgets.[38] Hence, as suggested in Figures 11.7 and 11.8, the military industrial complex is not a sufficient explanation for the widespread

---

[35]Christian Catrina, *Arms Transfers and Dependence* (New York: Taylor and Francis, 1988); Thomas Ohlson, ed., *Arms Transfer Limitations and Third World Security* (New York: Oxford University Press, 1988); Gary K. Bertsch, Heinrich Vogel, and Jan Zielonka, eds., *After the Revolution: East-West Trade and Technology Transfer in the 1990s* (Boulder, CO: Westview, 1991).

[36]Robert F. Coulam, *Illusions of Choice: The F-111 and the Problem of Weapons Acquisition Reform* (Princeton: Princeton University Press, 1977); Thomas L. McNaugher, *New Weapons, Old Politics: America's Military Procurement Muddle* (Washington, DC: Brookings Institution, 1989).

[37]International Institute for Strategic Studies, *The Military Balance, 1996–1997* (London: Brassey's, 1996).

[38]Johnnie E. Wilson, "Power Projection Logistics Now and in the 21st Century," in Association of the United States Army, *Army Greenbook* (Washington, DC: AUSA, 1994), pp. 137–43; Gordon Sullivan, "America's Army—Focusing on the Future," in Association of the United States Army, *Army Greenbook* (Washington, DC: AUSA, 1994), pp. 19–29; Winn Schwartau, *Information Warfare* (New York: Thunder's Mouth, 1994).

modernization drive across nations with varying percentages of their GDP based on industrial or service sectors.

Figure 11.9 indicates a wide range of absolute defense budgets for modernizers from $48 billion in FY95 to a low of $35 million. Similarly, 1995 gross domestic products (GDP) for modernizers range from $1.4 trillion down to $2.6 billion.

The strength of the claims of the respective national defense infrastructure on either the nation's income or the national government's expenditures shows a similarly broad spread, as shown in Figures 11.9, 11.11, and 11.12. Military demands on national income, seen in the defense budget's percentage of GDP, range from 15 percent to just over 0.25 percent for modernizers. Data in Figure 11.11 also shows that the power of the national defense institutions to control national expenditures, indicated by the defense percentage of government expenditures among modernizers, is also widely disparate, ranging from an estimated 63 percent to 1.3 percent.

Nonmodernizers have a similarly wide distribution of income, as shown in Figure 11.9; GDP ranges from $721 billion to $430 million, three orders of magnitude in difference. Absolute 1995 defense budgets range from a high of $7 billion down to $3.4 million. Similarly, as shown in Figures 11.11 and 11.12, the military's demand on their national budgets across these countries range from 69 percent to barely more than 1 percent. The military consumption of the national income in the nonmodernizing nations also shows a wide spread, from 33 percent to 0.1 percent.

In short, there is no clear national-level evidence that modernizers are clustered only among the most wealthy nations, nor among the nations whose militaries place strong demands on the national treasury or income. That there is indeed considerable dispersion within both populations is depicted graphically by Figure 11.3, which plots defense budgets as a percentage of government expenditures against defense budgets as a percentage of gross domestic product (GDP) for 1995.

C. Sociotechnical Systems and Technological Demands

Sociotechnical (STS) explanations focus on the ripple effects within using organizations. The logic is that, once modernizers acquire some IT-based equipment, in order to use it, they find they must buy more. Then they must massively restructure their organization to be successful. The reasons are the specialized and generally expensive knowledge requirements imbedded in the design and components of these extensive systems.[39]

The tenets emerging from the STS literature are rarely applied to militaries be-

---

[39]For a full discussion of the use of the term "knowledge" here, see Demchak, *Military Organizations, Complex Machines.*

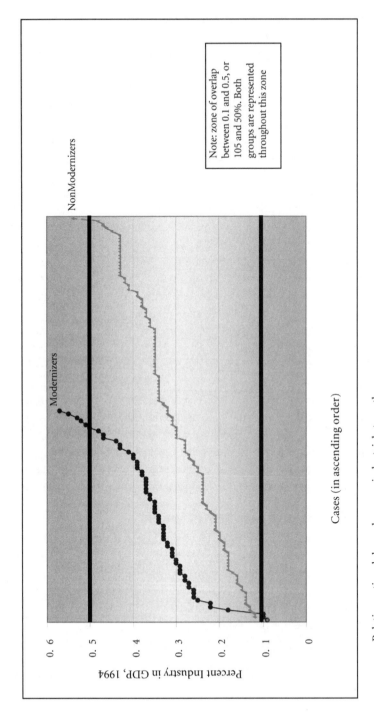

FIG. 11.7. Relative national dependence on industrial strength

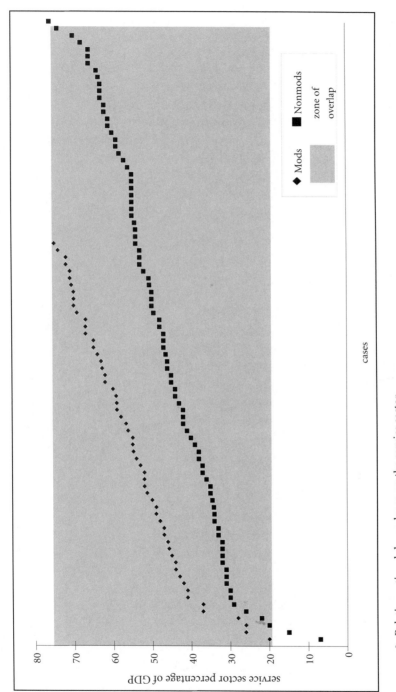

FIG. 11.8. Relative national dependence on the service sector

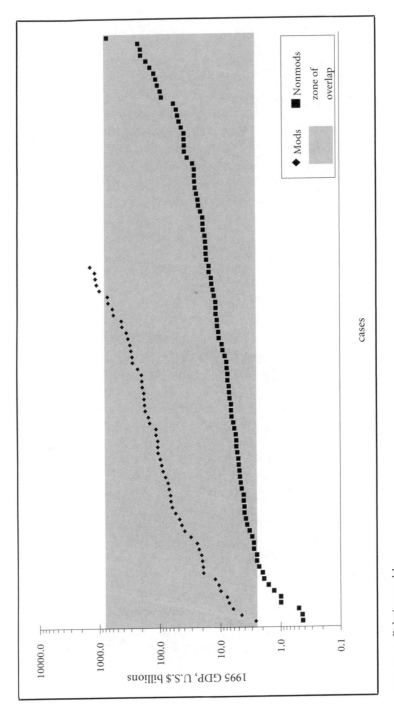

FIG. 11.9. Relative wealth

cause the field emerged from research on civilian organizations based on complex technologies. A number of authors have demonstrated the strong social channeling effect of the designs and subsequent implementation of systems of complex technologies.[40] As militaries become more like their civilian counterparts in their reliance on highly integrated complex technologies, this literature offers increasingly relevant tools and cautionary observations for future technology choices and evolutionary paths for military institutions.[41] This explanation suggests the following hypothesis.

Hypothesis 6: Sociotechnical Pull
   If nation A has acquired advanced military systems, it will be forced by the technical demands of the equipment to acquire more and more elements of complex systems simply to make the initial systems as effective as possible; this accumulation will result in the modernization program.

The difficulty with this explanation of the current trend is the lack of already purchased complex systems across all the modernizers. The social demands of complex computer systems force changes on using organizations only after the systems are first introduced.[42] Among seven militaries, modernized systems have yet to be bought, and sixteen others have not yet purchased the most complex command and control (C^2) systems (Figure 11.10). Hence, for 40 percent of the modernizing nations, the announced changes are not being driven by the real-world experiences with the social demands of complex technological systems. In the future, after these militaries acquire the equipment, a stronger case may be made that their organizations be driven to change in largely unexpected ways to successfully use the extant systems. However, the explanation of technological pull poorly explains the current trend.

[40]Urs Gattiker, *Technology Management in Organizations* (Newbury Park, CA: Sage, 1990); Luis R. Gomez-Mejia and Michael W. Lawless, eds., *Organizational Issues in High Technology Management*, Monographs in Organizational Behavior and Industrial Relations, vol. 11 (Greenwich, CT: JAI, 1990); Shoshana Zuboff, *In the Age of the Smart Machine* (New York: Basic, 1984); Jon Clark et al., *The Process of Technological Change: New Technology and Social Choice in the Workplace* (Cambridge: Cambridge University Press, 1990).

[41]Sheila Jasanoff et al., *Handbook of Science and Technology Studies* (Newbury Park, CA: Sage, 1995); Carl Mitcham, *Thinking through Technology: The Path between Engineering and Philosophy* (Chicago: University of Chicago Press, 1994).

[42]Kenneth F. McKenzie, Jr., "Beyond Luddites and Magicians: Examining the MTR," *Parameter* 25 (summer 1995): 15–21; Todd R. LaPorte and Paula Consolini, "Working in Practice but Not in Theory: Theoretical Challenges of 'High Reliability' Organizations," *Journal of Public Administration Research and Theory* 1 (Jan. 1991): 19–48.

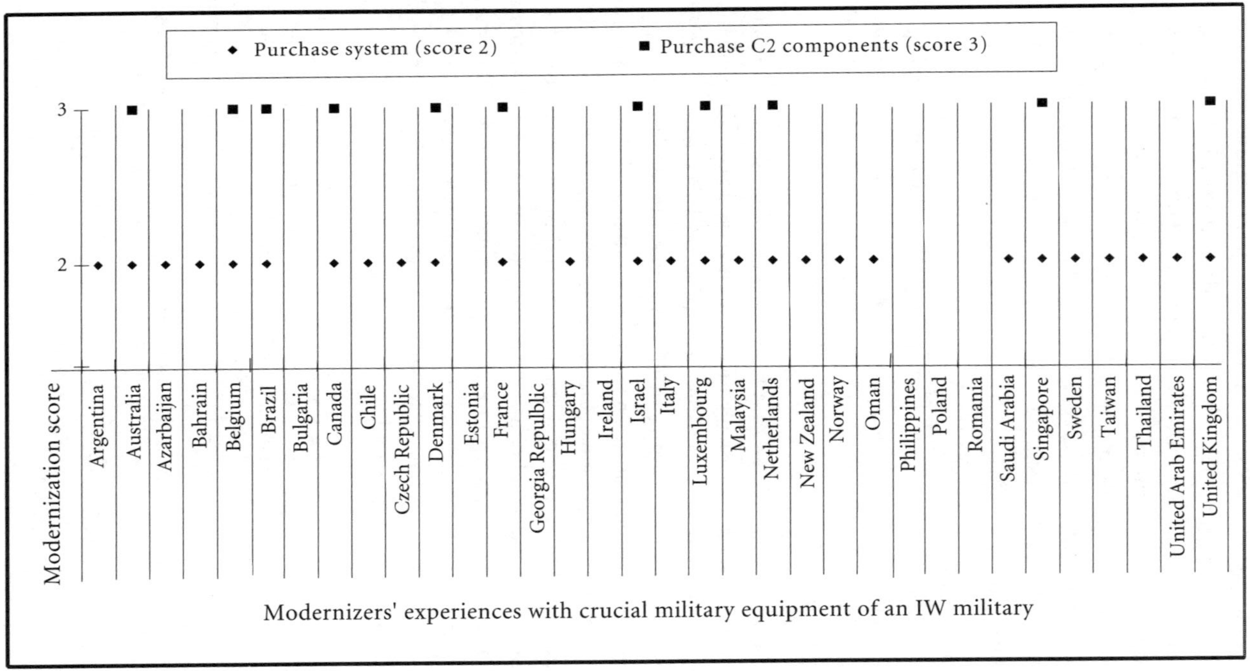

Modernization score

3

2

◆ Purchase system (score 2) ■ Purchase C2 components (score 3)

Argentina ◆
Australia ◆ ■
Azarbaijan ◆
Bahrain ◆
Belgium ◆ ■
Brazil ◆ ■
Bulgaria
Canada ◆ ■
Chile ◆
Czech Republic ◆
Denmark ◆ ■
Estonia
France ◆ ■
Georgia Republic
Hungary ◆
Ireland
Israel ◆ ■
Italy ◆
Luxembourg ◆ ■
Malaysia ◆
Netherlands ◆ ■
New Zealand ◆
Norway ◆
Oman ◆
Philippines
Poland
Romania
Saudi Arabia ◆
Singapore ◆ ■
Sweden ◆
Taiwan ◆
Thailand ◆
United Arab Emirates ◆
United Kingdom ◆ ■

Modernizers' experiences with crucial military equipment of an IW military

FIG. 11.10. "Technology pull" for states that have declared an intent to modernize their militaries. Score code: 3 = signing of contract or acquisition of computer-based command-and-control (C2) systems; 2 = signed contract or acquisition of computer-enabled major weapons systems ("platforms")

Institutional Mimetic Isomorphism of the Military Field

None of the previous explanations parsimoniously encompass the rapidity, breadth, and ambitiousness of the spread of modernization plans compared with historical patterns of the diffusion of military innovations. The explanation of institutional structuration, in contrast, captures the change in what is considered "legitimate" by a society, whether or not that change makes sense for a nation's security, economic, or technological circumstances. This section suggests that the apparent "irrationality" of the conditions surrounding this diffusion may best be explained by proposing a process of emergent institutional isomorphism. This section first presents a discussion of this explanation and then the logical inspection of its applicability.

A. Structuration through Mimetic Institutional Isomorphism

Structuration is a well-established concept in population ecology and organizational sociology.[43] Weber noted the tendency of organizations to compete with others in the same market, weeding out the weaker systems.[44] Over time the survivors would become more similar as they iterated onto the same basic set of attributes that ensured longevity. Until the next major change of their environment, then, that group of organizations, now called a "field," would be relatively homogeneous and successful. Weber called this inevitable cycle an "iron cage" of efficiency. Weber argued that this was bureaucratic rationality, and, now characteristic of all large organizations, it was a key attribute without which an organization was not competitive and would simply fail to thrive.[45] Similarly, structurated biological populations through competition develop a minimal and relatively similar set of attributes essential for survival in that environment.[46]

The difficulty with such market competition–based explanations is that they leave unexplained the modern phenomenon of rising isomorphism among organizations not routinely competing directly with one another. Where relevant task environments are not characterized by free and open competition, it is difficult to sustain the argument that, through competition, the most efficient form of structure becomes dominant.

In 1983, DiMaggio and Powell argued in their seminal "Iron Cage Revisited"

[43]A. Giddens, *The Constitution of Society: Outline of a Theory of Structuration* (Cambridge: Polity, 1984).

[44]W. G. Runciman, ed., *Max Weber* (Cambridge: Cambridge University Press, 1978).

[45]Walter W. Powell and Paul J. DiMaggio, eds., *The New Institutionalism in Organizational Analysis* (Chicago: University of Chicago Press, 1991).

[46]Michael T. Hannan and Glenn R. Carroll, "An Introduction to Organizational Ecology," in Glenn R. Carroll and Michael T. Hannan, eds., *Organizations in Industry: Strategy, Structure & Selection* (New York: Oxford University Press, 1995), pp. 17–32.

article that the character of this bureaucratic rationality—called "competitive isomorphism"—has altered in the intervening years, and the modern version of structuration did not necessarily emerge through competition or result in more efficient organizations. Four conditions could foster the development of a "field" without competition: "an increase in the extent of interaction among organizations in the field; the emergence of sharply defined interorganizational structures of domination and patterns of coalition; an increase in the information load with which organizations in a field must contend; and the development of a mutual awareness among participants in a set of organizations that they are involved in a common enterprise."[47] Built on noncompetitive inducements such as coercion, mimicry, or consensus, this newer form of structuration is distinguishable as "institutional isomorphism."[48]

The process leading to this new structuration can take one of three forms.[49] First, the organization can be forced into a structure in a process of "coercive isomorphism" in which culture, legal environment, or social expectations of key external actors can force isomorphism. For example, organizational change may be the direct result of a government mandate or other specific legal requirements.

Second, the organization can adopt or borrow increasingly standard responses to uncertainty in a process of "mimetic isomorphism." Copying successful others offers a less expensive and possibly viable solution, one that requires the mimicking organization to extend less effort searching for alternative options.

Finally, the organization can demonstrate "normative isomorphism" by which the rise of professionalism within its ranks produces a consensus among similarly trained and powerful professions about the best structure of the organization. Two mechanisms promote this congruency of images of organizations: common sources of formal education, and the strengthening and widening of professional networks across organizations. Professional and trade associations, careful filtering of individuals hired into these organizations, and intense on-the-job socialization strongly contribute to homogenization of outlooks among members of organizations.

Thus, for reasons not necessarily connected to efficiency or individual organizational goals, a set of structural forms can become the minimally legitimate set of options in front of the field's organizational members. For the world's militaries, this process means that over time the evolutionary path of this structurating field of military organizations can stray significantly far from optimal operations or configurations based on realistic security concerns. The dismantling of the former Soviet Union, for example, produced greater openness but also great in-

[47]DiMaggio and Powell, "The Iron Cage Revisited," p. 149.
[48]DiMaggio and Powell, "The Iron Cage Revisited."
[49]Ibid.

terest in Western military styles, equipment, and structures by former Warsaw Pact militaries.[50] Across modernizers, what these institutions are doing may have little to do with either their individual security circumstances or with their societal needs.

B. Application of the Institutionalist Approach

Kraatz and Zajac have identified seven propositions implicit in the DiMaggio and Powell theoretical explanation. The following four hypotheses paraphrased from Kraatz and Zajac test the institutionalization thesis.[51]

Hypothesis 7: Technical Environmental Constraints
If nation A's military is in a strongly institutionalizing organizational environment, then contrary technical environmental conditions will not deter a desire among senior leaders to pursue a modernization program.

Hypothesis 8: Homogeneity Pull
If nation A's military is in a strongly institutionalizing organizational environment, then announced organizational change will involve increasing modernization, thereby encouraging homogeneity of the field.

Hypothesis 9: Prestige Leader Pull
If nation A's military is in a strongly institutionalizing organizational environment, then its modernization program will increasingly resemble that of the accepted prestige leader among the field of global militaries.

Hypothesis 10: Diverse Environment Impotence
If national militaries are in a strongly institutionalizing organizational environment, then the diversity of their individual technical environments will not negatively affect the decision of local senior leaders to pursue a modernization program.

These four hypotheses address the behavior of a number of organizations currently pursuing modernization. Hypotheses seven and ten argue that, if institutionalization is occurring strongly, the local environmental realities and the organization's extant experiences will not deter the senior leaders from pursuing modernization. Hypotheses eight and nine argue that, if there is a prestige leader

[50]Chris C. Demchak, "Modernizing Militaries and Political Control in Central Europe," *Journal of Public Policy* 15, no. 2 (1995): 111–52.

[51]Kraatz and Zajac's propositions one, six, and seven are not included. The first proposition is an overall definition of the institutionalization effect and not useful for validation purposes, and it is too soon to assess the key elements of six and seven: performance effects and early versus late adoption. See Matthew S. Kraatz and Edward J. Zajac, "Exploring the Limits of the New Institutionalism: The Causes and Consequences of Illegitimate Organizational Change," *American Sociological Review* 61 (Oct. 1996): 812–36.

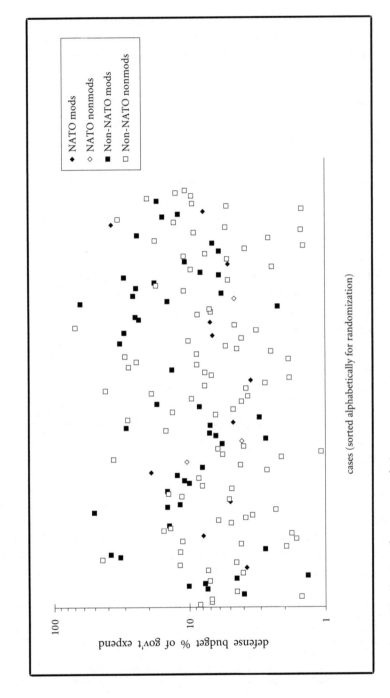

FIG. 11.11. Distribution of government choices

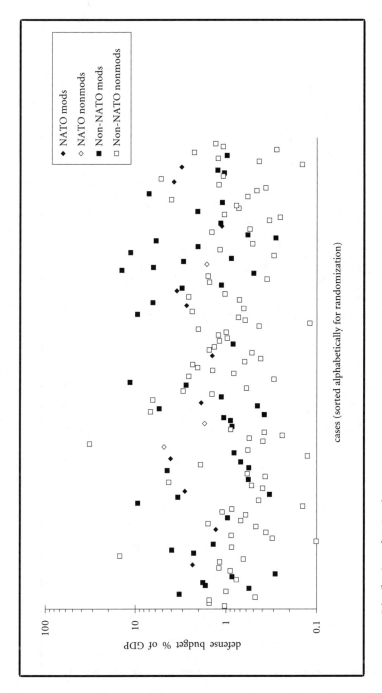

FIG. 11.12. Distribution of national resources

Legend:
- ◆ NATO mods
- ◇ NATO nonmods
- ■ Non-NATO mods
- □ Non-NATO nonmods

Y-axis: defense budget % of GDP (100, 10, 1, 0.1)

X-axis: cases (sorted alphabetically for randomization)

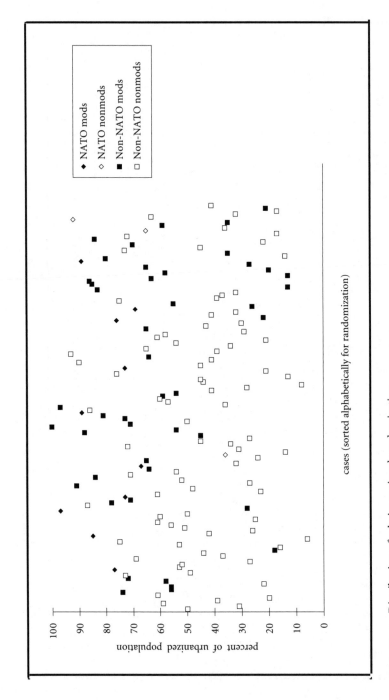

FIG. 11.13. Distribution of relative national modernization

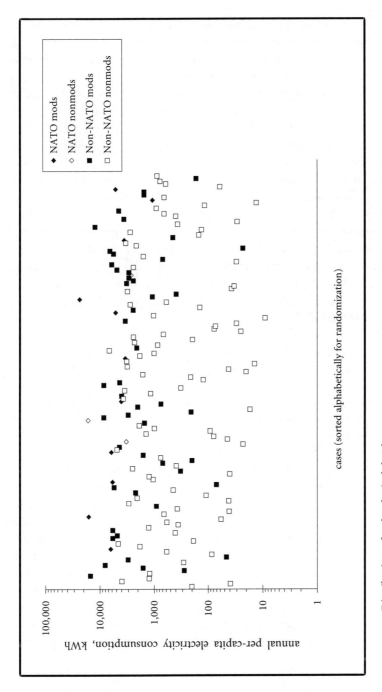

annual per-capita electricity consumption, kWh

cases (sorted alphabetically for randomization)

- ◆ NATO mods
- ◇ NATO nonmods
- ■ Non-NATO mods
- □ Non-NATO nonmods

FIG. 11.14. Distribution of technological development

and any organizational change, national militaries will be demonstrating strong institutionalization by mimicking the prestige leader, and the changes will advance homogenization. All of the relevant indicators support the interpretation that institutionalization is occurring globally among militaries.

To make this case in summary fashion, the otherwise inexplicable dispersion of these nations across selected indicators are shown in Figures 11.11–11.14. The patterns suggest a random diffusion not resolved by the previous explanations and consistent with the implications of hypotheses 7–10. Neither government choices (indicated by the defense budget percentage of government budget, Figure 11.11), available government resources (indicated by government budget percentage of GDP, Figure 11.12), relative level of national development (indicated by percentage of the population that is urbanized, Figure 11.13), nor technological modernization (indicated by per capita electricity consumption, Figure 11.14) suggest a material reason for the distribution of modernizing versus nonmodernizing states. Only the unpredictable reach of legitimizing interpretations explains the lack of consistent concentrations of modernizers along these variables. In short, only the process of institutionalization is a plausible explanation for the breadth of this trend.

Conclusion: Cycle of Structuration, Autopoiesis, and (Then) Adaptation

This analysis suggests that the future path of the world's modernizing nations is being led by the field-defining actions of the U.S. military's ambitious plans, budget size, defense industry support, international visibility, and the speed and ease of global commercial communications. Under such circumstances, institutional isomorphism is revealing a largely autopoiesis process that is producing a dangerous incongruity between the actual security dilemmas of these nations and the dampening capabilities of the international system in times of crisis. *Autopoiesis* is a term coined for the ecology literature in the early 1980s by Humberto Maturana, a Chilean physicist. It captures the process by which organizations "enact" their environment unintentionally through the cumulative effects of their internal operational decisions.[52] Autopoiesis postulates that actors in a system make dispersed random changes that automatically result in the same actors having to deal with a new reality emerging directly from their own unwitting constructions. A good analogy is the anthill, which ants inadvertently build by their normal operations and then have to operate around or over or through.

[52]Gareth Morgan, *Images of Organization* (Beverly Hills, CA: Sage, 1986), p. 139.

Such structuration is not necessarily neutral or positive. As homogeneity spreads, militaries will increasingly be similarly equipped with opaque technologically complex capabilities.[53] As the autopoietic electronic modernization process progresses, actors in neighboring countries will find it increasingly difficult to accurately gauge intentions or capabilities of states physically adjacent or reachable by long-range weapons and forces. Pursued initially to resolve a sea of unweighted uncertainties, modernization's isomorphism process can have paradoxical results, producing greater fears for national security and survival among nations not generally viewed as threatened or threatening.

Furthermore, technologically induced organizational changes will tend to establish a field of policy choices as well as condition the way that military options are selected by insiders and viewed by outsiders. When military forces and operations are difficult for leaders to understand, control moves away from political authorities and even from senior military officers.[54] It takes extraordinary efforts, technical expertise, and self-confidence for political leaders to overrule entrenched organizational procedures.[55] Worse, senior political and military leaders may not understand the consequences of decisions involving their forces.[56] Under any circumstances, institutional (mimetic or normative) isomorphisms and their potentially autopoietic outcomes are undesirable outcomes among purely public-good organizations because they are intrinsically difficult to test, and display a tendency to remain unchecked by actual experience. Such a scenario is particularly troubling among a coalescing community of militaries, since the only truly persuasive test of performance will be actual use in hostilities. The creation of an international community of militaries reputed to have great speed and lethality is not likely to enhance peaceful longer-term relations.

In its extreme form, autopoietic isomorphism would produce exactly the same military proportionately for each nation. These could produce conditions of perfect transparency, if this extreme were achieved or achievable. However, in-

---

[53]Robert Art and Robert Jervis, *International Politics: Anarchy, Force, Political Economy and Decision-Making* (Boston: Little, Brown, 1985), p. 521; Arden Bucholz, "Armies, Railroads, and Information: The Birth of Industrial Mass War," in Jane Summerton, ed., *Changing Large Technical Systems* (Boulder, CO: Westview, 1994), pp. 53–71.

[54]William Seymour, *Decisive Factors in Twenty Great Battles of the World* (London: Sidgwick and Jackson, 1988); Kenneth Macksey, *Military Errors of World War II* (London: Brassey's, 1987); Kenneth Macksey, *For Want of a Nail: The Impact on War of Logistics and Communications* (London: Brassey's, 1989).

[55]McNaugher, *New Weapons, Old Politics*; Hew Strachan, *European Armies and the Conduct of War* (London: Unwin Hyman, 1983).

[56]Paul Bracken, *The Command and Control of Nuclear Forces* (New Haven: Yale University Press, 1983); Scott D. Sagan, *Moving Targets: Nuclear Strategy and National Security* (Princeton: Princeton University Press, 1989).

stitutionalization must have its limits, and adaptive responses must re-emerge. With every military engaged in being or becoming small but putatively extremely rapid and lethal, national leaders will turn to devising responses to the new environment that has been autopoietically constructed. No matter how much Argentina's leaders feel that Chile is currently not a threat, they will find it difficult to ignore the Chileans' functional highly technologically enhanced, rapid reaction air mobile brigades. Chile's neighbors will find it necessary to focus once again on their local environment and respond not to what Chile does, but to what Chile is capable of doing. When structuration reaches its logical extreme, the weighting of uncertainties will change dramatically for military decision-makers.

Once the structuration and accompanying autopoiesis are nearly complete, the limits of the institutional isomorphism argument will become apparent. Then, traditional adaptation explanations will again come into play as each nation, now on a new evolutionary path defined by the structuration process, turns to responding in the traditional ways described in the extant literature. Military organizations will change according to the needs of security, defined by neighbors' actions. They will respond to the demands of the technology they own and the industry that has grown up around these new technologies. Their decisions will be less ideationally derived and more experientially driven, making the responses more congruent with local conditions. Over time, then, the organizations on the new path begin to differ from each other, and the structurated isomorphism begins to decay. For a period, then, traditionally adaptive theories will better explain variation among global militaries. When the competition among militaries again becomes quiescent and no natural tests are available, structuration begins anew and the cycle continues.

It is this cycle that fuels the continuing debate about which approach has more explanatory power. At some point in the future, as technological advances become dramatic or world relations more and more intricately interdependent, senior leaders will again face a sea of unweighted uncertainties, no standard response that seems to meet the needs, and the presence of a demonstration leader whose actions seem to define a successful response. At that point, an institutional isomorphic process is likely to re-emerge and shift the community onto a new evolutionary path. If the wider global community is both wise and fortunate, the new path will be one less likely to encourage misperception, crises, and the use of rapid-response, long-distance, closely targeted systems of highly lethal weapon systems.

Appendix: The Basic Data—Modernizers and Nonmodernizers

TABLE A1

Modernizers

Country Name	Defense Budget 1995 in $U.S. billions	Government Expenditures 1995 in $U.S. billions	Percent Defense Budget of Govt. Exp. 1995	Percent Defense Budget of Gross Domestic Product 1995	Gross Domestic Product 1995
Argentina	4.7	46.5	10.11	1.69	278.5
Australia	7.3	95.15	7.67	1.80	405.4
Austria	2.1	75.8	2.77	1.38	152
Azerbaijan	0.245	0.488	50.20	9.42	2.6
Bahrain	0.247	1.7	14.53	3.38	7.3
Bangladesh	0.481	4.1	11.73	0.33	144.5
Belgium	4.6	121.444	3.79	2.34	197
Botswana	0.199	1.99	10.00	4.42	4.5
Brazil	6.736	54.9	12.27	0.69	976.8
Bulgaria	0.352	4.4	8.00	0.81	43.2
Canada	9	114.1	7.89	1.30	694
Chile	0.97	17	5.71	0.86	113.2
China	8.4375			0.24	3500
Croatia	1.089	3.72	29.27	5.42	20.1
Cyprus	0.493	3.4	14.50	6.32	7.8
Czech Republic	0.931	16.2	5.75	0.88	106.2
Denmark	3.2	64.4	4.97	2.84	112.8
Egypt	3.5	19.4	18.04	2.05	171
Estonia	0.035	0.582	6.01	0.28	12.3
Ethiopia	0.14	1.7	8.24	0.58	24.2
Finland	1.9	31.7	5.99	2.06	92.4
France	47.7	249.1	19.15	4.07	1173
Georgia Republic	0.065	0.414	15.70	1.05	6.2
Germany	42.8	780	5.49	2.95	1452.2
Hungary	0.62	13.8	4.49	0.86	72.5
India	8	54.9	14.57	0.57	1408.7
Indonesia	2.7	38.1	7.09	0.38	710.9
Iran	1.459	20.84286	7.00	0.45	323.5
Ireland	0.618	20.3	3.04	1.13	54.6
Israel	9.2	53	17.36	11.49	80.1
Italy	20.4	431	4.73	1.87	1088.6
Japan	50.2	829	6.06	1.87	2679.2
Jordan	0.589	2.5	23.56	3.05	19.3
Korea, South	17.4	67	25.97	2.95	590.7
Kuwait	3.5	14.2	24.65	11.36	30.8
Luxembourg	0.142	4.05	3.51	1.42	10
Malaysia	2.4	19.9	12.06	1.24	193.6
Micronesia	0.0001	0.031			0.205
Monaco	0.0001	0.586			0.788
Myanmar	0.135	10	1.35	0.29	47
Netherlands	8.2	122.1	6.72	2.72	301.9
New Zealand	0.556	20.28	2.74	0.89	62.3
Norway	3.7	53	6.98	3.48	106.2
Oman	1.82	5.6	32.50	9.53	19.1
Pakistan	3.1	12.4	25.00	1.13	274.2
Peru	0.998	9.3	10.73	1.15	87

TABLE A1—*cont.*

Country Name	Defense Budget 1995 in $U.S. billions	Government Expenditures 1995 in $U.S. billions	Percent Defense Budget of Govt. Exp. 1995	Percent Defense Budget of Gross Domestic Product 1995	Gross Domestic Product 1995
Philippines	1	13.6	7.35	0.56	179.7
Poland	2.4	37.8	6.35	1.06	226.7
Qatar	0.294	3.5	8.40	2.75	10.7
Romania	0.885	6.6	13.41	0.84	105.7
Russia	106.927	545.5459184	19.60	13.43	796
Saudi Arabia	12.1	40	30.25	6.39	189.3
Serbia	2.927			14.21	20.6
Singapore	3.9	12.9	30.23	5.90	66.1
Slovakia	0.43	6.4	6.72	1.10	39
Slovenia	1.6	6.6	24.24	7.08	22.6
Spain	6.3	122.5	5.14	1.12	565
Sri Lanka	0.64	3.7	17.30	0.98	65.6
Sweden	5.8	146.1	3.97	3.27	177.3
Taiwan	11.5	30.1	38.21	3.96	290.5
Thailand	4	28.4	14.08	0.96	416.7
Turkey	13	35	37.14	3.76	345.7
Ukraine	0.868			0.50	174.6
United Arab Emirates	1.59	4.9	32.45	2.27	70.1
United Kingdom	35.1	447.6	7.84	3.08	1138.4
United States of America	272.2	1461	18.63	3.76	7247.7
Vietnam	0.544	5	10.88	0.56	97

TABLE A2

Nations with No Evidence of Modernization Intent on Record

Country Name	Defense Budget 1995 in $U.S. billions	Government Expenditures 1995 in $U.S. billions	Percent Defense Budget of Govt. Exp. 1995	Percent Defense Budget of Gross Domestic Product 1995	Gross Domestic Product 1995
Afghanistan	0	0	0.00	0.00	12.8
Albania	0.041			1.00	4.1
Algeria	1.3	17.9	7.26	1.20	108.7
Andorra	0.0001				1
Angola	1.1	2.5	44.00	14.86	7.4
Anguilla	0.0001	0.0176	0.57	0.19	0.053
Antigua	0.0014	0.135	1.04	0.33	0.425
Armenia	0.075			0.82	9.1
Bahamas	0.02	0.725	2.76	0.42	4.8
Barbados	0.013	0.71	1.83	0.52	2.5
Belarus	0.479	5.47	8.76	0.97	49.2
Belize	0.008	0.123	6.50	1.39	0.575
Benin	0.033	0.375	8.80	0.43	7.6
Bhutan		0.15	0.00	0.00	1.3
Bolivia	0.145	3.75	3.87	0.72	20
Bosnia-Herzegovina	0.878				1

TABLE A2—*cont.*

Country Name	Defense Budget 1995 in $U.S. billions	Government Expenditures 1995 in $U.S. billions	Percent Defense Budget of Govt. Exp. 1995	Percent Defense Budget of Gross Domestic Product 1995	Gross Domestic Product 1995
Brunei	0.312	2.1	14.86	6.78	4.6
Burkina Faso	0.104	0.548	18.98	1.41	7.4
Burundi	0.025	0.326	7.67	0.62	4
Cambodia	0.085	0.346	24.57	1.21	7
Cameroon	0.102	2.3	4.43	0.62	16.5
Cape Verde	0.0034	0.235	1.45	0.77	0.44
Cayman Islands	0.0001	0.161			0.75
Central African Republic	0.03			1.20	2.5
Chad	0.074	0.363	20.39	2.24	3.3
Colombia	2	24	8.33	1.04	192.5
Comoros		0.092	0.00	0.00	0.37
Congo	0.11			1.43	7.7
Costa Rica	0.055	1.34	4.10	0.30	18.4
Cote d'Ivoire	0.14	3.4	4.12	0.64	21.9
Cuba	0.588			4.00	14.7
Djibouti	0.026	0.201	12.94	5.20	0.5
Dominica		0.096	0.00	0.00	0.2
Dominican Republic	0.116	2.2	5.27	0.43	26.8
Ecuador	0.386	3.3	11.70	0.87	44.6
El Salvador	0.1	0.89	11.24	0.88	11.4
Equatorial Guinea	0.0025	0.0359	6.96	0.77	0.325
Eritrea	0.038			1.90	2
Fiji	0.028	0.591	4.74	0.60	4.7
Gabon	0.154	1.6	9.62	2.57	6
Gambia	0.014	0.09	15.56	1.27	1.1
Ghana	0.03	1.2	2.50	0.12	25.1
Greece	4.9	47	10.43	4.82	101.7
Grenada		0.127	0.00	0.00	0.284
Guatemala	0.13	1.88	6.91	0.35	36.7
Guinea	0.05	0.708	7.06	0.77	6.5
Guinea-Bissau	0.009			0.90	1
Guyana	0.007	0.303	2.31	0.44	1.6
Haiti	0.034	0.299	11.37	0.52	6.5
Honduras	0.041	0.668	6.14	0.38	10.8
Iceland	0.086	2.1	4.10	1.72	5
Iraq	2.628			6.39	41.1
Jamaica	0.03	2	1.50	0.37	8.2
Kazakhstan	0.404			0.86	46.9
Kenya	0.136	2.8	4.86	0.37	36.8
Kiribati		0.054	0.00	0.00	0.068
Korea, North	7	19.3	36.27	32.56	21.5
Kyrgyzstan	0.048			0.89	5.4
Laos	0.105	0.351	29.91	2.02	5.2
Latvia	0.105			0.71	14.7
Lebanon	0.278	3.2	8.69	1.52	18.3
Lesotho	0.018	0.4	4.50	0.64	2.8
Liberia	0.014	0.285	4.91	0.61	2.3
Libya	1.4	9.8	14.29	4.26	32.9

TABLE A2—*cont.*

Country Name	Defense Budget 1995 in $U.S. billions	Government Expenditures 1995 in $U.S. billions	Percent Defense Budget of Govt. Exp. 1995	Percent Defense Budget of Gross Domestic Product 1995	Gross Domestic Product 1995
Liechtenstein	0.0001	0.442			0.63
Lithuania	0.032			0.24	13.3
Macedonia	0.03			1.58	1.9
Madagascar	0.029	0.265	10.94	0.25	11.4
Malawi	0.01	0.674	1.48	0.14	6.9
Maldives		0.141	0.00	0.00	0.39
Mali	0.066	0.697	9.47	1.22	5.4
Malta	0.021	1.4	1.50	0.48	4.4
Mauritania	0.033	0.28	11.79	1.18	2.8
Mauritius	0.011	0.567	1.94	0.10	10.9
Mexico	2.24	54	4.15	0.31	721.4
Moldova	0.038			0.37	10.4
Mongolia	0.023	1.3	1.77	0.47	4.9
Morocco	1.38	8.9	15.51	1.58	87.4
Mozambique	0.084	0.607	13.84	0.69	12.2
Namibia	0.064	1.05	6.10	1.10	5.8
Nauru		0.069	0.00	0.00	0.1
Nepal	0.036	1.05	3.43	0.14	25.2
Nicaragua	0.028	0.551	5.08	0.39	7.1
Niger	0.032	0.4	8.00	0.58	5.5
Nigeria	0.172	6.4	2.69	0.13	135.9
Palau	0.0001	0.057			0.082
Panama	0.078	1.86	4.19	0.57	13.6
Papua New Guinea	0.04	1.9	2.11	0.39	10.2
Paraguay	0.094	1.66	5.66	0.55	17
Portugal	1.9	41	4.63	1.64	116.2
Rwanda	0.113			2.97	3.8
Saint Kitts and Nevis		0.1	0.00	0.00	0.22
Saint Lucia	0.005	0.127	3.94	0.78	0.64
Saint Vincent and the Grenadines		0.118	0.00	0.00	0.24
San Marino	0.0037	0.32	1.16	0.97	0.38
Sao Tome and Principe		0.114	0.00	0.00	0.138
Senegal	0.082	0.198	41.41	0.06	145
Seychelles	0.01	0.263	3.80	2.33	0.43
Sierra Leone	0.014	0.128	10.94	0.32	4.4
Solomon Islands		0.102	0.00	0.00	1
Somalia			0.00	0.00	3.6
South Africa	2.9	38	7.63	1.35	215
Sudan	0.298	1.06	28.11	1.19	25
Suriname	0.013	0.7	1.86	1.00	1.3
Swaziland	0.022	0.41	5.37	0.73	3
Switzerland	3.74	36.9	10.14	2.36	158.5
Syria	2.358	3.4	69.35	2.59	91.2
Tajikistan	0.066			1.03	6.4
Tanzania	0.069	0.631	10.94	0.30	23.1
Togo	0.048	0.274	17.52	1.17	4.1

TABLE A2—*cont.*

Country Name	Defense Budget 1995 in $U.S. billions	Government Expenditures 1995 in $U.S. billions	Percent Defense Budget of Govt. Exp. 1995	Percent Defense Budget of Gross Domestic Product 1995	Gross Domestic Product 1995
Tonga		0.086	0.00	0.00	0.228
Trinidad and Tobago	0.083	1.61	5.16	0.51	16.2
Tunisia	0.535	5.5	9.73	1.44	37.1
Turkmenistan	0.063			0.55	11.5
Tuvalu		0.0043	0.00	0.00	0.0078
Uganda	0.056	1.07	5.23	0.33	16.8
Uruguay	0.256	3.37	7.60	1.05	24.4
Uzbekistan	0.317			0.58	54.7
Vanuatu		0.076	0.00	0.00	0.21
Vatican	0.0001	0.175			
Venezuela	0.902	9.8	9.20	0.46	195.5
Yemen	0.401	1.2	33.42	1.08	37.1
Zaire	0.046	0.479	9.60	0.28	16.5
Zambia	0.096	0.767	12.52	1.08	8.9
Zimbabwe	0.236	2.2	10.73	1.30	18.1

CHAPTER TWELVE

Patterns of Commercial Diffusion

JOHN ARQUILLA

William Gladstone, one of the great prime ministers of the nineteenth-century *Pax Britannica,* viewed the spread of open markets and the sharing of advanced technologies as "great equalizers," bringing both peace and prosperity to the world. Benjamin Disraeli, his long-time rival, denounced the idea of benefits arising from untrammeled free trade and technology transfers as "pernicious nonsense." Their views form opposite ends of a spectrum of thought about what one might call "commercial diffusion." Between these extremes, though, lies a wide range of ambivalent opinions. While we might recognize the benefits of diffusing commercial products, concepts, technologies, and forms of organization, their spread presents dilemmas for state security interests. These concerns lie at the heart of contemporary U.S. policy debates in a number of key information technology areas, especially high-performance computing and encryption.

In the military realm, a guarded attitude toward diffusion seems requisite, and is apparent from early history to the present. For example, the ancient Hittites strove to keep the secret of their iron weapons, which helped launch them on the path of empire.[1] They failed and were destroyed. On the other hand, the Byzantines kept the secret of their incendiary "Greek fire" so well that the Eastern empire outlasted the fall of Rome by a thousand years.[2] More recently, the Germans

---

[1] John Keegan, *A History of Warfare* (New York: Vintage Books, 1993), p. 238.

[2] Edward Gibbon, *The Decline and Fall of the Roman Empire* (London: J. B. Bury, 1905), held that Greek fire "saved Constantinople on more than one occasion" (vol. 2, p. 673). He also noted that the later failure to prevent the diffusion of information about heavy artillery was what brought Constantinople to ruin.

retained their monopoly on blitzkrieg for only a few short years, and were soon defeated by adversaries that had learned to employ, just as skillfully, the tanks and planes that dominated maneuver warfare. More recently, the U.S. military-industrial complex has prevented the diffusion of much information about stealth weaponry, aiding the cause of U.S. primacy.

To think usefully about commercial diffusion, it is necessary first to develop a working definition, and then a sense of how diffusion processes work. Fortunately, there is an excellent body of work that studies diffusion—in all but name—referring instead to "growth," "backwardness," and "development."[3] In this chapter, hypotheses are drawn from these literatures, and then used to help understand the enduring quandaries that policy-makers have always faced in the commercial realm—especially the tension between the pursuit of economic gain and the security risks often engendered by commercial activity.

This is a particular dilemma today, as advanced information technologies clearly improve productivity and profitability, but they are almost all inherently "dual use"—that is, they can generally be used for a wide variety of military activities, from improving long-range missile guidance systems to facilitating the design and manufacture of better hardware, such as tanks, planes, and even submarines. A further concern is the risk that diffusion of advanced information technology easily grants potential adversaries a first-class ability to wage cyberwar or information warfare. A better understanding of the nature of commercial diffusion, and its relationship with security concerns, will thus help to chart a much needed policy that reconciles an economic need for openness with a military need to be guarded. To inform policy, though, it is first necessary to consider commercial diffusion's origins, scope, and character.

Early Concepts of Commercial Diffusion

Extensive trading networks emerged in the Mediterranean world more than three thousand years ago. Commerce was first stimulated by the diffusion of the concept and use of money. The Lydians began stamping coins with specific valuations (as opposed to simple weight measures of worth), and they were soon widely imitated. The emergence of money was wildly popular, as it enabled commerce to move well beyond the limits of barter. Soon the concept of money expanded to include letters of credit, or what were referred to as "leather money" (as the terms of finance were often written on cowhide). If anything, this nonspe-

---

[3]Examples of these literatures include: Paul Baran, *The Political Economy of Growth* (New York: Monthly Review, 1957); Alexander Gerschenkron, *Economic Backwardness in Historical Perspective* (Cambridge: Harvard University Press, 1962); and Peter Evans, *Dependent Development* (Princeton: Princeton University Press, 1979).

cie form of money was even more useful commercially, spurring an explosion of economic activity across the ancient world.[4]

Money quickly proved of benefit to all transacting parties—and was eagerly and rapidly diffused by all. The growth of commerce next spurred the rise of banks, which flourished under the aegis of Athenian maritime empire. The fall of Athens at the end of the war with Sparta, however, saw banking, and trade in general, stagnate. It was only with the rise of a stable empire, Rome, that trade, money, and banking came to flourish again. By 50 B.C., standard money had diffused throughout the empire, and loan rates were set at 1 percent per month. When the Western empire fell, five hundred years later, money and banking suffered severely, counterfeiting was rampant, and trade declined precipitously. The only reliable coinage could be found in the Eastern empire, whose mines in the Crimea became the sole significant source of gold coinage throughout the Middle Ages; and the feudal system in the West became characterized by autarky and only limited use of local coinage minted with baser metals.[5]

The financial and economic malaise of the Middle Ages came to an end, and money and banking began to rediffuse, this time without falling back, by a process that can only be viewed as coming from the "bottom up." Instead of relying on a leading state or empire to grant the stability and security upon which commercial progress could be made, the late medieval resurgence of banking and trade came from small merchants, banks run by individuals, and the largesse of the Catholic church. Their efforts were reinforced by the sack of Constantinople by Venetian and other raiders, whose Fourth Crusade was sidetracked by visions of plunder. The robbers went home with enormous amounts of Byzantine gold, which gave a "jump start" to Venetian banking. This in turn encouraged the reopening of gold mines in the West dating from the old Roman days. Money and banking were back to stay, aiding the process of nation-building in the West, and priming the world for the rise of transnational markets of vastly increased size. And gold would diffuse as the standard for all monetary assets, giving way only thirty-odd years ago at Bretton Woods.[6]

As feudalism waned and international commerce began to take hold once again, a tendency toward guarded practices also began to manifest itself. For ex-

---

[4]The clearest classical analysis of the spread of money can be found in Aristotle's *Ethics,* particularly Book 5. A more modern examination of the diffusion of the concept of money in the ancient world can be found in Joseph A. Schumpeter, *History of Economic Analysis* (New York: Oxford University Press, 1954), esp. pp. 62–65. See also Jack Weatherford, *A History of Money* (New York: Crown, 1997).

[5]In addition to previously cited sources, see Clive Day, *A History of Commerce* (London: J. B. Bury, 1926).

[6]See Henri Pirenne's *Economic and Social History of Medieval Europe* (New York: Everyman's Library, 1936), and *Medieval Cities* (Princeton: Princeton University Press, 1939). On the persistence of gold, see Weatherford, *A History of Money,* esp. ch. 3.

ample, the skills of the artisan were generally seen as quite advantageous to the possessor of some process, creating incentives to maintain the position of what would today be called the "sole source." Therefore, when it came to gaining and retaining the benefits of preindustrial skills, individuals, but more often their rulers, saw fit to control the diffusion of information about their processes. For example, in the fifteenth and sixteenth centuries, the great trading city-state of Venice went to extremes to protect its perceived intellectual property (e.g., advanced pottery techniques), driven by what cyberneticist Norbert Wiener depicted as "a national jealousy of secrets, exaggerated to such an extent that the state ordered the private assassination [!] of emigrant artisans, to maintain the monopoly of certain chosen arts and crafts."[7]

Another guarded aspect of commerce revolved around the idea of maintaining control of trade routes—either by force majeure or by simply keeping navigational secrets. The Portuguese empire in the fifteenth century, for example, sought to maintain its control over trade by monopolizing information about ocean navigation. Prince Henry mandated that only incomplete portions of nautical charts could be exported, so that all foreign merchant ships would still have a need for his pilots. Navigation, of course, had commercial and military applications, and the preclusive strategy of the Portuguese was soon overcome by the efforts of navigators in other countries. Thus the short run of Portuguese mastery, and the diffusion of knowledge about navigation, no doubt resulted from the fact that there was little of a proprietary nature in play. Instead, diffusion in this realm was dependent on developing an understanding of the stars, measuring the passage of time, and mathematical skill.[8]

Technological Aspects of Commercial Diffusion

The "dual use" phenomenon is endemic to the technological realm; thus industrial advances may often benefit both commerce and military capability. A good example of this can be seen in early-nineteenth-century advances in ship design and propulsion. The age of steam and steel ships portended both a revolution in seaborne commerce as well as in naval warfare. For Britain, by far the leading naval power of the day,[9] this posed the prospect of losing the overwhelming advantages it possessed in sailing ships of the line. This security-

[7]See Norbert Wiener, *The Human Use of Human Beings: Cybernetics and Society* (New York: Doubleday, 1950), p. 112.

[8]C. R. Boxer, *The Portuguese Seaborne Empire* (New York: Knopf, 1969), pp. 17–29. On the diffusion of skills relating to navigation, see Dava Sobel, *Longitude* (New York: Walker, 1995).

[9]George Modelski and William R. Thompson, *Seapower in Global Politics, 1494–1993* (Seattle: University of Washington Press, 1989), p. 143, contend that Britain possessed about two-thirds of global naval strength after the fall of Napoleon in 1815.

oriented issue provided incentives, therefore, to slow the diffusion of technologies that were reshaping the merchant and naval fleets of the world. And that was
exactly what the Royal Navy strove to do—to stretch out the useful lives of its
sailing vessels, and to manage the pace of change in naval propulsion and archi-
tecture.[10]

The key point in thinking about the commercial diffusion of new technology
is that innovations can occur in several realms: science, academia, the military,
industry, or commerce. Also, innovations often have multiple applications,
though this is seldom clear from the outset. Last, the process of diffusion of
commercial innovations is not always controlled by state-to-state interactions.
That is particularly true of the re-emergence of money and banking at the end of
the medieval period.

Commercial diffusion with military applications may also extend to the spread
of ideas and forms of organization, the subject of the next section.

The Diffusion of Commercial Ideas and Forms of Organization

Beyond defining the types of diffusion issues, though, one must also strive to
hypothesize about how commercial diffusion occurs. In the case of innovations
that are largely ideational, such as the concept of "money," the historical record
suggests that demonstration effects are often sufficient to encourage its widespread adoption, both within and between states. Commercial organizational
forms also appear to diffuse via imitative behavior.

A very good example of imitative diffusion of ideas and forms of organization
can be seen in the spread of mercantilism, beginning with the emergence of the
modern nation-state in the early 1500s. Mercantilism sought to harness the accumulation of wealth and gains from trade to the maximization of state power.
Indeed, mercantilism was one of the principal means by which vestigial feudal
actors were tamed by young central governments. It also became a central mode
of international competition, with nations striving to control routes of trade and
resources—often through trading firms that had monopoly status within the
spheres of influence of their nation-state masters.[11]

This first modern example of commercial diffusion highlights the roles of both
ideas (mercantilism) and forms of organization (the trading company)—each of

[10]Paul Kennedy, *The Rise and Fall of British Naval Mastery* (London: Ashfield, 1976), examines this
strategy. See also Bernard Brodie, *Sea Power in the Machine Age* (Princeton: Princeton University Press,
1943), pp. 138–39.

[11]For comprehensive studies of mercantilism in historical perspective, see Eli Heckscher, *Mercantilism*, trans. M. Shapiro, 2 vols. (London: Longman's, 1935); and Gustav Schmoller (trans. W. Ashley),
The Mercantile System and Its Historical Significance (New York: Oxford University Press, 1896). On the
trading company phenomenon, see John Keay, *The Honourable Company* (New York: Macmillan, 1991).

which were rapidly imitated by the great powers with overseas holdings (i.e., Spain, Britain, France), as well as by the trading states with lesser military power (principally Portugal and Holland). Mercantilism itself, however, was transformed as it spread. First practiced by Spain in the form of a "metals policy," the aim of which was to extract surplus from conquered lands in the New World, it soon shifted to an emphasis upon the accumulation of specie through gains from trade—a practice led by the British. Any country with a merchant fleet and control of sea routes or overseas holdings was thus able to pursue this British form of mercantilism.[12] Indeed the "new" approach to mercantilism proved to be far superior to the old metals-policy version.

The state capabilities required to implement mercantilist practices raised "entry costs" for those who later pursued this innovation. In the absence of a large commercial fleet, the navy to protect it, and the overseas holdings upon which to draw, a state could not engage effectively in mercantilist practices.[13] Further, the essentially conflictual view of international relations posited by the mercantilists was borne out by vicious, extended conflicts aimed at weakening rivals—the best example being the century-long struggle between Britain and France (and each of their East India Companies) for control of South Asia. Thus this form of commercial diffusion was limited both by high barriers to entry and by the risks inherent in international politico-military competition. Nevertheless, to the extent possible, many states strove to achieve the mercantilist ideal of "power and plenty."[14]

Aside from these constraints on the diffusion of mercantilism—which were of a principally "structural" nature (i.e., they related to state capabilities and resources)—another great check was imposed by the rise of a competing idea: lais-

[12]This is, essentially, maritime mercantilism as classically formulated. It is the version of mercantilism considered in this chapter. However, one should also note that mercantilist practices evolved further in the ensuing centuries so that even states without overseas holdings would be able to tie economic and governmental means and ends quite closely (e.g., see the nineteenth-century German "coalition of iron and rye").

[13]Thus the Dutch effort to compete as a mercantilist great power was thwarted by the British in three wars, from the 1650s to the 1670s, whose aims were largely commercial. On the "structural" inferiority of the Dutch, and the basic necessities of power to undergird mercantilism, see A. T. Mahan, *The Influence of Sea Power upon History, 1660–1783* (Boston: Little, Brown, 1890), pp. 165–72. See also G. J. Marcus, *A Naval History of England: The Formative Centuries* (Boston: Little, Brown, 1961), esp. pp. 123–50, and Jacques Barzun, *Introduction to Naval History* (New York: Lippincott, 1944), who noted of these mercantilist wars simply that "England took over Dutch trade and shipping, consolidating a large empire based on sea power" (p. 47).

[14]On the notion that mercantilism provided the dual benefit of advancing state power and enriching societal elites, see Jacob Viner, "Power versus Plenty as Objectives of Foreign Policy in the Seventeenth and Eighteenth Centuries," *World Politics* 1, no. 1 (Oct. 1948), reprinted in Jacob Viner, *The Long View and the Short: Studies in Economic Theory and Practice* (Glencoe, IL: Free Press, 1958).

sez-faire notions of free trade and of an international "harmony of interests."[15] Led by the essays of Hume and Smith, and by the desire to avoid a protracted economic war such as characterized the last decade of the Napoleonic era, the adherents of market liberalism grew in number and influence—for the most part in Britain. Thus, soon after the fall of Napoleon, when the Royal Navy was absolutely unchallengeable at sea, the British empire nonetheless shifted to advocacy of a pure free trade strategy. This process began in 1820 with the opening of parliamentary and public debates on the subject. By the 1840s it became the dominant British view on trade—and also swept along many other European states as well—at least for some decades.[16]

In the decades after the repeal of the Corn Laws, the diffusion of free trade areas was quite broad, and it appeared as though state-driven notions of mercantilism were in full retreat before the power and attractiveness of the idea of economic liberalism. Even on its "home turf" of maximizing state power, mercantilism suffered—as the free traders demonstrated that laissez faire practices made the imperial concept more humane and practicable. However, mercantile thought and practice were soon revived, its competition with the idea of free trade reshaped by the rise of industrialization—the second, more technological, dimension of commercial diffusion.

Explaining the Diffusion of Commercial Technologies

At the same time that British-inspired liberal trade practices were spreading, a British-led process of industrialization was underway. Powered by coal-driven engines, machines were now making the mass production of goods possible. The

[15]The intellectual foundations of liberal critiques of mercantilism can be seen in David Hume's essay "Jealousy of Trade" (1755), in which he avowed "not only as a man, but as a British subject, I pray for the flourishing commerce of Germany, Spain, Italy, and even France itself." Adam Smith's *An Inquiry into the Nature and Causes of the Wealth of Nations*, which appeared in 1776, is also generally viewed as expositing the benefits of free trade versus more guarded mercantilist practices—though Edward Mead Earle's "The Economic Foundations of Military Power," in Peter Paret, ed., *Makers of Modern Strategy* (Princeton: Princeton University Press, 1986), offers the view that Smith was not entirely hostile to mercantilist practices. Eventually, Hume's and Smith's ideas grew into what became known in the nineteenth century as the "Manchester Creed" for its strong belief that a harmony of interests is the natural state of affairs, rather than mercantilism's assumption of constant competition and conflict.

[16]Charles P. Kindleberger, "The Rise of Free Trade in Western Europe," *Journal of Economic History* 35, no. 1 (spring 1975)—reprinted in J. A. Frieden and D. A. Lake, eds., *International Political Economy* (New York: St. Martin's, 1987)—argues strongly that one can only understand the British shift, which resulted in repeal of the protectionist Corn Laws, as a response to the power of an attractive idea—and that the spread of free trade must be seen in these terms also. On this point, see the argument of Karl Polanyi, in *The Great Transformation* (New York: Farrar and Rinehart, 1944), who saw the key role played by ideas, but also noted the fifty-year-long intellectual struggle that ensued prior to Smith's free trade views rising to pre-eminence. As Polanyi notes: "Not until the 1830s did economic liberalism burst forth as a crusading passion" (p. 152).

same power was also being applied to revolutionize the movement of goods, as seen in the rise of railroads and steamships. In both of these areas Britain was a clear leader,[17] and it had many imitators. As the second half of the nineteenth century unfolded, all of the other great powers,[18] and a good number of other nations—including the United States and Japan, which had not as yet risen to great power status—were rapidly industrializing.

The role of Britain suggests that individual actors (i.e., nation-states in this case) do matter; as Britain's adherence to openness in the trade realm generally supported the diffusion of very advanced industrial technologies to others. Indeed, Gerschenkron has argued that British willingness to export sophisticated machinery was a boon to "late modernizers"—enabling them to leapfrog over the early steps in industrialization.[19] However, the British were guarded in some areas, especially those having to do with steam propulsion and ship design. Also, the availability of the latest technologies did not, by itself, allow industrialization, as labor, capital, and markets had to be in place in order to catalyze the process.

Other theories that explain the diffusion of industrialization fall into three categories. First, there are those that keep the state at the center of the process. Gerschenkron is the best exemplar of this perspective, as he strives to demonstrate, in case after case, the crucial role of state intervention in the process of industrialization. From the formation of state-run or -sponsored "industrial banks," to the return of mercantilist/protectionist policies aimed at nurturing infant industries, Gerschenkron chronicles the state's importance to the process of industrial modernization.[20] His studies also reflect a difference between the initial process of industrialization in Britain—which was led by the private sector—and the pattern of its diffusion to other states, whose governments sometimes directly funded (e.g., see the Berlin-Koenigsberg and Trans-Siberian rail links) or enabled industrial advances by means of a variety of protective policies (e.g., tariffs and other barriers to competition).

The second explanation of the diffusion of industry moves from a focus on the

---

[17]However, Britain's limited ability to manage the pace of change in commercial ocean transport could not be duplicated when it came to railroading, where nearly all states sponsored the rapid development of rail transport, financing the process through state-run industrial banks—and only Portugal, Spain, and Russia, out of security concerns, did so with gauges that differed in size from the rest of European trackage.

[18]Which, at this point, included Britain, France, Germany, Italy, and Russia, according to the general view of which nations were great, and when. See Jack Levy, *War in the Modern Great Power System* (Lexington: University of Kentucky Press, 1981).

[19]Gerschenkron, *Economic Backwardness in Historical Perspective.*

[20]Indeed, in one telling case, in which states succeed despite commercial opposition from within, we see the Rothschild banking empire acting to destroy a state-sponsored industrial bank in Austria—but then the Rothschilds themselves are pressed by the state to undertake the same tasks as those envisioned for the defunct bank that the Rothschilds had driven under.

state to consider social and cultural factors. According to this point of view, the characteristics of a particular society matter most. The best-known study of this "sociocultural" factor is Max Weber's *The Protestant Ethic and the Spirit of Capitalism.* Berger adds to this perspective his argument that Taoist and Confucian cultures also offer fertile ground for the diffusion of commercial innovations.[21] It may also be that Rogers falls into this category (see his Appendix A, "Generalizations about the Diffusion of Innovations").[22] And even Gerschenkron implicitly accepts the notion that characteristics of particular states and their leading actors do matter, as he notes, in each of his country studies, the importance of governmental attitudes toward entrepreneurs.[23]

The third explanation of industrial diffusion holds that its processes and patterns are very strongly influenced by interstate interactions. At this level of analysis a darker view of diffusion patterns may be developed—or at least a perspective that considers very seriously the possibility that leading states will work or tend to act in ways that prevent the modernization of others, to the extent that this serves their strategic interests. Certainly, many classical socialist views of capitalist exploitation of the underdeveloped fall into this category.[24] The tenets of dependency theory also support the view that states may attempt to control diffusion of commercial technologies.[25] Finally, Wallerstein points to a de facto consortium of leading states that act together to control and extract surplus from the peripheral, lesser developed areas of the world.[26] The view that predominates here is of a generally exploitative system.

What do the patterns of industrialization over the past two centuries have to say about commercial diffusion processes? First, there is a clear sense of breadth regarding the spread of industrial machines. They emerge virtually everywhere. Yet when one looks at the issue of the "depth" of diffusion, sharp variances emerge. As with mercantilism, industrialism has witnessed a concentration of capability in the hands of relatively few leading states. In 1900, for example, those states that constitute what is today called the G7 accounted for three-fourths of world manufacturing output. That figure is virtually the same today for the G7.

[21]See Peter Berger, *The Capitalist Revolution* (New York: Basic, 1987).

[22]Everett Rogers, *Communication of Innovations* (New York: Free Press, 1971).

[23]Gerschenkron, *Economic Backwardness in Historical Perspective,* esp. pp. 52–71.

[24]See esp. V. I. Lenin, *Imperialism: The Highest Stage of Capitalism* (New York: International, 1939), pp. 76–92. Another important study supporting this point of view is J. A. Hobson, *Imperialism: A Study* (London: Allen and Unwin, 1902).

[25]The clearest exposition of this view is in Fernando Henrique Cardoso and Enzo Faletto, *Dependency and Development in Latin America,* trans. Marjory Mattingly Urquidi (Berkeley: University of California Press, 1979).

[26]Immanuel Wallerstein, *The Modern World System: Capitalist Agriculture and the Origins of the European World-Economy in the Sixteenth Century* (New York: Academic, 1974). Wallerstein's "world system" theory, it might be argued, is a blend of notions of imperialism and dependent development.

In 1830, at the outset of the industrial era, the same countries accounted for only one-fourth of world output.[27] This pattern seems, at a minimum, to be supportive of Wallerstein's ideas about core-periphery patterns of development.

As far as the role of government is concerned, the rise and diffusion of industrial processes seems to have been very dependent upon private actors, much as the late medieval return of money and banking had been. That said, there were some important initiatives fostered by government, some of which directed investment—but the more important ones revolved around Gerschenkron's notion of the government's necessary role in providing a "favorable climate" for entrepreneurship. Both Britain and the United States seemed able to do this with government's hand light, though not invisible. In Japan and Germany, on the other hand, the role of government was more pronounced. Interestingly, at the level of lesser developed countries, there is much more direct governmental involvement, and a much less favorable environment for entrepreneurship. Thus, it seems that the issue of commercial diffusion is somewhat conditioned by the ability of countries to form effective public-private partnerships.

The next section considers the issue of commercial diffusion in the information age. The link between commercial technologies and security concerns will be explored—with special attention given to: the "public good" of communications systems of the most advanced sort; encryption systems that ensure the privacy of commercial (and other) transactions; and the high-performance computers that are, in effect, information-age analogs to the steam engines and smelting machines of the nineteenth century that drove industrialization forward. Currently, the United States is an undisputed leader in each of these three key aspects of the process of "informatization." Thus it should prove both analytically useful and policy-relevant to examine U.S. policies in each area.

Commercial Diffusion and "Informatization"

If the steam engine powered the Industrial Revolution, then most certainly it can be seen that the computer drives the Information Revolution. And by "computer" one must explicitly include all types of computational engines, from the tiny, onboard devices that now control fuel injection on most vehicles, to the supercomputers that help to manage and govern the most advanced, complex industrial processes. The pervasive nature of "informatization" can be seen to extend also to modern military weaponry—helping propel what has been called a revolution in military affairs to match the one going on in business.

[27]Data is drawn from Paul Bairoch, "International Industrialization Levels from 1750 to 1980," *Journal of European Economic History* 11 (spring 1982), pp. 294–96; and from Paul Kennedy, *The Rise and Fall of the Great Powers* (New York: Random House, 1987).

In the realm of infrastructure building, the United States has taken a stance in recent years that is quite similar to Britain's in the nineteenth century. Britain willingly provided and maintained freedom of the seas as a "public good," to help foster the growth of international trade. Today the United States is helping to interconnect the world to ensure free flow of information. The U.S.-created global positioning system (GPS), for example, is having profound effects on commerce, industry, and even agriculture. The FLAG Project (which features much private investment and support) is linking the world through fiber optics.[28] And this sort of activity has been undertaken despite the fact that some advanced information systems, such as GPS, can be used by adversaries as a tool for guiding their own weapons against U.S. forces. Further, the increasing dependence of a wide variety of commercial enterprises upon GPS will make it ever more difficult to arrange for even a temporary wartime shutdown of the system.[29]

In a related area, ensuring the privacy of commercial and other communications across the many links under construction around the world, the United States is proving a bit less forthcoming. Privacy and "information security" have become key concerns as the current debate over encryption technology demonstrates.[30] Policy-makers have been reluctant to allow the diffusion of the most sophisticated forms of encryption. Law enforcement officials, for example, fear losing their ability to intercept or monitor the communications of criminal organizations. The military would find its technical intelligence-gathering capabilities compromised if other state and nonstate actors had access to virtually unbreakable code-making tools.[31] However, commercial pressures are inexorable, and, at this writing (summer 2001), the U.S. government has significantly relaxed its export controls on the sale of strong encryption tools.[32] In addition to the

[28]See Neal Stephenson, "Wiring the Planet," Wired 4 no. 12 (Dec. 1996): 98–160. "FLAG" stands for "fiber-optic links around the globe." How different this is to nationalistic attitudes toward linking the world telegraphically in the nineteenth century. On this point, see Daniel R. Headrick, The Invisible Weapon: Telecommunications and International Politics, 1851–1945 (New York: Oxford University Press, 1991), esp. chs. 3–5.

[29]On the implications of the GPS system—including its defensive capabilities against illicit uses—see Dorothy Denning and Peter MacDoran, "Grounding Cyberspace in the Physical World," in A. Campen, D. Dearth, and R. T. Goodden, eds., Cyberwar: Security Strategy & Conflict in the Information Age (Fairfax, VA: AFCEA International, 1996).

[30]Codes can be broken only when their "keys" (i.e., substitution patterns) have been solved; while high-performance (and many lesser-powered) computers can quickly unravel keys of shorter length (say, "64-bit" key lengths), it is nearly impossible to crack codes with far longer keys. It is encryption of this latter sort whose diffusion the U.S. government seeks to control. For more on cryptography, see David Kahn, The Codebreakers (New York: Macmillan, 1967).

[31]Such unbreakable communications could also foster increased growth in the subterranean economy. Indeed, it is entirely possible that a kind of "virtual" currency could arise and be protected from detection by strong encryption. This could pose a serious threat to the taxation authority of the government.

[32]There is also a growing acceptance of the fact that the capability to generate strong encryption will

commercial pressure to allow diffusion of strong encryption, there is also a substantial legal challenge to government policy, based on the notion that encryption is a form of protected free speech.[33]

The spread of informatization is also evident in the large-scale adoption of networked forms of commercial organization. Just as, in the heyday of mercantilism, leading states each formed trading firms of great size and strength, so today the commercial sectors of leading states are seeing a broad shift away from classic "line organizations" toward networks. It is generally accepted that networks, in which power and decision-making authority are decentralized, are the sine qua non of advanced corporate organization.[34] As in the case of the mercantile era, there is little impediment to imitation, as an idea can be replicated more easily than an innovation based on the acquisition of an advanced technology. Thus leading corporations around the world are decentralizing authority to "business nodes," and loose-knit "strategic partnerships" are starting to supplant the proclivity to pursue winner-take-all acquisition and control policies.

U.S. "commercial diffusion policy" (if one can refer to it as such) encourages openness in fostering the emergence of a global information infrastructure. In the area of encryption, despite some governmental tendencies to remain guarded, the trend is clearly toward easing efforts to maintain proprietary control over advanced cryptographic technologies. No effort is being made—nor could there be a successful effort—to preclude or control the spread of networked organizations. In each of these areas, there is a tendency to allow commercial diffusion, and thus the spread of informatization. Even in a leading sector such as high-performance computing technology—which features both commercial and security-oriented factors—there is a considerable willingness to allow diffusion. Thus the result of "pulling and hauling" between governmental departments (State versus Defense versus Justice—arbitrated by the executive) and between private actors (led by electronics and telecommunications industries, but also by NGOs

---

grow, regardless of the U.S. stance toward its diffusion. The policy decision has been made, however, to avoid hastening the process of diffusion by decontrolling encryption completely. On this point, see the President's Commission on Critical Infrastructure Protection, a digest of whose findings is available at www.pccip.gov.

[33]See *Bernstein v. United States Department of Justice,* Case Number 97–16686; *Karn v. Department of State,* Civil Action Number 95–1812-LFO; and *Junger v. Department of State and the National Security Agency,* Case Number 96CV1723. In May 1999, Bernstein won a ruling in the Ninth U.S. Circuit Court of Appeals.

[34]On the strength of and attractiveness of networks to business organizations, see Arno Penzias, *Information and Ideas: Managing in a High-Tech World* (New York: Free Press, 1990); Ikujiro Nonaka and Hirotaka Takeuchi, *The Knowledge-Creating Company: How Japanese Companies Create the Dynamics of Innovation* (New York: Oxford University Press, 1995); and Fritjof Capra, *The Web of Life: A New Scientific Understanding of Living Systems* (New York: Doubleday Anchor, 1996), which lauds the network as an optimal form of organization for commerce—as well as most other forms of activity.

such as the Electronic Frontier Foundation) and government, have seen prodiffu-
sion forces consistently triumph. Perhaps this is happening because of the powerful pull of potential commercial gain, given that most grave threats—with the
exception of terrorism—have waned since the Cold War's end.

The case of high-performance computers, though replete with the aforemen-
tioned tendencies toward openness, nonetheless also features an effort to provide
some meaningful controls on diffusion. This makes it a case worth examining in
more detail, and it thus forms the focus of the next section's analysis.

Supercomputers: A Case of Selective
Commercial Diffusion Control

In an era in which information—and information technology—have emerged
as increasingly important dimensions of national power, U.S. export control
policy regulating trade in supercomputers turned sharply away from traditionally
(i.e., Cold War–era) restrictive strategies in 1995.[35] Instead, policy-makers favored
an approach intended to foster the diffusion of this technology—which is a key to
commercial modernization across almost every major industry. The shift toward
openness has had salutary effects on the supercomputer industry, helping U.S.
companies to maintain and extend their foreign market dominance.[36] However,
in the security sphere, concerns have arisen regarding the possibility that other
nations, or even nonstate actors, might use the vast power of these computational
engines to enable them to develop and secretly test a wide range of weapons.
While the use of supercomputers in support of nuclear proliferation activities
remains a paramount concern, the ability to design and produce advanced conventional weapons, and to engage in some forms of "information warfare,"[37] also
flows from the acquisition of high-performance computing (HPC) capabilities.
Thus, in this issue area, proliferation concerns continue to grow in their diversity
and complexity.

The current "open" policy received presidential approval after only a very
limited public debate. A built-in requirement for annual review of the national
security consequences of the policy, though, does allow considerable room for

[35]In 1993, supercomputers were defined as that class of computational engines capable of 1.5 billion
theoretical operations per second (TOPS).

[36]A government estimate of the immediate effects of the policy shift toward openness suggests that
computer exports grew by nearly an additional 10 percent annually, in the years immediately following
the policy shift.

[37]Generally speaking, "information warfare" (IW) is most associated with the use of cyberspace-
based weaponry (computer viruses, logic bombs, etc.) to disrupt or destroy military or civilian communications and infrastructure and automated system controls. For a more detailed discussion, see Martin
Libicki, *What Is Information Warfare?* (Washington, DC: National Defense University Press, 1995).

continuing reappraisal. Also, there is a 1984 agreement between the United States and Japan to coordinate their policies on the sale of technical systems that have military applications. This means that Japanese assent is necessary if the more open HPC export policy is to have any useful effect. Since Japan is the only other significant producer of HPC, Japanese acquiescence can, at least in the short run, pose the prospect of successful diffusion control.

The range of possible diffusion control strategies constitutes a spectrum of choices, ranging from a preclusive, closed stance at one end, to complete openness at the other. Current U.S. policy, which has elements of both, falls in the middle of the continuum. On the more proprietary side, the policy prohibits sales of any HPC to countries that pose national security concerns to the United States.[38] These include such "usual suspects" as Iran, Iraq, Libya, and North Korea. Cuba, already excluded from such purchases by virtue of the overall U.S. economic embargo, fills out the membership roster of this "pariahs club." At the other end of the spectrum, the countries of Western Europe, Japan, Canada, Mexico, Argentina, Australia, and New Zealand ("Tier I") have no restrictions whatsoever—in terms of governmental review—on their purchases.

Between these extremes lie two middle layers. As of the fall of 2000, machines capable of forty-five billion theoretical operations per second (TOPS) could be sold to South American countries, most of the former Soviet satellites, South Korea, and South Africa ("Tier II")—if certain record-keeping requirements were met. Individual licenses had to be issued to support their purchases above this level. The last class of purchasers ("Tier III") includes old Cold War adversaries Russia, China, Vietnam, and Syria. States of the Middle East, the Maghreb, and South Asia also fall into this more preclusive policy. All may purchase already "widely available" computational engines without any regulation up to two billion TOPS. Purchases between two and twenty-eight billion TOPS require licenses, with record-keeping by the exporter (usually the U.S. manufacturer). In cases of machines with speeds over twenty-eight billion TOPS, monitoring is considered on a case-by-case basis, at the discretion of the U.S. government. The policy states only that "additional safeguards *may* be required at the end-user location."[39]

Sometimes, precautions against diversion to military uses extend to having U.S. technicians in the employ of the manufacturer operate the supercomputer in

---

[38] As determined by the State and Defense Departments, primarily.

[39] Emphasis added. For further details about the classes of purchasers and the various levels of monitoring, see the Presidential Press Release entitled "Export Controls on Computers," 6 Oct. 1995. The rules are often quite vague regarding exactly who is responsible for monitoring, though the clear intent of the export control regime is to place the onus of compliance on the exporter. For a full discussion of the HPC monitoring regime, see the U.S. Dept. of Commerce, *Foreign Policy Report* (Washington, DC: Government Printing Office, 2001), esp. ch. 9, "High Performance Computers."

a segregated facility. This site must be sufficiently secure so as to prevent access or control by the purchaser, who enjoys only the outputs of the computational engine. However, the HPC export control policy articulates no mandatory requirement to take such precautionary measures, even concerning sales of the most advanced machines—a seemingly glaring gap in the regulations, given the risk that these HPC or their computational outputs might be passed on or sold to the rogue states that are not intended to have any access to supercomputers.

This policy resembles two earlier, well-known examples of the hybrid approach to coping with diffusion: efforts to control the spread of nuclear and missile technologies. The spread of nuclear power as an energy source has been allowed, for example, while the creation of nuclear weapons of mass destruction has been strongly discouraged. In the second instance, the principally military uses of missiles have led to a much more proprietary Missile Technology Control Regime (MTCR), although "friendly" states clearly have substantial access to this technology. The HPC export control strategy flows from these two examples, drawing from both the intent to allow peaceful, commercial uses while at the same time discouraging more military applications, and fostering the creation of admittedly arbitrary classes of "tame" and "rogue" states to help in identifying the most appropriate customer base.

Alternatives to the current policy would logically consist of shifting the emphasis of export controls toward either more guarded or more open regimes. Commercial concerns revolve around notions of short-term market share gains, and longer-term concerns about industry viability and competitiveness. If these concerns form the paramount interests in this case, then it would seem appropriate to pursue a strategy of greater openness toward export of HPC. On the guarded side, concerns about the diffusion of HPC extend to potential shifts in relative state power, the effects on weapons (nuclear and advanced conventional) proliferation and war-marking capabilities, and the prospects for arms racing. Also, it is important to factor into the policy analysis some sense of whether manufacturers beyond Japan and the United States can emerge to challenge their duopoly.

Depending on the state of the technology, a very preclusive strategy maximizes security while paying little heed to commercial effects. Conversely, an open strategy fosters commercial benefits but may have deleterious consequences in the security sphere. A hybrid approach, however, tries to serve security and business interests. But in an industry characterized by explosive growth, advances in HPC technology are virtually uncontrollable in the long run. In the nearer term, though, there are clear "thresholds" that competitors outside the United States and Japan cannot surmount on their own. This suggests the feasibility of a strategy of allowing unrestricted sales of supercomputers whose strength does not cross that (admittedly shifting) threshold.

The current diffusion control strategy fosters a very robust commercial environment for the U.S. HPC industry, as evidenced by its dominant share of the global market. Unlike a pure "open" strategy, current U.S. policy provides some hedge against the erosion of power advantages over potential rivals. Or does it? The answer to this question depends both upon the setting of the control threshold and policies regulating sales to specific countries. A government-sponsored analysis of the HPC issue—one that formed the basis of current HPC export control policy—developed a comprehensive methodology for identifying the appropriate thresholds on a class-by-class basis. However, the report avoids discussion of the risks entailed in selling supercomputers to potential politico-military rivals.[40] In this regard, the current HPC export control policy derives principally from notions of technological determinism,[41] with little attention given to possible shifts in relative military power. Herein lie some problems.

Concerns about the risks of a too-open policy should revolve around its effects on non- and counterproliferation policy, both in the nuclear and advanced conventional sphere. With this in mind, one can see that current HPC policy has less to do with new nuclear proliferators (the first nuclear weapons were made with the aid of only very minimal computing),[42] but may allow those already in possession of nuclear weapons to increase their production and refine capabilities quite significantly—and to do so without detection. As Gary Milhollin, of the Wisconsin Project on Arms Control, once observed: "Russia and China will be able to develop their nuclear arsenals [further] without having to conduct underground tests and we will not be able to monitor them."[43] This possibility could foster both qualitative and quantitative arms racing.

As nettlesome as the prospect of increasing opacity in nuclear development

[40]Seymour Goodman, Peter Wolcott, and Grey Burkhart, *Building on Basics: An Examination of High-Performance Computing Export Control Policy in the 1990s* (Stanford: Center for International Security and Arms Control, 1995), p. 14, notes, on this point: "This consideration is beyond the scope of this study." The assistant secretary of commerce for export administration and the assistant secretary of defense for counterproliferation policy sponsored the study.

[41]As Bill Archey, president of the American Electronics Association, has put the matter: "[It] may be difficult to restrict the export of a high performance computer, [but] it is virtually impossible to restrict the export of high performance computing." Cited in Gary H. Anthes, "Restrictions Lifted on Export of High Performance Computers," *Computerworld* (16 Oct. 1995): 32. William Reinsch, commerce undersecretary for export administration, has argued along the same lines, about the problematic nature of control efforts: "This is not a happy reality, but it is the technological reality." From Pat Cooper and Theresa Hitchens, "New U.S. Computer Export Rules Spark Optimism, Arms Control Fear," *Defense News* (22 Oct. 1995), p. 26.

[42]Brian Deagon, "Why Tech Export Curbs May Be a Futile Exercise," *Investor's Business Daily*, 23 Oct. 1995, p. A8, notes succinctly, citing a 1990 Los Alamos National Laboratory study, that "the U.S. nuclear arsenal in the 1960s was designed using computing power equal to today's hand calculator."

[43]Cited in Jonathan S. Landay, "Easier High-Tech Controls Reopen Proliferation Debate," *Christian Science Monitor*, 16 Oct. 1995.

seems, though, this problem could be eased if the U.S. government emphasized other nonproliferation policies, ranging from the control of very advanced communications technology to enacting serious penalties for detected violations of existing nonproliferation and arms control agreements. But the far greater long-term risk posed by the current diffusion control strategy may be that potential adversaries will now have the ability to bring their militaries into the information age. High-performance computing capabilities will allow them to develop the communications and battle management systems needed to fight in a fast-paced future battlespace devoid of definable "fronts." Currently, sensors and image analysis capabilities needed to wage this kind of war fall in a range of tens of billions of TOPS. However, even in this area one may note important nuances. For example, advances in the design of composite or explosive reactive armor rely upon HPC capabilities well above thirty billion TOPS. Similarly, requirements for the advanced antisubmarine sensing devices begin at twenty billion and extend to forty billion TOPS.[44] Many military advances lie even further out, at the highest reaches of HPC capability—currently available only to the United States and its allies.

Today, the capability to manufacture HPC, beyond the United States and Japan, remains in very few hands, and the power of computational engines made by those outside the duopoly would not likely exceed twenty billion TOPS—with India being the most likely candidate to achieve such a success. However, very soon major advances are likely to undermine the duopoly's control, putting even the hundred-billion TOPS level in sight by the year 2010.[45] This implies a strong need to monitor the "controllability threshold" carefully, and to determine whether the current level of controls might provide potential rivals with opportunities to enhance their military potential significantly. Further, even if a particular TOPS level ends up being "uncontrollable," the issue of whether to sell to or forgo sales and compel the other party to undertake independent HPC development requires close analysis. The basic question that underlies this line of reasoning is: "Why make an information proliferator's path to the production of very advanced conventional weaponry easier?"

Britain asked this sort of question during its mid-nineteenth-century naval rivalry with France, when ships converted from sail to steam as their mode of propulsion. A Royal Navy–Parliamentary panel concluded that, by controlling the export of steam engines, French development costs would rise by more than 35 percent. France, however, viewed the creation of its own steam engine industry as a prerequisite of great power, and it took up the cudgels to compete with the

---

[44]For some details on the HPC requirements for these and many other military systems, see Goodman et al., *Building on Basics*, pp. 44–58. Other key military applications include cryptography and ballistic missile defensive systems.

[45]Ibid., p. 27, projects the likely progress of HPC in the coming years.

British anyway.[46] But the French could compete only for a while, as the cost of the naval arms race became too heavy to sustain, and the French navy decided to shift away from direct competition in capital ships, replacing it with an emphasis on light, swift vessels, mines, and torpedoes instead—a policy espoused by the so-called *jeune école.*

The foregoing discussion raises the possibility that tighter controls may help to maintain relative advantages over potential competitors. However, preclusive approaches encourage, sometimes require, others to develop indigenous designs and production capabilities. Two potential "peer competitors" of the United States, Russia and China, both have computer industries barely capable of producing at the ten billion TOPS level—a minimum level of high-performance computing. Current HPC export control guidelines, though, will allow them to advance rapidly and quite cost effectively, even though both countries fall into the category of states subject to severe export controls. On the other hand, India, also subject to a restrictive HPC export control policy, has developed capabilities on its own of around twenty billion TOPS, confirming the point that exclusion may breed internal efforts at information proliferation.[47]

In summary, this discussion describes two tensions. First, efforts to safeguard national power advantages over others, which are associated with preclusiveness, have commercial costs, at least in terms of lost market share. Second, a policy of openness features immediate economic benefits, but tends to erode advantages in relative power over actual or potential rivals. Reconciling these deep tensions, to the extent possible, and selecting an appropriate export control strategy require knowledge of two types. First, a precise sense of the pace of advances in computing power must inform any preclusive efforts. The United States does not hold a monopoly over HPC technology, so a completely closed strategy is futile. However, if the rate of HPC advance follows predictable patterns—as it has thus far—a "moving preclusive limit" on HPC exports could emerge.

The other factor that should guide strategic decision-making about commercial diffusion of HPC is the economic impact of any policy. In 2001, the HPC export market had annual sales of roughly $70 billion, of which the United States held an approximately 80 percent share. Presumably, a fully preclusive strategy would exert strong downward pressure on U.S. market share, while openness would contribute to keeping a dominant position. An open attitude allows U.S.

---

[46]See Brodie, *Sea Power in the Machine Age,* p. 39. For a good survey of the many vigorous efforts to achieve commercial diffusion control from the start of the industrial era to the outbreak of World War I, see David J. Jeremy, ed., *International Technology Transfer: Europe, Japan and the USA, 1700–1914* (Hants, UK: Edward Elgar, 1991); and Clive Trebilcock, *The Industrialization of the Continental Powers, 1780–1914* (London: Longman, 1981).

[47]For some early details on Russia, China, and India, see Goodman et al., *Building on Basics,* pp. 17–21.

HPC vendors to sell at will, while preclusiveness can only encourage others to develop their own HPC, sooner or later. The preclusive approach hinders general growth of the size of the market, while openness could enlarge it considerably. Yet economic considerations and forecasts of possible outcomes regarding market size and share always have an element of uncertainty. Therefore the negative effects of preclusiveness might actually prove modest, as could the positive results from openness. Studies of these potential effects, hitherto not undertaken, should be designed and pursued.

Given the hard-to-control nature of technical progress in HPC, and the uncertainties about economic effects, a hybrid strategy that is weighted a bit more than current policy toward guardedness does seem better suited to the needs of the situation. Pure preclusiveness is impossible. But a somewhat more guarded approach, calibrated to an estimate (the best that can be done) of the rate of "uncontrollable" advances in HPC, might work. By measuring results from year to year, resetting limits based on market data, and assessing the ever-changing threshold of technological controllability, the hybrid approach, which combines preclusive and open elements, offers an optimal strategy. The current HPC export control strategy is clearly a hybrid; but security issues remain, suggesting that the degree of guardedness currently employed is insufficient. For in areas where the information revolution is having its greatest impact on security affairs, from refining arsenals of weapons of mass destruction, to operational warfighting, to command and control systems, HPC have become an essential element of national power.[48]

Finally, a strategic analysis of the effort to control the diffusion of advanced informational capabilities must grapple with the serious problem of HPC falling into the hands of pariahs, rogue states, and terrorists—through illicit resales or other forms of "downstreaming." Coping with what is euphemistically called the "resale dilemma" will require very close postcontract relations, accompanied by a resolve to impose stiff sanctions or embargoes on further sales as punitive measures in the face of violations. Indeed, after-sale HPC monitoring of this sort may become a major intelligence mission. A substantial body of practice of this sort, both in the nuclear realm and in foreign military sales of more advanced types of arms, should provide good guidance for the development of such a "policing" strategy.

[48]Indeed, the Select Committee on U.S. National Security and Military-Commercial Concerns (aka the Cox Committee) reported on Chinese efforts to gain access to high-performance computers as an element of their nuclear modernization strategy, and made a specific point of calling for much greater guardedness in the export control regime covering HPC. While some of the committee report remains classified, the part of it recommending tighter HPC export controls was made public on 25 May 1999. See "Public Remarks of Congressman Christopher Cox," *C-SPAN*, 25 May 1999.

Conclusion

Thinking about the concept of commercial diffusion provides new insights into the process by which innovations are disseminated—and replicated, or not. As in the cases of military- or security-related phenomena, the realm of commercial diffusion offers innovations that range from the largely ideational (e.g., money, mercantilism, free trade, networking) to those that are principally technological (e.g., steam engines, railroads, high-performance computers). Almost all commercial innovations have some relationship with security concerns. Despite this, in many cases, innovators have either done little to prevent the imitation of their innovations, or have actively encouraged diffusion (e.g., Britain's missionary zeal for free trade in the nineteenth century, and the U.S. passion for "wiring the world" today). Occasionally, though, efforts have been made to control the pace of industrial development, as the British did in the nineteenth century with steam propulsion technology, and as the United States is now attempting to do, however fitfully, with high-performance computers.

How is one to sort through all these different dimensions of the realm of commercial diffusion? Is there any strategic approach that can be elucidated for dealing with possible threats to prosperity—and in some cases, security? Based on the findings of this study, the best answers to the questions are equivocal. For example, in the area of the diffusion of ideas that have commercial value, as noted above, there is very little hope that active control measures will stop their spread. Instead, "entry costs" will govern the path of diffusion. For example, the use of money (metallic and otherwise) spread widely and very quickly throughout the ancient world; its creation was not an inherently costly process. On the other hand, the idea of linking economic policy to the growth of national power, in the form of mercantilism, was also widely attractive but required a basis of national military power before this economic strategy could be used. This led to a very controlled diffusion of mercantilist practices, with the great powers in charge.

However, in the case of the most advanced computers, it is not clear that the entry cost factor will ultimately impose a limit on their diffusion. Currently, supercomputers are about the size of a large office copying machine—albeit much more costly. It is hard to imagine contemporary limits on capabilities and resources, like those that precluded so many states from industrializing over the past century and a half. In the absence of a clear strategy for diffusion control, advanced information technologies will spread broadly to many nations, having ever-more-serious commercial and military consequences.

Commercial innovations, by their very nature, seem to have a bias toward spreading. But those to whom they spread often add their own innovations and modifications. For example, the Spanish empire began the process of mercantil-

ism with its "metals policy," based on the extraction of gold and silver from conquests in the New World. Other states lacked this same access, but they were
drawn to the idea of using economic policy to enhance state power. Thus, as
mercantilism diffused to others, it also came to be driven by gains from trade that
allowed the accumulation of specie. Another good example of "second tier" innovation in the diffusion process is the case of industrialization. Britain was the
leader in this instance, allowing the private sector to finance the creation of
manufacturing industries and the railroads that took products to market. Most of
the later industrializers (e.g., Continental Europe) veered from the British path
by forming state-run banks specifically chartered to finance the process of industrialization—and some states even owned railroads outright. The United States,
during this period, innovated even further, allowing the private sector to raise
capital and build rail infrastructure—but the federal government also often acted
decisively to protect and nurture its burgeoning industrial sector, with winning
results.

Today, the United States finds itself in the position of a leader in the process of
"informatization." But this leadership is precarious in nature, and highly dependent upon the diffusion control strategy that emerges in the coming years. In economic terms, allowing, or encouraging, sales of advanced technologies implies
increasing profits for U.S. firms, and good employment security for the emerging
class of "knowledge workers." In the security realm, however, sales of such inherently dual-use products will undoubtedly enhance the power of potential rivals
and adversaries. These factors seem locked in a tight, zero-sum embrace, as more
attention to security affairs will impose economic costs, and vice versa.

This dilemma that confronts the United States at the outset of the information
age is quite different from the situation in the early years of the Cold War. In the
1950s, helping former enemies from World War II, as well as many of the countries that they had victimized, both fed a U.S. economic boom and created robust
security partners for the duration of the long confrontation with the Soviet Union. Doing good while doing well was feasible then. Is it now? Possibly not. This
is one of those periods in which security considerations may have to supersede
the pursuit of profits.

The implication, then, is that the United States should cultivate a somewhat
more preclusive turn of mind regarding the diffusion of advanced information
technologies, heightening awareness of the military and security threats inherent
in their uncontrolled spread.

But this turn of mind toward preclusiveness must not become reflexive, for
there are a number of areas of information technology that simply cannot be
controlled. In the realm of encryption, for example, there is little hope of waging
anything more than a rearguard action against the diffusion of unbreakable code-

making capabilities. U.S. policy should therefore emphasize identifying the strategic paths that might be taken to cope with this fact, and its dire, profound implications for the intelligence-gathering process. For, as earlier cases of efforts at commercial diffusion control have shown, in the long run, things get out and around. There will sometimes be benefit in controlling the pace of diffusion (as in the case of HPC), but trying to control the fact of diffusion merely conjures up pixilated images of Canute.

CHAPTER THIRTEEN

Conclusion

The Diffusion of Military Technology
and Ideas—Theory and Practice

EMILY O. GOLDMAN AND ANDREW L. ROSS

Technological and organizational changes of the magnitude of the information revolution are usually accompanied by fundamental changes in warfare. For the United States, standoff precision weapons aided by information superiority may be the preferred policy option of political leaders who need to make the use of military force palatable to a casualty-sensitive public. Information superiority is key to achieving the operational concepts of dominant maneuver, precision engagement, focused logistics, and full-dimensional force protection laid out in *Joint Vision 2010* and now *Joint Vision 2020*.[1]

But the information revolution is diffusing. It will redistribute power to those once considered lesser actors, be they weaker nation-states or nonstate actors such as terrorists and organized crime syndicates.[2] The pace of diffusion may be more rapid than was the case with nuclear or aerospace weapons because IT is more affordable, has broad commercial applications, and faces fewer restrictions to its spread. Defense practitioners need to understand which actors are empowered by the information revolution and how these actors are likely to integrate and exploit their new capacity. As the cases presented in this volume demonstrate, the diffusion of military innovations depends on the capabilities of states, organizations,

[1]Chairman of the Joint Chiefs of Staff, *Joint Vision 2010* (Washington, DC: The Pentagon, 1996).

[2]Jessica T. Mathews, "Power Shift," *Foreign Affairs* 76, no. 1 (Jan.–Feb. 1997): 50–66; John F. Sopko, "The Changing Proliferation Threat," *Foreign Policy* 105 (winter 1996/97): 3–20; Kevin Soo Hoo, Seymour Goodman, and Lawrence Freedman, "Information Technology and the Terrorist Threat," *Survival* 39, no. 3 (autumn 1997): 135–55.

cultures, and individuals to mobilize a variety of key resources to utilize the new object or practice. In the process, institutions, organizations, cultures, and even social structures can change, and with them the international system itself.

While there is no shortage of hypotheses about diffusion, little systematic empirical research, particularly in the military arena, is available to guide decision-makers. The military arena offers rich insights into the diffusion process and how adopting states and organizations respond to diffusion opportunities. Militaries are comparative institutions. Their international effectiveness can be calibrated only in relation to other militaries. Accordingly, militaries tend to monitor closely other militaries, both those in their immediate vicinity as well as the leading militaries of the day, performing "net assessments." Adversaries have both reasons to use foreign lessons—to gain advantage in conflicts—and opportunity to do so, since most wars are events that can be observed. Merely demonstrating a military capability can spur responses abroad (e.g., to emulate, offset, or innovate). The price of comparative ineffectiveness is dramatically high—death, destruction, defeat in war, and the consequent loss of independence, or even dissolution of the state. On the other hand, the lessons learned from other militaries are often filtered through the culture and organizations attempting to adopt, adapt, or respond to the innovations observed elsewhere. The impact of diffusion on any particular military is rarely one of perfect emulation. Culture and organizational norms often have a prismatic effect, deflecting or bending the original innovation to fit the demands and widely held beliefs of the new setting.

This volume seeks to integrate theory and practice, to act as a guide to the policy community and to inform academic researchers. We do not labor under the illusion that we have eliminated the gap between theory and practice. But the gap can be bridged. As Alexander George points out: "[T]heoretical-conceptual knowledge is critical for policymaking."[3] Practitioners need to describe, explain, predict, and cope with the diffusion of military knowledge, whether that knowledge is embedded in "hardware" (the artifacts or techne) or in "software" (the organizational or human application component of an innovation or technology, such as doctrine, tactics, and organizational form).[4] The interest of practitioners in description and explanation is chiefly instrumental. Description and explanation are prized not so much for their contributions to understanding and the advancement of knowledge per se, but for their practical utility in discerning future

[3]Alexander L. George, *Bridging the Gap: Theory & Practice in Foreign Policy* (Washington, DC: United States Institute of Peace Press, 1993), p. xviii.

[4]For a discussion of the distinction between "hardware" and "software," see Andrew L. Ross, "The Dynamics of Military Technology," in David Dewitt, David Haglund, and John Kirton, eds., *Building a New Global Order: Emerging Trends in International Security* (Toronto: Oxford University Press, 1993), pp. 106–40; and Barry Buzan and Eric Herring, *The Arms Dynamic in World Politics* (Boulder, CO: Lynne Rienner, 1998).

challenges and opportunities and developing options for dealing with alternative futures.[5] In describing and explaining various instances of military diffusion, we have sought to advance theoretical knowledge as well as to aid practitioners in their efforts to anticipate, prepare for, respond to, and perhaps even shape the future.

The remainder of this concluding chapter is divided into two parts. First, the key theoretical findings concerning the diffusion process are discussed in light of the book's empirical cases. This is followed by an analysis of the lessons that can be drawn from comparing across the cases. This is specifically written with practitioners in mind, both those interested in spreading innovations to allies in order to enhance interoperability, as well as those interested in preventing the spread of innovations to actual and potential adversaries. Our goals are also twofold. First, we strive to introduce greater theoretical rigor into the study of diffusion in the military sphere. Second, we attempt to aid practitioners by helping them to develop a methodology to diagnose problems accurately. Moreover, by highlighting counterintuitive insights, we hope to convince practitioners and academicians alike of the importance of re-examining core, and possibly flawed, assumptions about the diffusion process.

Diffusion Diagnostics

For practitioners, the diagnostic task presented by the diffusion of military knowledge has four key components: (1) identifying who has an incentive, or is motivated, to adopt new practices; (2) identifying the models that are likely to be targets of emulation or offsets; this requires knowledge of the transmission paths for the spread of information; (3) identifying the ease with which military knowledge is likely to spread across different environments in order to gauge the scope, speed, and extent of diffusion; and (4) capturing the results of military diffusion within states and organizations in order to understand indigenous patterns and the range of possible adaptations. The findings summarized below are organized around these four main diagnostic tasks.

Motivations

Policy-makers need to understand the range of motivations that drive military diffusion. Practitioners must know which states have strong incentives to transform their military capabilities. This diagnostic task is a crucial first step in charting the future security environment.

The extant literature posits four motivations for the diffusion of military

---

[5]What George in *Bridging the Gap* refers to as diagnosis and prescription.

knowledge: strategic necessity; economic pressures; technology-push dynamics; and institutional pressures. Strategic necessity explanations emphasize that competition in the international system creates a powerful incentive to adopt new military methods and emulate the military practices of the most successful actors in the system. Strategic necessity drives allies as well as adversaries to adopt innovations. Spheres of influence and alliance obligations require a level of interoperability that ensures diffusion.

Economic pressures from the military-industrial complex, including the national defense community and the industrial and commercial sectors that have institutional or financial interests in the dissemination of military knowledge and technology, encourage diffusion. Technology-push characteristics of an innovation, such as commercial or dual-use applications, can encourage adoption. The capital investment or supporting infrastructure required may impede diffusion.

Institutional explanations include bureaucratic explanations in which inter- and intraorganizational competition and infighting play a role. New practices are likely to be adopted if they enhance an organization's resources, autonomy, and essence. Adaptations often result from bureaucratic infighting or existing organizational preferences that are grounded in prior experience and tradition. Neoinstitutional approaches (so-called because institutions are defined as norms) focus on noncompetitive pressures that stimulate the spread of forms and practices across organizations in the same line of business as they come to share ideas about the most legitimate way to practice their profession.

Strategic considerations emerge as a powerful driver of diffusion in nearly all the case studies. Warlike confrontations with aggressive expansionist European states in the eighteenth and nineteenth centuries compelled ruling elites in extra-European countries to introduce superior European military practices into indigenous armed forces in order to preserve their countries both physically and culturally. The diffusion of Western weaponry, tactics, and regimental culture to South Asia was stimulated initially by the quest for military and commercial advantage on the part of French and British trading companies. Native rulers subsequently adopted Western practices once they saw their effectiveness, in order to gain advantage over their neighbors and to survive the onslaught of the East India Company. While the forces of nationalism and industrialization had to be reckoned with by military forces worldwide, the ability of some states early on to leverage these developments and impose catastrophic defeat on their enemies compelled the vanquished to try to emulate the victors' superior military practices. Not surprisingly, the high-threat environments of the late interwar and World War II years caused adversaries to monitor each other vigilantly and attempt to exploit developments in armored warfare and carrier air power. The defeat of the Arabs by Israel in 1967 provided strong impetus in Egypt, Syria, and Iraq to de-

velop a more professional and competent approach to waging war in order to regain lost territory. Each turned to Moscow for arms, technical instruction, advisory assistance, and training. Even when innovations originated in states that are
not among the most technologically or industrially advanced, their subsequent
adoption, whether by regional neighbors or the major industrialized powers, occurs once their utility in regional conflicts is demonstrated.

Alliance obligations promote diffusion, although cultural affinity can promote
diffusion even in the absence of explicit alliance requirements. According to
Young, military diffusion among the Anglo-Saxon nations had its roots during
World War II and was institutionalized and regularized in a series of collective
defense treaties after the war. The chief goal has been to attain interoperability (or
at least to "manage" instances where it is not possible), and close collaboration
has endured and even deepened, despite diplomatic disputes and differences in
threat perceptions.

Only to a limited extent can the diffusion of conventional capabilities in the
Warsaw Treaty Organization (WTO) be explained by traditional East-West security motives or alliance obligations. While perceived in the West to be driven
chiefly by the specific nuclear and conventional requirements for mounting an
offense against NATO, Jones argues that the distribution of the most advanced
capabilities within the WTO never met the technical requirements of an offensive
doctrine. The actual distribution of capabilities was driven by political and internal security considerations: to cope with the recurring crisis of Communist Party
legitimacy in Eastern Europe that threatened contagion throughout the bloc. It
was the threat from within the Warsaw Pact that drove the actual disposition of
forces and doctrinal developments.

Potter's analysis of the diffusion of nuclear weapons reveals no typical proliferator profile or simple general explanation for why states go nuclear. Most U.S.
policy-makers assume that military threats drive states to acquire nuclear weapons. However, Potter concludes, while international pressures are greater than
domestic, no single international pressure drives most proliferators. Traditional
security motives such as deterrence and war-fighting advantage coexist along with
motives including the quest for status and prestige, as well as autonomy and influence. The most prevalent primary determinants are deterrence and auton-
omy/influence. Interestingly, while the prevalence of deterrence and war-fighting
motivations displays no variance over time, autonomy/influence and status/pres-
tige motives have increased in importance since the 1960s.

Potter's observation about the increasing importance of status drivers over
time is consistent with institutional arguments that early and later adopters have
different motivations. Early adopters favor innovations with certain technical
characteristics and are driven by efficiency considerations. Later adopters are

drawn to practices considered legitimate within the relevant social system. Potter's findings in the nuclear case resonate with Demchak's analysis of the spread of the U.S. model of an information technology–based military. She attributes today's unprecedented global diffusion of the U.S. model of military modernization to a process of growing institutional convergence among militaries. Military organizations are more closely connected today because of the global reach of telecommunications and international forums such as NATO's Partnership for Peace, and so seek to acquire the characteristics of "modern" militaries, especially those of the leading military power, the United States. In information warfare as in nuclear weapons, legitimacy and prestige are increasingly important drivers, but they do not extinguish traditional security motivations. Once rapid, lethal, highly technological military innovations spread widely enough, they will have altered the international environment to such an extent that states will be forced to redirect their military efforts to the demands of their local environments. As actual capabilities become clearer, not global norms about what constitutes a "modern" military but rather what potential adversaries are actually capable of doing will become the focus of efforts to upgrade militaries.

Several chapters emphasize economic and technology push explanations. Potter notes the importance of technological momentum in the nuclear case, as well as the active promotion of civilian nuclear technologies, with spin-offs in the military sphere. The high-tech aspects of recent cases should not distract us from the parallels in earlier periods, as Herrera and Mahnken's analysis reminds us. Private entrepreneurs laid railroads to serve the interests of private industry and trade, and as the national rail net became denser, states reaped the greater military potential. Private interests can serve public aims, and Arquilla's analysis of commercial diffusion directly addresses these motivations in the current period. Commercial innovations by their very nature tend to spread, both because recipient states (and firms) seek the nonmilitary benefits and because sending states (and firms) have an economic incentive to encourage diffusion to gain and maintain market share. Industry can be a powerful lobby as the examples of supercomputers, satellite launch technology, and encryption software demonstrate. Moreover, a Washington consensus today favors openness and the free flow of information, even though the events of September 11 may have encouraged a tendency toward a somewhat more cautious approach. There remains a bias against control, with the implication that military applications of these dual-use capabilities will naturally diffuse along with their commercial counterparts.

The quest for security in a competitive international system is a powerful driver of diffusion, both among adversaries and allies. Nevertheless, competition is but one of several motivations. As noted above, status and legitimacy are important for later adopters. Economic incentives are also powerful, today perhaps even more than during the early industrial age, given the relative cheapness and accessibility of new technologies. Many new technologies do not require infrastructure to develop and operate more complex systems. Their myriad commercial applications make them all the more attractive. This suggests that global diffusion is not always driven by direct competition. So a state's strategic requirements do not necessarily dictate the model that will be chosen for emulation. This raises the important question of model selection, which is closely related to the analysis of transmission paths for diffusion.

Transmission Paths

A second diagnostic task for practitioners interested in charting the diffusion of new military knowledge is to identify which military model or models states are likely to emulate or offset, and this depends on how states learn about innovations and then can acquire the necessary knowledge to adopt them. While U.S. models are preeminent targets of emulation, there may also be alternative models within a state's region or cultural affinity group. In the economic sphere, the Japanese or "Asian" model of economic development was far more attractive to regional actors than the "Anglo-Saxon" model. To understand what those who are motivated to transform their militaries are responding to, practitioners must examine the process of transnational communication by which knowledge about new military technologies and practices diffuses.

Innovative technologies and practices spread via a variety of transmission belts. Policy analysts focus on diffusion among competitors as the greatest concern. But collaborative or voluntary processes also affect the international balance of power, and may be more problematic precisely because they are not driven by the survival imperative. Allies may adopt more familiar or accessible models rather than the most effective or suitable methods. Political considerations can overwhelm the dictates of military effectiveness, as interoperability problems within NATO recently have demonstrated and earlier Soviet control of Warsaw Pact diffusion also showed. Thus practitioners need to examine cross-national interactions among states and organizations to understand why a country adopts an innovation and whether certain transmission paths are more effective conduits for diffusion than are others. States may adopt the same innovation for different reasons, and they may acquire the innovation via different transmission paths. For practitioners seeking to inhibit diffusion, this may mean that eliminating one known or anticipated transmission path will not necessarily prevent diffusion altogether. For those trying to encourage the spread of certain innovations, the transmission path that works for one adopter may not be satisfactory to bring about adoption in another.

Many chapters emphasize the importance of demonstration effects as conduits for diffusion. The East India Company's successful new military units spurred indigenous rulers to adopt them. The spectacular battlefield accomplishments of the Napoleonic and Prussian military systems revolutionized warfare in Europe before the twentieth century. The rapid defeat and fall of France at the hands of the Germans and the successful British raid on the Italian fleet at Taranto are examples from the two world wars. Outside Europe, the efficacy of Israel's fast attack craft/antiship missile combination and its remote piloted vehicles and Iraq's effective use of chemical weapons and ballistic missiles also gained significant attention. And most recently, U.S. performance in the Persian Gulf War made strong impressions on observers and induced emulation.

But Demchak points out that interstate wars are relatively rare events. Militaries are only intermittently, unevenly, and incompletely aware of each other's abilities and practices. While global commercial communications networks are primary vehicles for the dissemination of information today, these are relatively recent developments associated with the information age. Moreover, several chapters emphasize the contestability of demonstration effects. It is not always clear what lessons should be drawn from battlefield experience, even by the parties engaged in the conflict. Demonstration effects are only one dimension of the transmission process, and the link between observation and adoption is tenuous.

Innovations may spread through collaboration and cooperation. French and British trading companies purposefully imported Western tactics and regimental culture into South Asia to protect their economic empires. The Soviets actively transmitted technologies, organizational forms, tactics, and operational concepts to their Arab allies through provision of large amounts of hardware, an advisory presence, and the training and education of Arab military personnel in the Soviet Union. Informal contacts at all organizational levels and continuous dialogue among the ABCA states (America, Britain, Canada, Australia and New Zealand) have allowed for an unprecedented level of "intellectual" exchange and diffusion, which has become institutionalized, even in the absence of one overarching collective defense treaty binding all five countries together. Europeans regularly sent military missions abroad with the expressed intent of diffusing their military knowledge to non-European militaries. Allies fighting side by side also facilitated the spread of military practices, and in some cases, such as U.S.-British carrier operations in the Pacific in 1945, were the only effective mechanism for the transfer of methods whose effectiveness had already been demonstrated in combat. In the nuclear arena, the United States and Soviet Union declassified and actively exported technical nuclear power information, experts, and research reactors by the mid-1950s, in an effort to enhance their respective influence among the international scientific community. Nor were they alone in actively disseminating nuclear knowledge. The French assisted the Israelis, the Chinese assisted the Pakistanis, the British planned to assist the Australians, and the Israelis may have cooperated with the South Africans. Finally, Great Britain's colonial conquests in earlier centuries, like America's passion to wire the world at the close of the twentieth century, reflects a liberal capitalist world view favoring broad access to markets for international trade that has enabled dissemination of commercial innovations—even those with military applications—and reinforced a bias against establishing controls on their spread.

New technologies and ideas can also be imposed. Once the French defeated and occupied an adversary, they attempted to impose the Napoleonic military system, with varying degrees of success. The most far-reaching experiment in imposed or "coercive" diffusion is the WTO. The Soviets were the sole suppliers of hardware. They coordinated tactics and doctrine through meetings, conferences, and joint exercises. By the early 1960s, the Soviets imposed an offensive military doctrine on their loyal members to prevent the spread of "heretical" doctrines from Albania, Romania, and Yugoslavia to other WTO states.

Diffusion occurs via commercial as well as military and political channels. As Arquilla's and Demchak's chapters document, globalization has transformed the U.S. defense industrial and informational base to a global one. Innovations originate from science, academia, the military, and industry. Innovations also feature multiple applications. Accordingly, military diffusion is not simply a state-to-state process, controlled and managed by central political decision-makers in the service of the national interest. Firms, organizations, educational institutions, and individuals all play important roles in the transmission of new knowledge and applications.

Finally, innovations can be transmitted through socialization processes, professional networks, and key individuals. Demchak attributes the spread of the information technology-based military model to the emergence of a global military community, fed by international trade shows, increasing openness and interaction among military organizations worldwide, and common sources of formal education. Professional networks are broader and denser than ever before, promoting mutual awareness of a common military enterprise. Demchak's analysis highlights the importance of access to information, as do several other chapters. The Napoleonic military system spread via proselytizing by such "product champions" as Henri Jomini, much as mobile mechanized warfare was prodded along by Liddell Hart and air power by Guilio Douhet. Thus key individuals have played pivotal roles in promoting diffusion of military innovations.

One implication of the socialization argument, born out by Demchak's analysis, is that militaries have been known to adopt new technologies and methods for waging war even when the innovations may not be the most rational or logical given their strategic situation. Malaysia, Myanmar, and Botswana are eager to modernize and adopt sophisticated weapon systems to model the attributes of a modern military despite a lack of technically advanced enemies. The legitimacy associated with a model can spur its diffusion. Other chapters also demonstrate how prestige (of the modeled state and the weapon system itself) and identity factors can subvert rationality and affect the transmission process. The Soviet Union enjoyed prestige among many developing countries during the Cold War as an alternative to the U.S. model. Ballistic missiles and weapons of mass destruction have emerged as symbols of prestige among many developing nations. Finally, Hoyt suggests that pariah states may identify with one another and adopt each other's military methods, particularly when those methods challenge international norms.

In addition, in contrast to arguments based on the "rational shopper" image, proximity and accessibility greatly influence model selection and transmission paths. The close relationship between Israel and the United States increased U.S. familiarity with Israeli innovations and encouraged diffusion in both directions. Arab militaries still bore the influence of their colonial masters well after independence. As Eisenstadt and Pollack argue, they later adopted Soviet technology, doctrine, and operational concepts to varying degrees because only the Soviets were willing to provide the necessary assistance to facilitate the transfer. Advanced arms on favorable terms, not an objective assessment of the relative merits or suitability of Soviet military systems, explain the model "choice."

Several chapters reveal that transmission is not a one-way street proceeding from more to less militarily advanced states. The East India Company learned from the South Asians and vice versa. As Napoleon's methods spread throughout Europe, the French learned from those they defeated (e.g., the Poles); the Prussians and French continually learned from each other. The Americans learned from British carrier methods and U.S. innovations subsequently diffused back to Britain. Hoyt goes the furthest in challenging conventional wisdom on the direction of transmission flows, disputing that lesser states merely imitate great powers. His cases show how technologies originating in the "core" can be applied in novel ways by states in the "periphery," and then those new approaches diffuse back to the core.

Collectively, the chapters show that transmission paths are multiple, and the most efficient model given strategic circumstances may not automatically be the target of emulation. Highly prestigious models will be prime targets for emulation if the adopter desires acceptance in the social system (e.g., Western society) within which the innovation is institutionalized. Professional networks are key avenues for transfer of information, implying that organizations will also pattern their practices after those with which they have greatest interaction (e.g., the most familiar model). The key issues are the recipient's relationship to various networks and how robust the transmission path is.

Are some transmission paths more effective than others? The transmission mode tells us about the source of information but not about how extensively a new model penetrates into a receiving country's state, organization, and society. The chapters demonstrate that states often emulate selectively, and that voluntary dissemination of information does not necessarily lead to successful assimilation of the model. Knowledge diffused among friends and allies where one would assume there was greater information sharing can encounter difficulties, so access to information alone does not guarantee an innovation's adoption. Despite Moscow's assistance in training and educating Arab military personnel, their methods were never fully adopted. In fact, the Syrians, who relied most heavily on Soviet organizational forms, tactics, and operational practices, had a less extensive and intensive relationship with the Soviets than did Egypt or Iraq. The significant Soviet advisory presence in Iraq throughout Desert Storm had no decisive impact on Iraqi thought and practice.

Diffusion researchers such as Blaut contend that diffusion is less likely when the innovation is incompatible with the existing cultural system as a whole.[6] Several of the cases presented here suggest that factors affecting the rate of diffusion may operate independently of the transmission path. One implication is that both adversaries and allies may have difficulty emulating innovations even when highly motivated.

Facilitators and Inhibitors

The next diagnostic task for practitioners is to prioritize how much of a threat or challenge a particular modernizing military represents. How great a threat the diffusion of an innovation represents depends in large part on whether the adopting state can readily absorb the new military technologies and implement the accompanying practices, so that its military rapidly becomes more effective. This diagnostic task requires identifying the relevant factors that facilitate or inhibit the integration of new knowledge into the receiving military's organization and practices. These factors include attributes of the innovation as well as the receiving state and military.

Diffusion requires interaction and sharing across national policy communities. Transmission alone does not ensure that an innovation will be adopted. Nor does the level of motivation exclusively determine the extent of emulation. Al-

[6]James M. Blaut, "Two Views of Diffusion," *Annals of the Association of American Geographers* 67, no. 3 (Sept. 1987): 343–49.

though Napoleon's adversaries were compelled to adopt his methods to avoid defeat, only Prussia was willing to risk political and social upheaval to transform its military. While the defeat of France by Prussia in 1870–71 shook the French military establishment to its foundations, only the form and not the substance of the Prussian model was adopted. Obstacles to emulation shape the scope, pace, and extent of diffusion, regardless of the particular transmission path.

In addition to the qualities of information sources and communication networks, facilitators and inhibitors stem from the qualities of the innovation, qualities of the adopters, and the compatibility between the innovation and the existing cultural system into which the innovation is being introduced.

Qualities of the Innovation

Most of the case studies support the contention that technology, or hardware, spreads more easily than software (e.g., doctrine, tactics, organizational form, and macrosocial change). Arquilla argues that leading-edge technologies today, including the most advanced computers, do not pose the same limits on capabilities and resources that precluded states from modernizing their militaries in the past. Yet even centuries earlier, as Lynn points out, native South Asian rulers had the resources to secure European-style weaponry and the arsenals to manufacture their own. We should not assume that peripheral states necessarily lack the resources to secure new technologies—or to develop their own, as Hoyt's chapter reminds us.

Many of the cases suggest that advantages in weaponry are superficial and fleeting. The ability to acquire, develop, or adapt the "software" necessary to take advantage of a new technology may be the more crucial factor in assessing the true capacity (as a threat or an ally) of a state's military. The British, Soviet, and U.S. armies all possessed tanks, yet each had difficulty developing the doctrine and organization to wage combined-arms armored warfare. The Soviet Union's Arab allies imported their weapons, but had great difficulty assimilating Soviet doctrine because they lacked the necessary skills of flexibility and adaptability. Arab militaries were mere caricatures of the Soviet military, with no evident correlation between a reliance on Soviet equipment and the degree to which that country's military adopted Soviet organizational forms, concepts, and practices.

Herrera and Mahnken similarly conclude that the acquisition of new technology is the first and easiest step, and that the successful importation or development of the tactics and the administrative and training apparatus that made the military practice effective was the harder step. They temper their observation with the example of the railroad. The expense and dependence on civilian investment prevented economically backward states from building a sufficient railway net. The case of the aircraft carrier reinforces the caveat that the expense and capital intensiveness of a technology may create significant barriers to its spread. In general, an innovation requiring large-scale capital investment will diffuse more slowly than one with incremental costs. Israel's low-cost innovations spread quickly, while the resource, manpower, and technology requirements of chemical weapons and missile production programs have posed greater constraints on their spread.

Again, however, generalizations must be qualified. Iraq demonstrated that even a developing state can produce chemical weapons and missiles. As Arquilla points out, government initiative can overcome what may seem to be excessively high capital barriers relative to the state's resources. More generally, Arquilla emphasizes "entry costs" related to the technology, idea, or organizational form. Some innovations, like money and coinage, were not particularly costly. Others, such as harnessing economic policy to promote national power (mercantilism) required a level of national military power that few states could match. In his judgment, dual-use technologies and ideas with commercial value are prone to spread. Entry costs will govern the extent and path of diffusion, and there is no a priori reason to assume that the entry costs of software will always exceed that of hardware.

Young's analysis provides an excellent example of these seemingly contradictory insights. For the ABCA states, with a high degree of cultural affinity, the software has been more readily transmitted than the hardware. Regular dialogue among the ABCA militaries fosters and sustains an ongoing process of transmission of ideas. Doctrinal and procedural standardization is easier, while contemporary defense procurement's linkages to domestic economic concerns complicate the diffusion of hardware. Equipment standardization is a highly sensitive issue because it requires the purchase of foreign systems or domestic manufacture under license, both of which threaten indigenous defense industries.

Qualities of the Adopters

Most chapters emphasize how an adopting state's characteristics shape and influence diffusion. Those qualities can be loosely grouped under two categories: situational factors and policy choices. Situational factors include geography, economic resource capacity, and the social, political, cultural, and organizational environment in the adopting state. Policy choices refer to the government's commitment to a certain policy direction, such as nonproliferation.

Not surprisingly, geography influences receptivity to new military practices. Of France's adversaries, only the British did not embrace Napoleonic warfare, principally because of their island location. Nor did the British adopt Prussian innovations. Britain was a maritime power, whose strategic requirements differed from those of the Continental opponents with which it was directly engaged. The strategic challenge of operating over the vast expanses of the Pacific made offensive carrier air power a particularly attractive option for the Americans and the Japanese. The requirement of operating in the confined waters of the Mediterranean, in proximity to land-based air, explains why the British approach to carrier air power privileged armor over airplanes.

Economic resources may constrain or enable diffusion. Soviet Middle East allies lacked the financial, material, and human resources to maintain Red Army replicas. Economic backwardness and an undeveloped railway net hampered Austrian and Russian attempts to emulate the Prussian military model. Structural problems in Britain's aircraft industry prevented full exploitation of carrier air power. By contrast, the 1943 arrival of new aircraft carriers made multicarrier task forces feasible for the United States. Technical know-how, the availability of fissile material, along with the resources to purchase the material and support weapons-fabrication facilities, have influenced the spread of nuclear weapons.

Economic capacity generally influences access to hardware, and the lack of hardware tends to impede the development of supporting software. But the Japanese experience with carriers in World War II demonstrates that while resources may affect implementation, they need not inhibit ideational adoption. The Japanese Navy operated under severe financial constraints and lacked the resources to sustain its carrier capacity throughout the war. Nevertheless, the organization assimilated the transformation in warfare heralded by air power. Japan's doctrine and organizational arrangements continued to reflect a commitment to the multicarrier task force concept, despite an increasing shortage of carriers, planes, and trained pilots.

The role played by the social environment in shaping diffusion is most clearly illustrated in Herrera and Mahnken's chapter, where the innovations emanate from broader and deeper socioeconomic transformations. Nationalism and industrialization shaped the diffusion pattern of the Napoleonic and Prussian military models. Across Europe, lingering aristocratic privilege prevented full implementation of universal service laws that were the heart of the Prussian model.

The political environment—regime type, political interests, and institutional capacity—also influences the course of diffusion. Austria avoided Napoleonic reforms because it aggravated nationality issues. A nation-in-arms policy risked civil war in the unstable multiethnic empire. Lynn emphasizes the role of institutional stability for military modernization. He attributes the greater effectiveness of the East India Company in introducing Western military practices to native troops, relative to the failure of indigenous rulers to emulate the Company model, to the Company's greater political stability.

Many chapters emphasize how the organizational environment—bureaucratic politics, intraorganizational infighting, and organizational culture—affects diffusion. Factionalism in the French military prevented emulation of Prussia's conscription policy and general staff. Stalin's purges weakened armor advocates in the Russian Army. The U.S. Army's organizational culture prevented it from adopting the concept of combined-arms armored warfare rapidly, while the Soviet Union's horse cavalry dominated the Red Army and similarly hindered the innovative use of tanks.

No explanation of the diffusion of carrier air power practices can overlook the role played by organizational facilitators and inhibitors. The Royal Navy's loss of control of the Fleet Air Arm from 1918 until 1939 removed all air power advocates from the navy. Similarly, the establishment of independent air forces in Germany and Italy meant their navies would have to depend on the largesse of land-based air advocates for resources and support. By contrast, the U.S. and Japanese navies retained control over their flyers and led the way in naval aviation reforms.

None of the facilitators or inhibitors mentioned thus far is decisive in all cases. Our focus on various dimensions of the adopter's capacity should sensitize practitioners to the range of forces that can influence the ability to assimilate an innovative set of technologies and practices. In addition, the willingness and ability of reformers to reallocate resources and take risks to alter existing institutions and practices should not be slighted as an important factor shaping the diffusion process. As Potter points out, in many countries the political decision to acquire nuclear arms precedes the development of the technological capacity to do so. In the nineteenth century, according to Herrera and Mahnken, among France's adversaries only Prussia was willing to risk upheaval to transform its military. Prussian reformers, such as Scharnhorst, recognized that military reform had to proceed in parallel with political reform, and that Prussia would have to implement sweeping political and military changes to transform its subjects into citizens. A sufficiently reform-minded regime can surmount immense political and social obstacles. Individuals can also exert decisive political influence. India's pre-1974 stance on nuclear weapons is widely attributed to Nehru's personal philosophical opposition.

Cultural Compatibility

Several authors focus on how cultural factors (e.g., shared values about how society should be structured and function, and about the purpose and limits of armed violence) affect diffusion, and since that element has received relatively little attention in the military diffusion literature to date, we single it out for special attention here. Imported practices must be integrated into the indigenous culture. The greater the compatibility between the innovation and the existing cultural system as a whole, the easier is their adoption and implementation. Quincy Wright long ago noted that technologies are not "superficial devices from which all cultures can benefit and which may originate anywhere and diffuse easily and rapidly. On the contrary, technologies are related to the culture as a whole and the origin, diffusion, and influence of a particular invention cannot be understood except in terms of the total culture which originated or utilizes it."[7]

The thesis is developed most explicitly by Lynn, who argues that a synergy existed between Indian (especially Hindu) qualities and the British regimental model that proved far more effective in mobilizing South Asian values than had any native military organization. The British were also willing to adapt the European model to India's indigenous culture. The combined result was to multiply the effectiveness of imported practices, rather than to erode them.

Eisenstadt and Pollack's analysis complements Lynn's chapter. They argue that a society's culture helps determine which skills and behavioral predilections the nation's manpower will bring to military service. Arab culture and social institutions valued conformity rather than individuality or independence of thought. Consequently, junior officers demonstrated little initiative or capacity for independent action. Whereas inherent synergies existed between British regimental culture and South Asian values, Soviet doctrine, which called for high degrees of tactical initiative, did not mesh with Arab cultural norms. Yet just as British regimental culture was more effective when adapted to South Asian proclivities, Arab armies functioned most effectively when they adapted imported models to their own styles of war. Whether an adopter can integrate (or adapt) an innovation to suit its culture will also be influenced by the context within which the innovation is transmitted. If the British had rigidly insisted on the isolation of sepoy soldiers from their families and home communities, regimental culture would probably not have been successfully incorporated into the South Asian military tradition.

Young also details how diffusion is facilitated when nations share common values and language. The cultural affinity and linguistic similarities among the English-speaking countries of the United States, Britain, Canada, Australia, and New Zealand has led to the transmission of military expertise and intellectual diffusion that far exceeds any identifiable security requirements.

An innovation developed and honed in one setting is rarely transplanted wholly to another without modification. Historically, states have either adapted innovations to make them functionally effective in their new setting, or selected certain aspects of the model to adopt. Few chapters identify instances of faithful emulation. Departures from original patterns occur because the environment and values in the importing society usually diverge from those of the source. As

---

[7]Quincy Wright, *The Study of International Relations* (New York: Appleton-Century-Crofts, 1955), pp. 375–76.

Westney argues, "Since the environment in which the organizational model was anchored in its original setting will inevitably differ from the one to which it is transplanted, even the most assiduous emulation will result in alterations of the original patterns to adjust them to their new context, and changes in the environment to make it a more favorable setting for the emerging organization."[8] Faithful emulation is rare and does not ensure military effectiveness. In fact, Lynn's intriguing study of the sepoy suggests that the very opposite can occur. Moreover, Eisenstadt and Pollack show that Arab military effectiveness depended not on the ability to fully assimilate Soviet or Western doctrine, but rather on the ability of Arab armies to adapt foreign doctrine to their own unique approaches.

What national or organizational leaders believe is "best practice" may be less important than what they can implement. Simply put, not every country could afford Mahan's big ship navy; nor can they afford to develop chemical weapons and missile production programs. A highly motivated regime can sometimes surmount such obstacles by shifting resources, but other deficiencies—such as literacy rates and skill levels, and even social class structure—may take much longer to surmount.

Diffusion Effects

The final diagnostic task is to anticipate how adopting states might incorporate innovations into their organizations and practices. When exposed to a new military practice, a state or military organization can respond in one of two ways: adoption or nonadoption. Nonadoption includes no response, the development of offsets (e.g., methods to counter the innovation), or development of yet another innovation. Adoption can occur in one of three ways: imitation, or replication of the innovation in its entirety; adaptation, or modification of the innovation in the process of its adoption and implementation; or selective emulation, faithful replication of part of the innovation.

Across the broad range of cases examined here, few cases of true imitation emerge. Syria most closely followed Soviet practices organizationally, tactically, and operationally. But close scrutiny reveals that the Syrians implemented only a caricature of the Soviet model. They never acquired or exhibited the flexibility and adaptability inherent in the Soviet approach.

Selective emulation and adaptation are more common among the cases examined in this volume. Because Soviet doctrine was based on a modern, mechanized military that only an industrial power could afford to raise and maintain, adoption of the Soviet model was necessarily selective and entailed adaptations.

[8]D. Eleanor Westney, *Imitation and Innovation: The Transfer of Western Organizational Patterns to Meiji Japan* (Cambridge, MA: Harvard University Press, 1987), p. 6.

Soviet influence was greatest at the organizational level while reliance on Soviet tactics varied widely, and Arab allies generally did not adhere to Soviet operational principles.

Adoption of Napoleonic methods was also selective. Most states adopted individual tactical innovations, such as skirmishing and the attack column, while eschewing fundamental political, social, and military reform. Practices that reflected broader social change, such as universal conscription and open access to commissions, were incompatible with ancien regime values. Austria never seriously attempted to institute the *levée en masse*, while Russian officers continued to be recruited from the upper class. Only Prussia was willing to risk social and political upheaval to respond to the Napoleonic revolution. The Prussians also extended the innovation cycle by creating a general staff and instituting an organized reserve system. Responses to Prussian reforms were also selective and adaptive. Many adopted the Prussian general staff in form but not substance. Similarly, the demonstrated effectiveness of German armor spurred the British, Soviet, and U.S. armies to expand their armored forces considerably. Although each adopted variants of Germany's doctrine and organization, each army's approach contained unique doctrinal and organizational elements.

Many of the chapters focus on emulation, a major concern for great powers interested in anticipating the likelihood that competitors will arise. Historically, inferior powers have sought ways to degrade the capabilities of their superiors, not by emulating superior strength but by exploiting vulnerabilities. Today, the accessibility of new relatively inexpensive technologies provides less than great powers with a broadening array of means for offsetting superior strength.

Offsets, however, are not only pursued by inferior powers in response to the innovations of superior powers. As Hoyt demonstrates, innovations originating in the periphery may be imitated by other peripheral states first, but core states also absorb the niche technology, as was the case with Israel's remotely piloted vehicles, or offset the innovation. Offsetting responses include attempts to prevent, counter, or limit the utility of an innovation. Strategies of prevention have included export controls, arms control treaties, and other multilateral regimes. In response to Israel's fast missile-armed attack craft, which provided its possessors with a sea denial capability, the core developed automated point defense systems, upgraded its air defenses, and developed new methods to counter the threat, such as missiles mounted on helicopters. In response to Iraq's use of chemical weapons, core states sought to prevent development and diffusion via the Chemical Weapons Convention, and prioritized defensive countermeasures including improved detection and protective gear. Peripheral states responded in several ways. Many did nothing. Others adopted countermeasures and civil defense routines. Some, such as Syria, Iran, and Myanmar, attempted to emulate. Responses to

Iraq's ballistic missile conceptual innovation, included technology control regimes, missile defenses, development of missile forces as a deterrent, and preparation for preemptive strikes. Several states entered into cooperative missile programs with Iraq. Others obtained their own missile production capabilities. The West pursued a missile technology control regime.

Despite the attention rightly paid in policy circles to potential offsets to superior U.S. capabilities, Demchak's analysis reveals that inferior states do not always resort to offsets. Global signs of emulation are evident today. Focusing on ground forces, she identifies trends around the world similar to the internal organizational adaptation of the U.S. Army's modernization program. Many countries have been negotiating the purchase of advanced communication and control systems, and are acquiring stand-alone major weapon systems with advanced computer-mediated and networked internal systems. Motivations such as strategic necessity may stimulate offsetting responses, while the desire for prestige and legitimacy may stimulate emulation. "Core" powers are as likely to respond with offsets to innovations by "peripheral" states as peripheral states are to respond with emulation attempts to innovations originating in the core.

Collectively, the chapters provide some initial insights into the scope, rate, and patterns of diffusion of military knowledge over time. In tracking diffusion trajectories, we learn that innovations are composed of multiple facets—technological, tactical, doctrinal, organizational, administrative—and they often include social or cultural components as well. To understand how innovations diffuse, and therefore how a military revolution will unfold in time and space, it is necessary to disaggregate the "model."

For example, the conventional wisdom is that the Napoleonic and Prussian models spread widely. But disaggregating the technology, tactics, organization, administration, operations, officer corps composition, and conscription yields a more nuanced depiction of how each one spread. Similarly, Potter's ladder of nuclear weapons capability, which segments the diffusion process rather than relying on one indicator (the detonation of a nuclear device), provides a more nuanced and analytically useful picture of the diffusion pattern. Diffusion can proceed at different rates and along different paths, depending on the facet examined. The challenge today is to identify and track a manageable number of key dimensions that most accurately capture the essence of the IT-RMA.

Full assimilation or adaptation also takes time. The East India Company was on the Indian subcontinent for well over a century before it created sepoy infantry. The extensive intellectual diffusion among the ABCA states has been nurtured for nearly a half-century. Although it is conventional wisdom that the IT-RMA is likely to spread rapidly because it relies on commercially based, dual-use, and inexpensive technologies, many thought that nuclear weapons would spread more than they have. But as Potter points out, the scope of nuclear weapons diffusion has been small and the rate slow. Although information and nuclear technologies differ in many respects, a key finding of this analysis is that for military diffusion, the spread of ideas, or software, has throughout time been the crucial dimension that accounts for military effectiveness.

From Theory to Practice

One of the central contributions of this volume is to alert practitioners to be cautious in their expectations that the spread of new military knowledge is easy or straightforward. It cannot be easily controlled, nor held back indefinitely. This is so for several key reasons. First, culture will continue to shape the development and diffusion of military knowledge, producing indigenous adaptations that will be difficult to predict. True emulation is rare, implying that others will probably not leverage the IT-RMA the same way as the United States.

Second, the key technologies underlying the current revolution in military affairs are driven by the civilian commercial economy. There is tremendous commercial pressure for them to spread because they are a competitive advantage in the global economy. The ability to monitor and control their dissemination, assuming we could develop a normative consensus on the desirability of control, is highly questionable, because limiting technology transfers has impacts well beyond the military realm; it can affect modernization and the eradication of social problems in civil society.

Third, the current RMA promises to empower traditionally weaker states and even nonstates that can leverage the technologies to develop niche capabilities that strike at the vulnerabilities of more powerful states. Technological and conceptual breakthroughs that originate in the periphery will be difficult to anticipate and control.

But practitioners are not powerless. We must move from a conceptual analysis of military diffusion to a specific set of strategies that are useful for policy practitioners, to identify the variables over which policy-makers have some control as distinct from those that they cannot change. Policy-makers must specify:

1. The modernization motivations that can be successfully influenced. Potter's suggestion that external security incentives are more amenable to control than domestic drivers may help practitioners focus their efforts on countries where they can hope to have some influence.

2. Strategies to strengthen knowledge transfers. Access to information is not enough; friends cannot assume that diffusion will be easy. Our finding that an established relationship between the sending and receiving parties did not neces-

sarily predict the success of diffusion was surprising, and has important implications for allied relations today. Here, Young's analysis of sustained collaboration among the ABCA states is instructive. Also, Lynn's analysis of the sensitivity of the British to the particular context of South Asia is insightful for successful diffusion, as is Eisenstadt and Pollack's conclusion that Soviet arrogance inhibited diffusion to their Arab clients.

3. The indicators one must track to assess the capacity for others to leverage the current RMA, in order to map the future strategic environment. A reduced capacity to assimilate can hinder diffusion even when the level of motivation is high, implying the need to accurately identify and track the factors that facilitate or hinder diffusion, which vary with the adopter and the innovation.

These themes are explored in greater detail below in an effort to offer practical guidance for policy-makers.

Culture and Diffusion

Despite the convergence tendencies ascribed to modernization, Westernization, and globalization, the cultural dimension of the diffusion process remains significant.[9] Young's examination of intra-Western diffusion supports Huntington's contention that the diffusion of military methods will occur more rapidly and completely within civilizations, and face more impediments across the boundaries between civilizations. A concerted effort to broaden and deepen "intellectual interoperability" within the context of NATO and other institutions of the growing democratic security community,[10] while beneficial, may require finding common ground culturally as well as politically and militarily.

Lynn's account of the introduction of European military practices into South Asia by the British East India Company and the synergy generated by the fusion of European and indigenous cultures embedded in the sepoy illustrates how cross-cultural diffusion can be effective if imported practices and indigenous culture are compatible.

The "fit" between European military culture and South Asian culture described by Lynn contrasts with the less than successful diffusion of Soviet military organization and doctrine to Arab countries such as Syria, Egypt, and Iraq. As Eisenstadt and Pollack show, even though Soviet military knowledge had much to offer Syria, Egypt, and Iraq, and Soviet military hardware was acquired by all

[9]For useful discussions of the dimensions of culture, see Peter J. Katzenstein, ed., *The Culture of National Security: Norms and Identity in World Politics* (New York: Columbia University Press, 1996).

[10]On security communities, see Emanuel Adler and Michael Barnett, eds., *Security Communities* (Cambridge: Cambridge University Press, 1998).

three countries, an inability to absorb, or at least adequately adapt, the requisite "software"—organization and doctrine—negated the potential value of imported hardware. For Eisenstadt and Pollack, the critical failure in these three cases was at the tactical level because of "Arab cultural proclivities." British and U.S. doctrines have proved no more compatible with Arab culture than the Soviet. Even when alien military knowledge is eagerly sought rather than imposed, culture continues to get in the way.[11]

That the transmission of military knowledge is facilitated by shared culture and hindered by dissimilar cultures would not surprise proponents of a cultural interpretation of international dynamics.[12] The "cultural affinities" that, according to Samuel Huntington, facilitate political and economic cooperation among states within civilizations should also facilitate military cooperation, including the transmission of military knowledge. Conversely, the cultural clashes between states of different civilizations may impede the diffusion of military knowledge. Thus fears of an emerging Chinese-Islamic coalition, featuring the transfer of military technology from East Asia to the Middle East, are probably exaggerated. While not impossible, this would require a high degree of adaptation. China and North Korea have helped Islamic states develop ballistic missiles and (in the case of Pakistan) nuclear weapons.

It is tempting to conclude that the United States and other potential purveyors of military knowledge should, when purposely seeking to diffuse their military knowledge, adopt the British East India Company model rather than the Soviet model, and that those seeking to import foreign military hardware and software should emulate the South Asian behavior described by Lynn rather than that of the Arabs described by Eisenstadt and Pollack. The Soviets, as Eisenstadt and Pollack note, tended to adopt an arrogant, patronizing attitude toward their Arab "clients." The East India Company model, however, is an imperial model. Few importers of military hardware and software are likely to find that model attractive. Moreover, even if the East India Company does offer something of a model for diffusion across cultures, firms that provide military-related technologies and know-how today are less closely tied to the state than was the East India Company. The Loral and Hughes companies, for instance, worked constructively with their Chinese clients, but in their efforts to please a customer, they acted to the detriment of U.S. security.

---

[11]In their chapter on the diffusion of the innovations underlying the military transformations of the nineteenth century, Herrera and Mahnken show that even among Western major powers, diffusion has been culturally conditioned.

[12]Samuel P. Huntington, "The Clash of Civilizations?" *Foreign Affairs* 72, no. 3 (summer 1993): 22–49; and Samuel P. Huntington, *The Clash of Civilizations and the Remaking of World Order* (New York: Simon and Schuster, 1996).

Yet the East India Company experience is instructive nonetheless for its conscious adaptation to South Asian culture. In Lynn's words: "In order to insure that its officers respected the men's language, customs, and religion, the Company tried to insure that its officers become knowledgeable about them." Practitioners must remain aware of and sensitive to the potential array of cultural obstacles that will work to hinder the successful absorption of imported military practices. Successful absorption of imported techniques may well require that the organization and doctrine thought by innovators to be integral to the effective use of hardware be adapted and molded to fit indigenous cultural traditions.

"Management" and "Control" of Diffusion

Instead of facilitating the diffusion of innovations, militaries often seek to limit, manage, control, or prevent diffusion, particularly to foes but sometimes to friends as well. Diplomacy, especially arms control, has figured prominently in post–World War II efforts to prevent the spread of weapons of mass destruction. Economic tools—unilateral and multilateral export controls and sanctions, for instance—and military means—the bombing of Iraq's nuclear-related facilities during the 1991 Gulf War, for example—have been employed to forestall diffusion.

Jones's examination of intra-alliance diffusion showcases a unique and apparently effective approach to the management and control of the diffusion of military hardware and software. Within the institutional framework of the WTO, the Soviet Union maintained monopoly control of alliance military technology. The Soviet Union's stance on sharing military knowledge of any kind with its allies was dictated by political rather than military requirements: maintain preeminence within the WTO and the need to provide a rival, political rather than military according to Jones, to the West German *Bundeswehr*. Air and naval assets acquired by the WTO served, essentially, as symbols of both "sovereign" military power and participation in the alliance. With the exception of East Germany, the ground forces capabilities possessed by its allies were distinctly inferior to those maintained by the Soviet Union.

Despite the obvious attractions, the United States should be wary of imitating the Soviet approach to managing and controlling diffusion. The United States also has a reputation for withholding its military knowledge from its allies, transferring less than state-of-the-art technologies to some. Some allies (and potential allies) perceive this policy as intended to enhance U.S. military advantage. U.S. encouragement of transatlantic consolidation of defense industries feeds European concerns about the "de-Europeanization" of defense capabilities.[13] The

---

[13]See Elizabeth Becker, "Defense Dept. Urges Industry to Cooperate with Europe," *New York Times*, 8 July 1999, p. A10.

United States must determine how to maintain meaningful alliance relationships while managing the spread of "critical" military innovations to its allies.

No military innovation has met with stronger control responses than nuclear weapons. An extensive and highly restrictive web of nuclear management and control measures, both unilateral and multilateral, has developed since World War II. Potter's systematic examination of the pattern of external and internal security, political incentives and constraints, and triggering events provides a sound basis for better measures to contain weapons of mass destruction.

The finding of the predominance of international incentives and constraints over domestic ones on decisions to go nuclear is noteworthy and fortuitous. It is noteworthy because it provides a focus for action. If international factors are more significant than domestic ones, resources should be concentrated on weakening the set of international security and political incentives that lead to decisions to exercise the nuclear option, and on strengthening the international constraints on decisions to do so. The potential deterrent and war-fighting advantages and prestige and influence derived from ascending to nuclear weapons status should be diminished or denied; states tempted to go nuclear should be assured of the adverse military consequences of yielding to that temptation; the bad neighborhoods that incline some to exercise the nuclear option should be the target of concerted, sustained international protection and clean-up efforts; and the sanctions imposed on states that deem themselves worthy of membership in the nuclear club should be swift and devastating (unlike the tepid response of the international community to the Indian and Pakistani nuclear tests of May 1998).

The finding that external factors outweigh internal factors is fortuitous because external agents find it difficult to manipulate domestic incentives and constraints. Domestic incentives and constraints should not be ignored, but greater leverage over decisions to go nuclear will be realized by attempting to influence and manipulate international rather than domestic incentives and disincentives.

Attempts to manage diffusion from the core may be undermined by "peripheral" states.[14] Timothy Hoyt's case studies of Israeli and Iraqi innovations demonstrate that states outside the circle of major powers generate conceptual and technological innovations, that innovations need not be revolutionary to be significant, that they need not be capital intensive or high cost, and that military innovation can flow upward from the periphery to the core as well as downward from the core to the periphery.

Defense planners in the world's leading powers must recognize that their countries do not monopolize military innovations. They must monitor and pre-

---

[14]On defense industrialization in the developing world, see Andrew L. Ross, "Developing Countries," in Andrew J. Pierre, ed., *Cascade of Arms: Managing Conventional Weapons Proliferation* (Washington, DC: Brookings Institution Press, 1997), pp. 89–127.

pare to respond to military innovations in the periphery. They can even learn from the likes of Israel and Iraq. Major powers, such as today's allegedly "indispensable nation," are not the sole sources of military innovation and diffusion. The periphery will continue to be a source of innovation, particularly asymmetrical innovations that can challenge and undermine the ascendancy of any state that regards itself as indispensable.

Peripheral states, whether roguish or not, and nonstate actors have a special incentive to pursue conceptual innovations that pose an asymmetrical challenge to status quo powers such as the United States. Asymmetrical challenges, including the violation of international norms and use of nuclear, biological, or chemical weapons, increase the costs and risks of major power military interventions. Developing countries are increasingly capable of generating technological breakthroughs, even if only in niches, especially as their economic, industrial, and scientific capabilities rise. Newly industrializing countries such as Brazil, India, Singapore, South Korea, and Taiwan are especially noteworthy potential sources of innovation (particularly those confronted, as is Israel, with compelling security challenges). Globalization and the diffusion of dual-use technologies will enhance the abilities of developing countries, and even nonstate actors, to pursue significant military innovations.[15] The conceptual and technological innovations that will emerge in the developing world will be shared with and replicated by other developing countries, especially when they can generate revenues and amortize development costs. As illustrated by U.S. interest in Israeli programs to produce remotely piloted and unmanned aerial vehicles, even major powers will seek to exploit—through collaborative research and development, licensed production, and even outright purchases—the military innovations of developing countries.

Military Transformation and Diffusion

Defense planners in the United States have been captured by the notion of a revolution in military affairs, planning for the military after next, by leaping ahead, embracing emerging technologies, and creating, quite consciously, a transformation in how militaries go about their business. The focus of attention is on the new technologies, organization, and doctrine required to realize a transformation. Too little attention is being devoted to the all-too-certain diffusion of any innovations that might be introduced.

The diffusion of the innovations at the heart of four nineteenth- and twentieth-century transformations—the Napoleonic revolution, the Prussian-led industrialization of warfare, combined-arms warfare, and carrier-based air power—

---

[15]For discussion of the globalization of the arms industry, see Ann Markusen, "The Rise of World Weapons," *Foreign Policy,* no. 114 (spring 1999): 40–51.

are instructive. Defense planners would do well to factor some of these lessons into their decisions about the pursuit, introduction, exploitation, and timing of military innovations.

First, transformation leaders do not long monopolize "their" transformations. Elements of the innovations upon which these nineteenth- and twentieth-century transformations rested soon made an appearance in other countries, particularly in rival powers. France's *levée en masse* was followed by Prussian conscription. The industrialization and bureaucratization of warfare, pioneered by Prussia, was, even though imperfectly replicated, widespread by 1914. The German approach to combined-arms armored warfare was soon adopted to various degrees by its competitors. Britain's was not long the only navy with aircraft carriers. Competitors can be expected to replicate, adapt, and improve advances introduced by innovation leaders and employ those advances against the very powers that introduced them.

Second, leaders are frequently surpassed by followers. Countries that introduced military innovations or initiated transformations seldom maintained their lead. The advantage often goes to followers. As Goldman noted, rarely have initial innovators been the ones to fully realize the potential of an innovation. The French nationalism that was exploited for military purposes sparked and was trumped by Prussian, then German, nationalism, with severe military consequences for France. The Prussian/German industrialization of war was improved upon by the U.S. industrialization of war. The British introduced the tank, but the Germans fully exploited it. And British leadership in the development of carrier air power disappeared during World War II.

Third, leadership in effecting a military transformation is no guarantee of victory. Napoleonic France went down to defeat. Prussia bested Austria and France, but its unified Germanic successor subsequently fell to coalitions whose members had exploited the opportunities presented by the industrialization of warfare. Germany lost World War II despite its vaunted leadership in the development of organization and doctrine for combined-arms armored warfare. The transformation of naval warfare evident in the displacement of the battleship by the carrier as capital ship is attributable more to competitive U.S. and Japanese interwar innovations than to earlier British innovations. While one of those competitive innovators won World War II, the other lost.

Fourth, the roots of military transformations are typically nonmilitary. This has implications for efforts to create a military revolution and for its diffusion. Neither the Napoleonic transformation nor the industrialization of warfare was "created" or "planned" in any conventional sense. Napoleon brought about a transformation in the art of war but not the political and social transformation that made it possible. Prussian military planners took advantage of industrialization but did not bring it about. The twentieth century's armored warfare and naval air power transformations were continuations of the industrialization of war pioneered by the Prussians. Today's military planners should harbor no illusions about controlling the forces of change.

Fifth, wholesale replication of the innovations of a transformation may be unnecessary. Limited, selective emulation and adaptation can be sufficient. France was defeated eventually by a coalition of countries that, save Prussia, avoided a sweeping transformation and that included Britain, a country that made no real effort to respond to the Napoleonic transformations. Germany was brought down by a coalition whose members had embraced the industrialization of war but not, fully, the Prussian conception of a general staff. It was U.S. materiel and resources as much as its slow and less than fully effective emulation of German organization and doctrine that enabled it to overcome the German armored warfare lead. British carrier air power technologies and practices were not swallowed whole by the United States and Japan.

Sixth, software generally does not travel as well as hardware. Napoleonic military art, which was founded on software advances—organization and doctrine—was not easily reproduced elsewhere. Whether because of geostrategic position, internal political dynamics, culture, economic backwardness and the consequent lack of resources, or lack of commitment and incompetence, the Prussian concept of a general staff and its emphasis on the education, training, and technical expertise that were central to Prussian military planning were seldom fully or adequately assimilated by its major power competitors. Major power armies built up inventories of tanks during the interwar period but resisted adopting the superior German organization and doctrine upon which effective armored warfare rested. The challenge to established naval practices posed by revolutionary carrier air power practices encountered resistance in all of the major power navies during the interwar period.

The seventh, and final, point: all of these nineteenth- and twentieth-century cases of military transformation illustrate the inadequacies of contemporary realism with its emphasis on power and position as determinants of state behavior and global power relations. Military innovation and diffusion are shaped by societal, cultural, institutional, organizational, bureaucratic, individual, doctrinal, and historical forces. Today's realists slight the complexities of diffusion in a misguided quest for parsimony. Practitioners ignore the complexities at their peril.

Diffusion in the Information Age

The transformation sought by contemporary advocates of a revolution in military affairs requires the exploitation of nonmilitary developments, primarily in information technology, but also in high-tech materials, micromechanics, and genetic engineering. This ongoing transformation is driven by commercial, economic dictates. At its heart is information technology—the ubiquitous computer. Computing technologies emerge as the focus of Arquilla's examination of commercial diffusion and the security implications of the diffusion of commercial, dual-use innovations.

Arquilla highlights the age-old tension that drives the dialectic between commerce and security: the desire for economic gain versus the desire to preserve national advantage. It is possible to have both but, apparently, not all of both. The pursuit of economic gain can erode security, and the pursuit of security can impose economic costs. Furthermore, the effort to preserve security by withholding innovations and forgoing the commercial benefits of dissemination may drive others to develop innovations independently. Thus the very security disadvantages that monopolization should have prevented occur without any of the advantages of commercial dissemination—and now with the additional burden of understanding the new innovations of others.

In high-performance computing (HPC) technologies, U.S. decision-makers, as Arquilla points out, have sought to balance security and commercial interests through a hybrid strategy that imposes restrictions on commercial diffusion in the interests of security while accepting a higher security risk. A substantial portion of the economic benefits of commercial diffusion might be realized using this strategy. Arquilla proposes "a somewhat more guarded approach," one that tilts more toward security interests.[16]

A calibrated approach that hedges somewhat more against security risks is eminently sensible. The diffusion of HPC and other information technologies such as software will, however, continue. Practitioners should seek to ensure that the process through which undesirables acquire HPC and other sensitive dual-use information technologies is as arduous, costly, and prolonged as possible. But more states will have HPC in the not so distant future. A much more restrictive, conservative stance by the United States on the commercial diffusion of HPC might actually encourage even more states to develop HPC technologies independently.

Defense planners, consequently, might be expected to begin preparing for a future in which HPC technologies are more widespread and have been employed to create information-age militaries and improve nuclear and advanced conventional capabilities. Nonstate actors have begun to exploit HPC for nefarious purposes. Planners may respond to the prospects of such a future in one of two ways.

---

[16]The Clinton administration, in early July 1999, tilted in the other direction. See David E. Sanger, "U.S. Loosens Rules that Limit Selling Fast Computers," *New York Times*, 2 July 1999, pp. A1, A3.

A "preventive defense" response would focus on reducing the demand for HPC and other information technologies by more effectively addressing the security challenges that lead states to seek such technologies.[17] Many defense planners will be more inclined to respond by developing the capabilities needed to ensure U.S. military superiority. These two responses are not entirely mutually exclusive. The dilemma is that the second response will serve to undermine the effectiveness of a preventive defense response. Along with the tension between security and commercial interests must be added the familiar security dilemma.[18] These defense planning dilemmas cannot be escaped or resolved. They can only be managed. Defense planners will find it impossible to avoid the equivocal responses discussed by Arquilla.

The ubiquity of these planning dilemmas is evident in the opportunity to initiate an information technology–led RMA. The United States is well positioned to bring about an IT-RMA that will maintain its current military superiority and transform future wars.[19] But just as the United States has adjusted to strategic transformations in the past, others will adjust to a U.S.-led transformation.[20] Some, undoubtedly, will seek to jump on the U.S. bandwagon. Close allies, such as Young's Anglo-Saxon security community perhaps, will sooner and more fully reap the benefits of U.S. military advances. Diffusing the innovations of a U.S.-led military transformation will be a significant issue in NATO. How this issue is dealt with will affect the U.S. position in NATO and the attitudes of NATO members toward the privileged position the United States enjoys in the alliance. The United States could use its position as the leader of an IT-RMA transformation to widen the gap between its military capabilities and those of others and thereby reinforce its dominant position in NATO, or it could facilitate adoption within NATO and reassure allies about the manner in which U.S. power and influence will be exercised. The challenges of the latter option are twofold. First, the U.S. government does not necessarily "own" the technology; private firms do. Giving it away will be expensive. More to the point, however, the current problem with the NATO allies is that they are generally unwilling to spend the money to obtain the technologies and capabilities, a problem that the U.S. Defense Capabilities Initiative has attempted to surmount.

[17]On the concept of preventive defense, see Ashton B. Carter and William J. Perry, *Preventive Defense: A New Security Strategy for America* (Washington, DC: Brookings Institution Press, 1999).

[18]See Robert Jervis, "Cooperation under the Security Dilemma," *World Politics* 30, no. 2 (Jan. 1978): 167–214.

[19]See Joseph S. Nye, Jr., and William A. Owens, "America's Information Edge," *Foreign Affairs* 75, no. 2 (Mar.–Apr. 1996): 20–36.

[20]On strategic adjustment, see Peter Trubowitz, Emily O. Goldman, and Edward Rhodes, eds., *The Politics of Strategic Adjustment: Ideas, Institutions, and Interests* (New York: Columbia University Press, 1999).

States not in a position to bandwagon with the United States will have a stake in challenging a U.S.-led transformation. At the least, those challenges will counter, erode, or over time undermine the military advantages that accrue to the United States and its allies. At the worst, they will expose the expected transformation as a house of cards.

Commercially based, dual-use technologies underlie the current transformation. Thus innovations and counterinnovations will probably diffuse rapidly to both state and nonstate actors.[21] Readily available, low-cost information technologies ensure that innovators will not long retain their exclusive advantages. Lower entry costs will enable nonstate actors as well as states to challenge transformation leaders. Demchak's examination of global military modernization should eliminate any lingering doubts about whether other militaries will seek to emulate a U.S.-initiated transformation. Many militaries have already adopted the IT-RMA military as the modern model. Even if the level of resources required to create a "system of systems,"[22] "net-centric warfare,"[23] or otherwise fully exploit the IT-RMA will initially be available to only a few states, niche players will emerge with the ability to erode the capabilities of transformation leaders.[24] Indeed, nonstate actors' ability to challenge states may be a distinguishing feature of an IT-RMA transformation. Defense planners in the United States and elsewhere should expect the emergence of cyber-terrorists, cyber-criminals, cyber-mercenaries, and the like.

The history of nineteenth- and twentieth-century military transformations suggests that the United States and other innovators will seldom have a choice about whether new military practices are disseminated. They will spread. The pace of diffusion may be slowed for a time. The cost of emulation can be rendered more dear. And the extent of initial diffusion can be influenced. But, unfortunately for defense planners, diffusion happens.

Given the inevitability of efforts to emulate, modify, enhance, and counter an IT-RMA transformation, U.S. planners should entertain the notion of emulating

---

[21]At work here is the ascendancy of spin-on over spin-off. For discussions of the shift from spin-off to spin-on, see Wayne Sandholtz, et al., *The Highest Stakes: The Economic Foundations of the Next Security System* (New York: Oxford University Press, 1992).

[22]William A. Owens, "The Emerging System of Systems," *Proceedings* 121, no. 5 (May 1995): 36–39.

[23]Mark Tempestilli, "The Network Force," *Proceedings* 122, no. 6 (June 1996): 42–46; Arthur K. Cebrowski and John J. Garstka, "Network-Centric Warfare: Its Origin and Future," *Proceedings* 124, no. 1 (Jan. 1998): 28–35; and David S. Alberts, John J. Garstka, and Frederick P. Stein, *Network Centric Warfare: Developing and Leveraging Information Superiority* (Washington, DC: C⁴ISR Cooperative Research Program, Department of Defense, 1999).

[24]As Keohane and Nye put it: "The off-the-shelf commercial availability of what used to be costly military technologies benefits small states and nonstate actors and increases the vulnerability of large states." Robert O. Keohane and Joseph S. Nye, Jr., "Power and Interdependence in the Information Age," *Foreign Affairs* 77, no. 5 (Sept.–Oct. 1998): 88.

a British strategy for adjusting to a strategic transformation. As Arquilla notes, Britain attempted to preserve its naval supremacy by managing the shift from ships of wood and sail to ships of steel and steam, slowing the introduction of these innovations and, thereby, their consequent diffusion. A calculated, nuanced, even equivocal approach to an IT-RMA would enable the United States to exploit its present unrivaled military position more fully, and for a longer period of time, than an unrestrained rush to transformation. As an example, a restrained approach to the exploitation of space, which is central to visions of an IT-RMA, may be particularly appropriate.[25] The intelligence, surveillance, and reconnaissance (ISR) component of an IT-RMA is heavily dependent on space-based assets for support and force enhancement.[26] But calls for moving beyond the current militarization of space to the weaponization of space in order to be the first with the ability to "fight in space, from space, and into space" should be resisted.[27] The United States is indeed well positioned to lead in weaponizing space. But the U.S. lead will be followed, eventually, by others. The high level, and higher future level, of U.S. dependence on space-based assets for ISR means that the United States has the most to lose from the competitive weaponization of space. Not every competitive advantage should necessarily be exploited.[28] More may be lost than gained by extending the terrestrial security dilemma into space.

Conclusion

Defense planners today are caught in the throes of a military revolution. It has been thrust upon them by the information revolution. Their response has focused, appropriately and unsurprisingly, on the ways in which the military might exploit the information revolution. The call for military transformation was not quieted by the events of September 11 and their aftermath. Indeed, the attacks on New York City and the Pentagon and the subsequent U.S. war on terror have

[25]According to one advocate of the exploitation of space, "Space power is catalyzing a true transformation of war" and is "the leading edge of a transformation of war." Colin S. Gray, "The Influence of Space Power upon History," *Comparative Strategy* 15, no. 4 (Oct.–Dec. 1996): 293.

[26]See Owens, "The Emerging System of Systems." Space support consists of "launching . . . controlling, and repairing space assets." Force enhancement includes "operating satellites providing surveillance, reconnaissance, missile warning, tactical intelligence, navigation, communication, and weather information to warfighters on land, at sea, and in the air." Steven Lambakis, "Exploiting Space Control," *Armed Forces Journal International* 134, no. 11 (June 1997): 42.

[27]Lambakis, "Exploiting Space Control," p. 42. Lambakis here defines space control as "protecting and operating survivable space systems to ensure freedom of action in space, and maintaining capabilities to deny the enemy access to space for military advantage." Force application requires "maintaining capabilities to attack enemy targets on earth and in space using space-based combat power."

[28]See Bruce M. DeBlois, "Space Sanctuary: A Viable National Strategy," *Airpower Journal* 12, no. 4 (winter 1998): 41–57.

been cited as evidence of the need for transformation[29]—despite the fact that distinctly untransformed U.S. forces were used to oust the Taliban regime and deprive Al Qaeda of its Afghan refuge.[30] The advanced U.S. military capabilities—sensors, communications networks, targeting technologies, and unarmed and armed UAVs, for instance—on display in the Afghan phase of the new war served to demonstrate once again the gap between the capabilities of the United States and the rest of the world. Transformation will only widen that gap.

Diffusion dynamics, however, ensure that that gap will be bridged. The U.S. advantage will not be fleeting, but it will not be permanent. The United States, in the interests of interoperability, will encourage its allies to adopt transformational doctrines, organizations, and technologies. Others will attempt to emulate new U.S. practices. New technologies will be replicated. Counters will be developed. Most important, perhaps, the information technologies at the heart of the current transformation may diffuse more rapidly than did earlier transformational technologies—to state and nonstate actors alike. Nonstate actors without the cultural resistance to innovation evident in the U.S. and other militaries may well exploit new opportunities more quickly than their limited resources would lead us to expect. One of Al Qaeda's most distinctive characteristics, for instance, is its networked organizational structure.

Isaiah Berlin's foxes, who employ new information technologies to do many little things better, and hedgehogs, who institute a new order, are both at play in the defense planning establishment today.[31] Neither, however, has fully factored into their calculations the implications of the diffusion of the military transformation they embrace. Practitioners must make a concerted effort to anticipate how others will exploit, respond to, adapt, and counter the emerging transformation. The historical cases presented here provide a basis for anticipating the range of possible responses.

The contributors to this volume have highlighted the complexity of diffusion dynamics. Few sweeping, unqualified generalizations have been advanced here; most chapters refute structural realism's simplified and deterministic view that diffusion will be rapid and highly imitative. Diffusion is highly contingent. Its scope, speed, and extent depend upon the interaction of strategic necessity, culture and doctrine, and institutional and organizational mindsets. Motivations, transmission paths, facilitators and inhibitors, and effects are similarly interactive

[29]See, for instance, Andrew F. Krepinevich, "Arming Soldiers for a New Kind of War," *New York Times*, 26 Oct. 2001, p. A23.

[30]Michael O'Hanlon, "Winning with the Military Clinton Left Behind," *New York Times*, 1 Jan. 2002, p. A21.

[31]Isaiah Berlin, *The Hedgehog and the Fox: An Essay on Tolstoy's View of History* (London: Weidenfeld and Nicolson, 1967).

and, consequently, subject to considerable variation. This complexity does not make life easy for practitioners (just as it frustrates theory-builders). Whether they seek to facilitate the adoption of new hardware and software by friends and allies or to prevent its diffusion to potential adversaries, recognition of the complexities of diffusion dynamics is a prerequisite for effective action. Complexity and the contingent nature of diffusion are not necessarily obstacles to effective action. Understanding the nature and impact of contingency is indispensable as defense planners attempt to prepare for, respond to, and shape the future.

Index

In this index an "f" after a number indicates a separate reference on the next page, and an "ff" indicates separate references on the next two pages. A continuous discussion over two or more pages is indicated by a span of page numbers, e.g., "57–59." *Passim* is used for a cluster of references in close but not consecutive sequence.